murach's
Android
programming

2ND EDITION

Joel Murach

TRAINING & REFERENCE

murach's
Android
programming

2ND EDITION

Joel Murach

MIKE MURACH & ASSOCIATES, INC.

4340 N. Knoll Ave. • Fresno, CA 93722
www.murach.com • murachbooks@murach.com

Author: Joel Murach

Editors: Mike Urban
Ray Halliday

Android Consultant: Leo Landau

Production: Maria Spera

Books for Java programmers

Murach's Java Servlets and JSP (3rd Edition)

Murach's Android Programming (2nd Edition)

Murach's Beginning Java with NetBeans

Murach's Beginning Java with Eclipse

Murach's Java Programming (4th Edition)

Books for database programmers

Murach's MySQL (2nd Edition)

Murach's Oracle SQL and PL/SQL for Developers (2nd Edition)

Murach's SQL Server 2012 for Developers

Books for web developers

Murach's HTML5 and CSS3 (3rd Edition)

Murach's JavaScript (2nd Edition)

Murach's jQuery (2nd Edition)

Murach's PHP and MySQL (2nd Edition)

Books for .NET programmers

Murach's C# 2012

Murach's ASP.NET 4.5 Web Programming with C# 2012

Murach's Visual Basic 2012

Murach's ASP.NET 4.5 Web Programming with VB 2012

For more on Murach books, please visit us at www.murach.com

Printed in the United States of America

10 9 8 7 6 5 4 3 2
ISBN: 978-1-890774-93-6

Content

Expanded contents

Chapter 7   How to work with themes and styles

Chapter 8   How to work with menus and preferences

Section 5 Advanced Android skills

Chapter 17   How to deploy an app

Chapter 18   How to work with locations and maps

Introduction

Android is the world's most widely used operating system for mobile devices such as smartphones and tablets. In the coming years, Android's popularity is likely to continue for a couple reasons. First, Google releases Android under an open-source license that allows it to be modified and freely distributed by device manufacturers. Second, since Android is open-source with a large community of developers, it is able to evolve according to the needs of the developers who use it.

Who this book is for

This book is for anyone who wants to learn professional skills for developing Android apps. The only prerequisite is a basic understanding of the Java programming language, roughly equivalent to chapters 1 through 18 of our book, *Murach's Beginning Java*. Once you have the necessary Java skills, this book should work for you even if you have absolutely no experience developing mobile apps.

What you'll learn in this book

As I wrote this book, I did my best to focus on the practical features that you'll need for developing professional Android applications. Here's a quick tour:

- In section 1, you'll quickly master the basics of Android programming. In chapter 1, you'll learn some background terms and concepts, and you'll use Android Studio to open and run some existing apps. In chapters 2 and 3, you'll learn how to develop your first Android app, a simple but complete Tip Calculator app. And in chapter 4, you'll learn how to test and debug your apps.

- In section 2, you'll learn some essential Android skills by enhancing the Tip Calculator app that you developed in chapter 3. To start, in chapter 5, you'll learn how to use different layouts and widgets to develop a sophisticated user interface. In chapter 6, you'll learn several techniques for handling high- and low-level events. In chapter 7, you'll learn how to use themes and styles to improve the appearance of your app. In chapter 8, you'll learn how to use menus and preferences. And in chapter 9, you'll learn how to use fragments to allow your app to take advantage of large screens that are available from some mobile devices such as tablets.

- In section 3, you'll learn how to develop a News Reader app. Along the way, you'll learn more essential Android skills. In chapter 10, for example, you'll learn how to read an RSS feed from the Internet, save that data in a file, and display it on the user interface. In chapter 11, you'll learn how to use a service to download data for an app even when the app isn't running, and to notify a user when new data is available. And in chapter 12, you'll learn how to respond to actions that are broadcast by Android and its apps.

- In section 4, you'll learn how develop a Task List app that stores one or more to-do lists. In chapter 13, you'll learn how to create a database to store these lists. In chapter 14, you'll learn how to use tabs and a custom adapter to display these tasks on the user interface. In chapter 15, you'll learn how to use a content provider to allow other apps to work with the same data as this app. And in chapter 16, you'll learn how to create an app widget that can display some of this app's data on a device's Home screen.

- In section 5, you'll learn how to deploy apps to the Google Play store. Then, you'll learn how to create a Run Tracker app that uses the Google Maps API.

Why you'll learn faster and better with this book

Like all our books, this one has features to help you learn as quickly and easily as possible. Here are just three of those features.

- To help you develop applications at a professional level, this book presents complete, non-trivial apps. For example, chapter 12 presents the final version of the News Reader app, chapter 16 presents the final Task List app, and chapter 18 presents a Run Tracker app. Studying these apps is a great way to master Android development.

- All of the information in this book is presented in our unique paired-pages format, with the essential syntax, guidelines, and examples on the right page and the perspective and extra explanation on the left page. This helps you learn more while reading less, and it helps you quickly find the information that you need when you use this book for reference.

- The exercises at the end of each chapter give you a chance to try out what you've just learned and to gain valuable, hands-on experience with Android programming. They guide you through the development of some of the book's apps, and they challenge you to apply what you've learned in new ways. And because most of these exercises start from code that you can download from our website, you'll spend your time practicing new skills instead of doing busywork.

How our downloadable files make learning easier

To make learning easier, you can download the source code for all the apps presented in this book from our website (www.murach.com). Then, you can view the complete code for these apps as you read each chapter, you can run these apps to see how they work, and you can copy portions of code for use in your own apps.

You can also download the starting points and solutions for the exercises in this book. That way, you don't have to start every exercise from scratch. This takes the busywork out of doing these exercises. As a result, you get more practice in less time. In addition, if you encounter a problem, you can easily check the solution. This helps you to keep moving forward with less chance that you'll get hung up on a minor issue.

What version of Android this book supports

Most of the apps presented in this book have been designed to work with Android 4.0 (API 15), also known as Ice Cream Sandwich, through Android 6.0 (API 23), also known as Marshmallow. Since Android is backwards-compatible, the apps presented in this book should continue to work for future versions of Android too, though they won't be able to take advantage of the new features available from those versions of Android.

What operating systems this book supports

This book supports the Windows, Mac OS X, and Linux operating systems. If you're using Windows, you can use Appendix A to set up your computer for this book. If you're using Mac OS X, you can use appendix B. And if you're using Linux, you can use appendix C.

What IDE this book supports

This book shows how to use Android Studio to code, test, and debug applications. This IDE is available for free and runs on all modern operating systems. Although it's possible to use other IDEs, Android Studio is now the official IDE for Android, replacing Eclipse with ADT. As a result, most Android developers use Android Studio for new development.

How to get the software you need

You can download all of the software that you need for this book for free from the Internet. This software includes the Java SDK (Software Development Kit), Android Studio, and the Android SDK, as well as the source code for the apps that are presented throughout the book. To find out how to download and install this software, please see the appendixes.

Support materials for trainers and instructors

If you're a corporate trainer or a college instructor who would like to use this book for a course, we offer these supporting materials: (1) a complete set of PowerPoint slides, (2) test banks, (3) additional chapter exercises that aren't in this book, (4) projects that the students start from scratch, and (5) solutions to the additional exercises and projects.

To learn more about these materials, please go to our website at www.murachforinstructors.com if you're an instructor. Or if you're a trainer, please go to www.murach.com and click on the *Courseware for Trainers* link, or contact Kelly at 1-800-221-5528 or kelly@murach.com.

A companion book

As you read this book, you may discover that your Java skills aren't as strong as they ought to be. In that case, I recommend that you get a copy of one of our beginning Java books (*Murach's Beginning Java with NetBeans* or *Murach's Beginning Java with Eclipse*). Either book gets you up to speed with Java and shows you how to use all of the skills you need for developing Android apps.

Let us know how this book works for you

When I started writing this book, my goal was (1) to teach you Android programming as quickly and easily as possible and (2) to teach you the practical Android concepts and skills that you need for developing professional apps. Now, I hope I have succeeded. If you have any comments about this book, I would appreciate hearing from you at murachbooks@murach.com.

Thanks for buying this book. I hope you enjoy reading it, and I wish you great success with your Android programming.

Joel Murach

Joel Murach
Author

Section 1

Get started fast with Android

This section gets you started quickly with Android programming. First, chapter 1 introduces you to some concepts and terms that apply to Android development. In addition, it shows you how to use Android Studio to open and run existing projects.

After that, chapter 2 shows you how to use Android Studio to develop the user interface for an app. Then, chapter 3 shows you how to write the Java code for an app. And chapter 4 shows you how to test and debug apps.

To illustrate these skills, this section uses a simple but complete Tip Calculator app. This app calculates the tip you would give based on the amount of a bill. When you complete these chapters, you'll be able to write, test, and debug simple applications of your own.

1

An introduction to Android and Android Studio

This chapter starts by presenting some background information about Android. This information isn't essential to developing Java applications, so you can skim it if you want.

This chapter finishes by showing how to use the Android Studio IDE to work with an existing project. This gives you some hands-on experience using Android Studio to work with projects such as the projects for this book that you can download from our website.

An overview of Android

Android is a Linux-based operating system designed primarily for touch-screen mobile devices such as smartphones and tablets. Before you begin developing apps that run on Android, it's helpful to take a moment to consider the types of Android devices and apps that are available today. In addition, it's helpful to understand the history, versions, and architecture of Android.

Types of devices

Figure 1-1 starts by showing two of the most popular Android devices, a smartphone and a tablet. However, the code for Android is open-source. As a result, it can be customized to work with other types of electronic devices such as eBook readers, cameras, home automation systems, home appliances, vehicle systems, and so on.

An Android phone and tablet

Other types of Android devices

- Readers
- Cameras
- Home automation systems
- Home appliances
- Vehicle systems

Description

- *Android* is a Linux-based operating system designed primarily for touchscreen mobile devices such as smartphones and tablet computers.
- Since the code for Android is open-source, it can be customized to work with other types of electronic devices.

Figure 1-1   Types of Android devices

Types of apps

Android has a large community of developers writing *applications*, more commonly referred to as *apps*, that extend the functionality of Android devices. Figure 1-2 lists some categories of different types of apps and describes some of the functionality available from each category. As you review this list, keep in mind that these categories are only intended to give you a general idea of the various types of Android apps. More categories exist, and the functionality that's available from Android apps is constantly expanding.

If you have a smartphone or tablet, you should be familiar with some of the apps listed in this figure. For example, you probably use apps to send and receive text messages and email. You probably use apps to take pictures, listen to music, and watch video. You probably use apps to get directions and navigate to a location. You probably use apps to check the weather, read news, and browse the web.

In addition, you may use apps from social media companies like Facebook and Twitter to work with social media. You may use apps to play games like Angry Birds. The main point here is that there are many different types of apps and that application developers are constantly creating new apps that use the capabilities of Android phones in new ways.

Some apps come preinstalled on the device. For example, most phones include apps for managing contacts, using the phone, sending and receiving text messages and email, working with photos and video, and web browsing. Other apps are available for download through Google Play or third-party sites. Some apps are free. Other apps are expensive. All apps were created by somebody like you who learned how to develop Android apps.

Types of Android apps

Category	Functionality
Communications	Send and receive text messages, send and receive email, make and receive phone calls, manage contacts, browse the web.
Photography	Take photos, edit photos, manage photos.
Audio	Play audio, record audio, edit audio.
Video	Play video, record video, edit video.
Weather	View weather reports.
News	Read news and blogs.
Personalization	Organize home screen, customize ringtones, customize wallpaper.
Productivity	Manage calendar, manage task list, take notes, make calculations.
Finance	Manage bank accounts, make payments, manage insurance policies, manage taxes, manage investment portfolio.
Business	Read documents, edit documents, track packages.
Books	Read and search eBooks.
Reference	Get info from a dictionary, thesaurus, or wiki.
Education	Prepare for exams, learn foreign languages, improve vocabulary.
Shopping	Buy items online, use electronic coupons, compare prices, keep grocery lists, read product reviews.
Social	Use social networking apps such as Facebook and Twitter.
Fitness	Monitor and document workouts.
Sports	Track sport scores, manage fantasy teams.
Travel	Get directions, use GPS to navigate to a location, get information about nearby places to visit.
Games	Play games such as arcade games, action games, puzzles, card games, casino games, sports games.

Description

- Android has a large community of developers writing *applications*, more commonly referred to as *apps*, that extend the functionality of Android devices.
- Android apps are available for download through Google Play or third-party sites.

Figure 1-2   Types of Android apps

A brief history

Figure 1-3 summarizes the history of Android. In 2003, a handful of entrepreneurs in Palo Alto, California, founded Android, Inc. In 2005, Google bought Android, Inc. Then, in 2007, the *Open Handset Alliance* was announced. This alliance consists of a consortium of hardware, software, and telecommunication companies devoted to advancing open standards for mobile devices. That same year, Google released Android code as open source under the Apache License. In addition, Google helped to create the *Android Open Source Project* (*AOSP*) and put it in charge of the maintenance and further development of Android.

In 2008, the first version of the Android *Software Development Kit* (*SDK*) was released. This SDK contains all of the tools that Android developers need to develop apps. Later in 2008, the first Android phones became available.

Since 2008, new versions of the Android SDK have continued to be released, and Android devices have continued to proliferate. During that time, millions of apps have been developed, and billions of apps have been downloaded.

A brief history of Android

Year	Event
2003	Android, Inc. is founded in Palo Alto, California.
2005	Google buys Android, Inc.
2007	The *Open Handset Alliance* is announced. This alliance consists of a consortium of hardware, software, and telecommunication companies devoted to advancing open standards for mobile devices.
	Google releases Android code as open source under the Apache License.
	The *Android Open Source Project* (*AOSP*), led by Google, is tasked with the maintenance and further development of Android.
2008	Android *Software Development Kit* (*SDK*) 1.0 is released. This kit contains all of the tools needed to develop Android apps.
	The first Android phones become available.
2009-present	New versions of the Android SDK continue to be released.
	Android devices continue to proliferate.
	Millions of apps are developed.
	Billions of apps are downloaded.

Description

* Android has experienced tremendous growth since its release in 2008.

Figure 1-3   A brief history of Android

Versions

Figure 1-4 describes all major releases of Android starting with version 2.2 and ending with version 6.0. Here, each version of Android corresponds with an API (Application Programming Interface) number. For example, version 5.1 corresponds with API 22, and version 4.0.3 corresponds with API 15. In addition, each version has a code name that's based on something sweet. For example, version 5.1 uses the code name Lollipop, and version 6.0 uses the code name Marshmallow.

As you develop an Android app, you must decide the minimum API level that your app supports. As of October 2015, many developers choose 15 or 16 as the minimum API level to support since that covers a high percentage of all Android devices.

The distribution percentages shown here are from October 2015. As time progresses, more users will upgrade from older devices to newer ones. As a result, you should check the URL shown in this figure to get current percentages before you decide the minimum API level for your app.

As you review this figure, you may notice that it doesn't include some versions of Android such as 1.0, 2.0, 2.1, and 3.0. That's because there are virtually no devices left that run on these versions of Android. As time marches on, it's inevitable that the same fate will befall other older versions of Android too.

Android versions

Version	Code name	API	Distribution
2.2	Froyo	8	0.2%
2.3.3 - 2.3.7	Gingerbread	10	3.8%
4.0.3 - 4.0.4	Ice Cream Sandwich	15	3.4%
4.1.x	Jelly Bean	16	11.4%
4.2.x		17	14.5%
4.3		18	4.3%
4.4	KitKat	19	38.9%
5.0	Lollipop	21	15.6%
5.1		22	7.9%
6.0	Marshmallow	23	

A URL for current distribution percentages

`http://developer.android.com/about/dashboards/index.html`

Description

- The distribution percentages in this figure are from October 2015. To get current percentages, please visit the URL shown above.

- As you develop an Android app, you must decide the minimum API level that your app supports. As of October 2015, many developers choose Android 4.0.3 (API 15) as the minimum API level to support since that covers a high percentage of all Android devices.

Figure 1-4   Android versions

System architecture

Figure 1-5 shows the Android system architecture, which is also known as the *Android stack*. This stack has four layers.

The bottom layer of the stack is Linux, an open-source operating system that's portable and secure. This operating system provides low-level drivers for hardware, networking, file system access, and inter-process communication (IPC).

The second layer up in the stack contains the native libraries. These libraries are written in C or C++. They include the *Dalvik virtual machine* (*VM*), which works similarly to the *Java virtual machine* (*JVM*). However, the Dalvik VM was designed specifically for mobile devices and their inherent limitations, such as battery life and processing power.

The third layer up in the stack contains the application framework. This layer is written mostly in Java, and it provides libraries that can be used by the top layer of the stack. In this book, you'll learn how to use some of these libraries, such as the libraries for the notification manager, content providers, and the location manager.

The top layer of the stack contains Android apps. These apps include pre-installed apps such as the apps that you can use to manage the Home screen, manage your contacts, make and receive calls, browse the web, and so on. In addition, you can download and install other apps. These types of apps are written in Java, and they are the type of apps that you'll learn to develop in this book.

Android stack

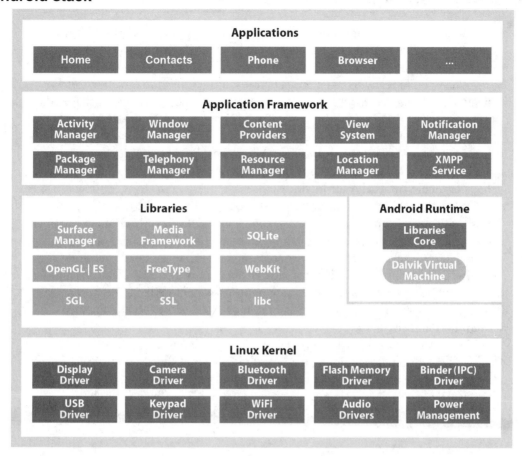

Description

- Linux is an open-source operating system that's portable and secure.
- The native libraries are written in C or C++. These libraries provide services to the Android application layer.
- The *Dalvik virtual machine* (*VM*) works similarly to the *Java virtual machine* (*JVM*). However, the Dalvik VM was designed specifically for mobile devices and their inherent limitations such as battery life and processing power.
- The application framework provides libraries written in Java that programmers can use to develop Android apps.

Figure 1-5   The Android system architecture

How apps are compiled and run

When you develop an Android app, you typically use an *IDE* (*Integrated Development Environment*) such as Android Studio to create a project. A *project* contains all of the files for the app including the files for the Java source code.

When you're ready to test a project, you can run it. Figure 1-6 shows how this works. When you run a project, the IDE typically compiles and packages the project automatically before running it. This is known as *building* the project.

When the IDE builds a project, it compiles the *Java source code* (.java files) into *Java bytecodes* (.class files). Then, it compiles the Java bytecodes into *Dalvik executable files* (.dex files) that can be run by the Dalvik virtual machine that's available from all Android devices.

When the IDE builds a project, it puts the .dex files and the rest of the files for the project into an *Android package* (.apk file). This file contains all of the files necessary to run your app including the .dex files and other compiled resources, uncompiled resources, and a binary version of the Android manifest. The *Android manifest* is a file that specifies some essential information about an app that the Android system must have before it can run the app. In its non-binary version, the Android manifest is stored in a file named AndroidManifest.xml.

For security reasons, all Android apps must be digitally *signed* with a certificate. During development, the IDE typically signs the app for you automatically using a special debug key. Then, it runs the app on the specified physical device such as a smartphone or tablet. Or, it runs the app on the specified *emulator*, which is a piece of software that runs on your computer and mimics an Android device. An Android emulator can also be called an *Android Virtual Device* (*AVD*).

The *Android Debug Bridge* (*ADB*) lets your IDE communicate with an emulator or a physical Android device. This is necessary to provide the debugging capabilities described in chapter 4.

When you are ready to release the app, you must sign the app in release mode, using your own private key. For more information about this, please see chapter 17.

Android system architecture

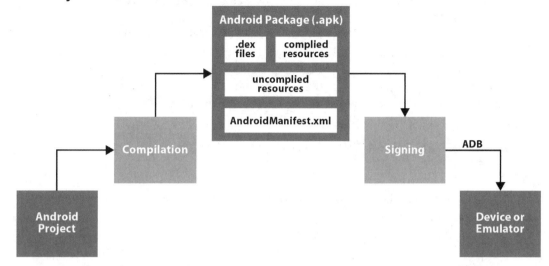

Description

- When you develop an Android app, you typically use an *IDE* (*Integrated Development Environment*) such as Android Studio to create a *project*, and you typically use Java as the programming language.

- When you develop an Android app, you can run it on a physical Android device, such as a smartphone or tablet. Or, you can run it on an *emulator*, which is a piece of software that runs on your computer and acts like an Android device. An Android emulator can also be called an *Android Virtual Device* (*AVD*).

- Before you can run a project, you must build the project. Typically, the IDE automatically builds a project before running it.

- When the IDE builds a project, it compiles the *Java source code* (.java files) into *Java bytecodes* (.class files) and then into *Dalvik executable files* (.dex files). Dalvik executable files can be run by the Dalvik virtual machine that's available from all Android devices.

- When the IDE builds a project, it puts the files for the project into an *Android package* (.apk file). This file contains all of the files necessary to run your app on a device or emulator including the .dex files, compiled resources (the resources.arsc file), uncompiled resources, and a binary version of the AndroidManifest.xml file.

- To run an app on an emulator or device, the app must be *signed* with a digital certificate that has a private key. During development, the IDE automatically signs the app for you in debug mode using a special debug key. Before you release an app, you must sign the app in release mode, using your own private key. For more information about this, please see chapter 17.

- The *Android Debug Bridge* (*ADB*) lets your IDE communicate with an emulator or a physical Android device.

Figure 1-6   How an Android app is compiled and run

An introduction to Android Studio

Android Studio is an IDE that you can use to develop Android apps. Android Studio is based on IntelliJ IDEA. It's open-source, available for free, and runs on all modern operating systems. Although it's possible to use other IDEs, Android Studio is currently the official IDE for Android, replacing Eclipse with ADT. As a result, most Android developers use Android Studio for new development.

How to work with the Welcome page

Figure 1-7 shows the Welcome page for Android Studio. This page is displayed the first time you start Android Studio. In addition, it's displayed if you close all open projects.

The right side of the Welcome page provides items that let you start new Android Studio projects from scratch and open existing Android Studio projects. In addition, it includes an item that allows you to import projects from other IDEs such as Eclipse with ADT.

The left side of the Welcome page provides a list of recent projects that have been opened. You can easily reopen any of these projects by clicking on them.

How to open an existing project

If you attempt to open an existing Android Studio project, you'll get a dialog box that lets you navigate to the directory that contains the project you want to open. For example, if you installed the source code for this book as described in the appendixes, all of the projects presented in this book are stored in subdirectories of this directory:

```
\murach\android\book_apps
```

The Welcome page

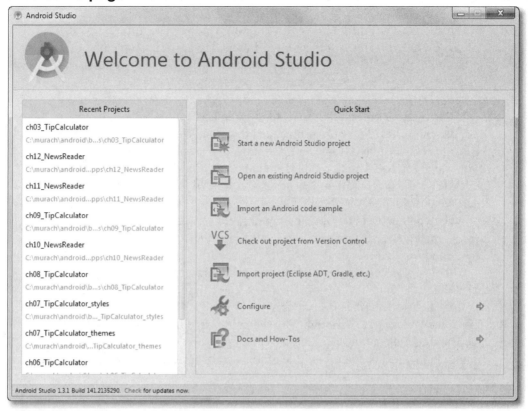

Description

- An Android Studio *project* consists of a top-level directory that contains the directories and files for an app.

- When you start Android Studio for the first time, it displays a Welcome page. Android Studio also displays this page if you close all open projects.

- The left side of the Welcome page displays recently opened projects. You can reopen them by clicking on them.

- The right side of the Welcome page displays options that you can use to start a new project, open an existing project, or import a project from another IDE such as Eclipse with the ADT bundle.

Figure 1-7   The Welcome page

How to view the user interface for an app

Figure 1-8 shows the main Android Studio window. The different parts of this window are known as *tool windows*. In this figure, for example, the Project tool window on the left side of the main window displays the directories and files that make up the project for the Tip Calculator app that's described in chapter 3.

In this figure, I opened the XML file for the user interface by expanding the res/layout directory for the project and double-clicking on the XML file for the layout. This displayed the user interface in the graphical editor. However, if you want, you could click on the Text tab below the graphical editor to view the XML for the layout. Then, whenever you wanted to return to the graphical editor, you could click the Design tab.

When you open a layout in the graphical editor, Android Studio typically displays the Palette, Component Tree, and Properties tool windows. You can arrange and resize these windows to fit your needs.

Although you can create a user interface by entering its XML, it's usually easier to use the graphical editor. Then, you can edit the XML that's generated to fine-tune the user interface if necessary.

For now, you can experiment with the graphical editor. If you have experience working with another GUI builder tool, you shouldn't have much trouble using it. In the next chapter, you'll learn the details for using the graphical editor to develop the user interface for an Android app.

The layout for the Tip Calculator activity in the graphical editor

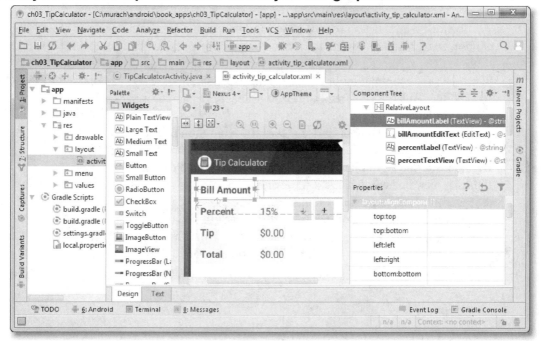

Description

- In Android development, an *activity* defines one screen of an app.

- In Android development, a *layout* stores the XML that defines the user interface for an activity.

- In Android Studio, the different parts of the main window are known as *tool windows*.

- The Project tool window displays the directories and files that make up a project. If this window isn't visible, you can display it by clicking the Project tab on the left side of the main window.

- You can expand and collapse the nodes of the Project window by clicking on the triangles to the left of each node.

- To work with the user interface, expand the res\layout directory for the project, and double-click on a layout file to open it. Then, you can click on the Design tab to use the graphical editor to work with the user interface. Or, you can click on the Text tab to work directly with the XML that defines the user interface.

- When using the graphical editor, the Palette, Component Tree, and Properties tool windows typically open to the left and the right of the user interface. These windows can be pinned to either side of the main window if you need more visual space.

Figure 1-8   How to view the user interface for an app

How to view the code for an app

To work with the Java source code for an app, you can use the Project window to expand the java directory and expand the package that contains the Java file. Then, you can double-click on that file to open it in the code editor. In figure 1-9, for example, the Java file for the Tip Calculator activity is open in the code editor. This file is stored in the com.murach.tipcalculator package of the java directory.

You can also rename or delete a Java file from the Project window. To delete a file, right-click on the file and select Delete from the resulting menu. To rename a file, right-click on the file and select Refactor→Rename from the resulting menu. If you rename a file, Android Studio automatically changes both the name of the Java file and the name of the class. Since the name of the Java file must match the name of the class, this is usually what you want.

For now, don't worry if you don't understand all of this code. In chapter 3, you'll learn the skills that you need to develop this kind of code.

The Java code for the Tip Calculator activity

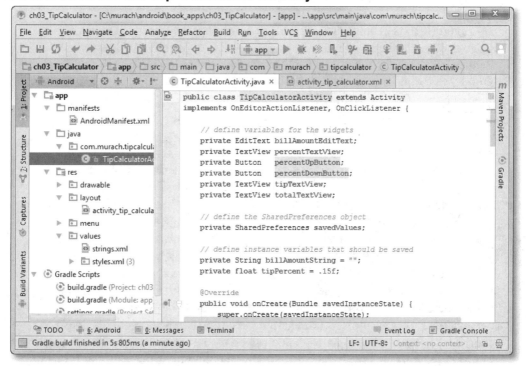

Description

- To work with the Java code for an app, use the Project window to expand the java directory, expand the package that contains the code, and double-click on the file.
- To rename a Java file, right-click on the file, select the Refactor→Rename item, and enter a new name for the file. This automatically renames the class that corresponds with the file.
- To delete a Java file, right-click on the file, select the Delete item, and use the resulting dialog to confirm the deletion.
- To delete a package, right-click on the package, select the Delete item, and use the resulting dialog to confirm the deletion.

Figure 1-9   How to view the Java code for an activity

How to run an app on a physical device

Figure 1-10 begins by showing how to run an app on a connected device. To do that, select the Run button in the toolbar. This typically displays a Choose Device dialog box. If it doesn't, you can display this dialog by clicking on the Run→Edit Configurations item from the menus. Then, you can use the resulting dialog box to select the "Show chooser dialog" option.

From the Choose Device dialog, you can select the device and click the OK button. In this figure, for example, this would allow you to select the Samsung Android 4.4.4 device.

If the Choose Device dialog doesn't display the physical device, make sure a compatible physical device is connected to your computer. Typically, you use a USB cable to connect your device to your computer.

Once your app is running on the physical device, you can test that app using any of the hardware that's available on that device. For example, you can rotate the device to test how it handles screen orientation changes.

By default, Android Studio automatically compiles an app before running it. Since this saves a step in the development process, this is usually what you want. Sometimes, though, Android Studio gets confused and doesn't compile a class that's needed by your project. For example, Android Studio sometimes doesn't compile the R class that contains the compiled resources needed by your project. In that case, you can usually fix this issue by selecting the Build→Clean Project item from the menu bar. This cleans the project and rebuilds it from scratch.

How to run an app on an emulator

This figure also shows how to run an app on an emulator. To do that, you use the same technique for running an app on a physical device. However, from the Choose Device dialog, you can select an emulator that's already running. Or, you can start a new emulator, by selecting the "Launch emulator" option and selecting the emulator you want.

Once your app is running in an emulator, you can use the emulator to test your app. To do that, you can use your mouse to interact with the touchscreen to test the app.

When you're done testing an app in an emulator, you can leave the emulator open. That way, the next time you run the app, Android Studio won't need to start the emulator, which can take a frustratingly long time on most systems. As a result, Android Studio will be able to run your app much more quickly the next time.

The Choose Device dialog

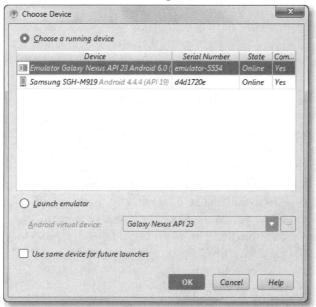

Description

- To run any app, click the Run button in the toolbar. This typically displays the Choose Device dialog to let you pick the device.
- To run on a physical device or emulator that's already running, select the device or emulator and click the OK button.
- To start an emulator, select the "Launch emulator" option, select the emulator from the drop-down list, and click the OK button.
- On a device or emulator, you need to unlock the screen to allow the app to run.
- By default, a project is compiled automatically before it is run, which is usually what you want.
- To clean and rebuild a project, select the Build→Clean Project item from the menu bar and respond to the resulting dialog.

Notes

- If the Choose Device dialog isn't displayed when you run an app, you can display it by clicking on the Run→Edit Configurations item from the menus. Then, you can use the resulting dialog to select the "Show chooser dialog" option.
- If your device isn't displayed in the Choose Device dialog, you can disconnect it and reconnect it. This should display the device.
- If you haven't authorized a device to work with the current computer, the device displays a dialog when you unlock the screen. You can use this dialog to authorize the device.

Figure 1-10   How to run an app

The user interface

Figure 1-11 shows the user interface for the Tip Calculator app after it has been displayed in an emulator for Android 6.0 (API 23). Of course, you can also run this app on emulators for other Android platforms. Similarly, you can run this app on a physical device.

The emulator shown in the figure displays an on-screen keyboard. As a result, you can use this on-screen keyboard, known as a *soft keyboard*, to enter text as shown in this figure. To do that, you click on the Bill Amount text box. When you do, the emulator should display a soft keyboard that's optimized for entering a number with a decimal point. When you're done using that keyboard to enter an amount for the bill, you can click the Done key (the one that looks like a check mark). When you do, the app calculates the tip and total for the bill.

By default, an emulator should allow you to use your computer's keyboard to enter text. This provides an easy way to enter text. However, it doesn't accurately emulate touchscreen devices. As a result, you may want to use your mouse to work with a soft keyboard to test your apps.

If your emulator doesn't display the soft keyboard when you click in a text box, you can change the input settings for the keyboard. The procedure for doing this varies depending on the version of Android. With API 16, you can use the Settings app to change the Language and Input→Keyboard and Input→Default→ Hardware (Physical Keyboard) option. With API 23, you can click the keyboard icon that's displayed in the emulator and use the resulting dialog to change the input method for the keyboard.

By default, this Tip Calculator app uses a tip percent of 15%. However, if you want to increase or decrease the tip amount, you can click the Increase (+) or Decrease (-) buttons. When you do, the app recalculates the tip and total amounts for the bill.

The Tip Calculator app with the soft keyboard displayed

Soft keyboard

Keyboard icon

Description

- To calculate a tip, click the Bill Amount text box and enter the bill amount. When you're done, press the Done key.

- To increase or decrease the tip amount, click the Increase (+) or Decrease (-) buttons.

- The app automatically recalculates the tip and total amounts whenever the user changes the bill amount or tip percent.

- On most emulators, you can enter text with your computer's keyboard or with the on-screen keyboard known as a *soft keyboard*.

- If your emulator doesn't display a soft keyboard when you click in a text box, you can change the input settings for the keyboard on your emulator. To do that, you may need to use your emulator's Settings app.

Figure 1-11   The Tip Calculator app running in an emulator

Perspective

In this chapter, you learned some background information about Android. In addition, you learned how to use Android Studio to open and run existing Android Studio projects. With that as background, you're ready to learn how to create your first Android app.

Terms

Android	Java source code
application	Java bytecodes
app	Dalvik executable files
Open Handset Alliance	Android package
Android Open Source Project (AOSP)	Android manifest
Software Development Kit (SDK)	signed app
Android stack	emulator
Dalvik virtual machine (VM)	Android Virtual Device (AVD)
Java virtual machine (JVM)	Android Debug Bridge (ADB)
Integrated Development Environment (IDE)	activity
	layout
project	tool window
building	soft keyboard

Summary

- *Android* is a Linux-based operating system designed primarily for touch-screen mobile devices such as smartphones and tablet computers. It was first released in 2008. Android code is open-source.

- *Applications*, more commonly referred to as *apps*, extend the functionality of Android devices.

- Android system architecture, known as the *Android stack*, consists of four layers: Linux, native libraries, the application framework, and Android apps.

- An Android app is typically developed using an *IDE (Integrated Development Environment)* like Android Studio, using Java as the programming language.

- Android apps can be run on a physical Android device or on an *emulator*, also called an *Android Virtual Device (AVD)*.

- An Android project must be built before it is run, compiling the *Java source code* (.java files) into *Java bytecodes* (.class files) and then into *Dalvik executable files* (.dex files).

- All of the files for an Android project are put into an *Android package* (.apk file), which includes a binary version of the AndroidManifest.xml file.

- To run an app on an emulator or device, it must be digitally *signed* with a certificate.

- The *Android Debug Bridge (ADB)* lets your IDE communicate with an emulator or a physical Android device.

- Android Studio is the official IDE for Android development. It's open-source, available for free, and runs on all modern operating systems.

- A *project* is a directory that contains all of the files for an app.

- In Android development, an *activity* defines one screen of an app.

- In Android development, a *layout* contains XML that defines the user interface.

- In Android Studio, the different parts of the main window are known as *tool windows*.

- A *soft keyboard* is an on-screen keyboard that you can use to enter text on touchscreen devices and emulators.

Before you do the exercises for this chapter

Before you do any of the exercises in this book, you need to install the JDK for Java SE, the Android SDK, and Android Studio. In addition, you need to install the source code for this book. This may take several hours, so plan accordingly. See the appendixes for details.

Exercise 1-1   Open an existing project and run it

Open a project and review its code

1. Start Android Studio. Then, open the project that's stored in this directory:
   `\murach\android\book_apps\ch03_TipCalculator`

2. Open the layout for the activity. This should open the layout in the graphical editor. If it doesn't, click the Design tab to display the graphical editor. Note the widgets displayed in this editor.

3. Click on the Text tab to view the XML for this layout. Note that the XML corresponds with the widgets displayed in the graphical editor.

4. Open the Java class for the activity. Review this code. Note how it works with the widgets displayed in the corresponding layout.

Run the app on a device

5. In the toolbar, click the Run button. This should display the Choose Device dialog. If it doesn't, edit your run configuration so it displays this dialog.

6. Connect the device you configured in the appendix. This should display the device in the Choose Device dialog, though it may indicate that the device is unauthorized.

7. Unlock the screen on the device. If you get a dialog that indicates that the device is unauthorized for the computer, use the dialog to authorize the device.

8. Use the Choose Device dialog to select the device.

9. Test the Tip Calculator app by using the soft keyboard to enter a bill amount and by clicking on the Increase (+) and Decrease (-) buttons to modify the tip percent.

Run the app on an emulator (optional)

10. Start the emulator. To do that, click on the AVD Manager button in the toolbar, and click on the Run button for the emulator that you created in the appendix. Depending on your system, it may take a long time for the emulator to launch, so be patient! After loading, you need to unlock the emulator screen by dragging the lock icon to the right or up.

11. In the Android Studio toolbar, click the Run button.

12. Use the Choose Device dialog to select the emulator. This should run the app on that emulator.

13. Test the Tip Calculator app by using the soft keyboard to enter a bill amount and by clicking on the Increase (+) and Decrease (-) buttons to modify the tip percent. If the soft keyboard isn't displayed, you can change the Hardware (Physical Keyboard) option. To do that, start the Settings app, click the Language and Input option, scroll down to the Keyboard and Input category, click the Default option, and turn off the Hardware (Physical Keyboard) option.

14. In the emulator, click the Home button to navigate away from the app.

15. Run the Tip Calculator app on the emulator again. This time, the app should run more quickly since the emulator is already running.

2

How to start your first Android app

In the previous chapter, you learned how to open and run an existing project for an Android app. In this chapter, you'll learn how to start developing such an app from scratch. First, this chapter shows how to use Android Studio to start a new project for an app. Then, it shows how to use Android Studio to develop the user interface for the app.

The Tip Calculator app

This chapter begins by reviewing the user interface of the Tip Calculator app. Then, it shows how to use Android Studio to create a new project for this app.

The user interface

Figure 2-1 shows the user interface for the Tip Calculator app. In Android development, the components that make up the user interface are known as *widgets*. Widgets can also be referred to as *controls*. The Tip Calculator contains ten widgets: seven TextView widgets, one EditText widget, and two Button widgets.

A TextView widget can be referred to as a *text view* or a *label*. It especially makes sense to refer to a TextView as a label when that widget provides text that labels another widget. For example, the TextView widget that displays "Bill Amount" on the left side of the Tip Calculator labels the EditText widget on the right side of the user interface. Similarly, the TextView widget that displays "Percent" labels the TextView widget that displays the value for the tip percent.

An EditText widget can be referred to as an *editable text view* or a *text box*. When the user touches the editable text view for the Tip Calculator, the user interface automatically displays a soft keyboard that allows the user to enter a total for the bill. After the user enters a total for the bill and touches the Done key, the app automatically calculates the tip.

The two Button widgets in the Tip Calculator allow the user to increase or decrease the tip percent. This automatically recalculates the tip.

The user interface for the Tip Calculator

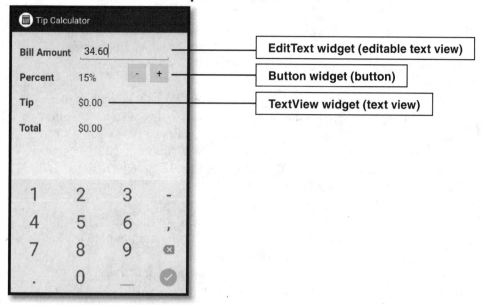

EditText widget (editable text view)

Button widget (button)

TextView widget (text view)

Description

- The user interface for the Tip Calculator contains ten *widgets*: seven TextView widgets, one EditText widget, and two Button widgets.

- A TextView widget can be referred to as a *text view* or a *label*.

- An EditText widget can be referred to an *editable text view* or a *text box*.

- When the user clicks on the editable text view for the bill amount, the user interface automatically displays a soft keyboard that allows the user to enter a total for the bill. After the user enters a total for the bill and touches the Done key (the check mark key), the app automatically calculates the tip.

- The user can click on the Increase (+) and Decrease (–) buttons to increase or decrease the default tip percent. This automatically recalculates the tip.

- A widget can be referred to as a *control*.

Figure 2-1   The user interface for the Tip Calculator app

How to create a new project

If Android Studio displays the Welcome page when you start it, you can create a new project for an app by selecting the "Start a new Android Studio project" item. Otherwise, you can select the File→New→New Project item from the menu system.

Either way, Android Studio displays a Create New Project dialog box like the one shown in figure 2-2. This dialog box displays a series of steps that allow you to create a new project for an app like the Tip Calculator.

In the first step, you can enter a name for the application as well as a name for the project and package. For the Tip Calculator app, I entered a name of "TipCalculator", and a company domain of "murach.com". This generated a package name of "com.murach.tipcalculator".

In the second step, you can select a minimum SDK for the application. For the Tip Calculator app, I accepted the default value for the minimum SDK. In this case, that happened to be Android 4.0.3 (API 15).

In the third step, you can select a template for the type of activity you want the project to start with. In Android development, an *activity* defines a screen. For the Tip Calculator app, I selected the Empty Activity template. This is the simplest template, and it's usually what you want when you're getting started.

In the fourth step, you can enter a name for the activity. For the Tip Calculator app, I entered a name of "TipCalculatorActivity" for the activity. This is the name of the Java class for the activity. When I did that, Android Studio automatically set the name of the corresponding layout to "activity_tip_calculator". This is the name of the file that stores the XML that defines the user interface for the activity. In Android development, a *layout* is a container that contains one or more child elements such as widgets and determines how they are displayed.

When you finish all four steps, you can click the Finish button. When you do, Android Studio creates a directory that corresponds with the project name, and it creates some additional directories and files that it uses to configure the project.

The New Android Application dialog box

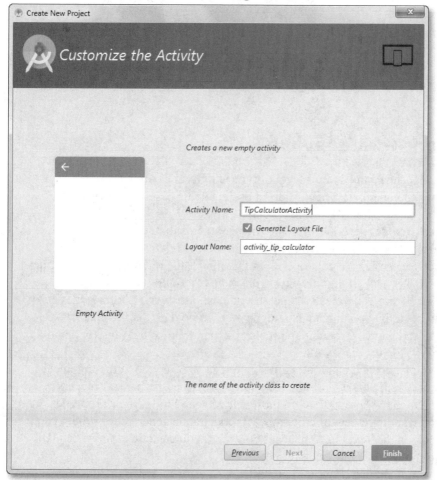

How to create a new project

1. Start Android Studio.
2. If the Welcome page is displayed, select the "Start a new Android Studio project" item. Otherwise, select the File→New→New Project item from the menu system.
3. Enter a name for the application, project, and package and click the Next button.
4. Select minimum SDK for your project and click the Next button.
5. Select the Empty Activity template and click the Next button.
6. Enter a name for the activity and click the Finish button.

Figure 2-2   How to create a new project

How to develop the user interface

The first step in developing an Android app is to develop the user interface. The easiest way to do this is to use the graphical layout editor that's provided by Android Studio to develop most of the user interface. This generates most of the XML for the user interface. Then, if necessary, you can review and modify this code.

How to work with a layout

Figure 2-3 shows the default layout of the activity that Android Studio generates when you create a new Android project that's based on the Blank Activity template as described in the previous figure.

Android provides several different types of layouts that you'll learn about later in this book. For now, you'll learn how to use the relative layout to create the interface for the Tip Calculator app. In this figure, the relative layout contains a single TextView widget that displays a message of "Hello world!"

Since a new project typically generates at least one layout file for you, you can usually begin by opening an existing layout. To do that, display the Project window, expand the res (resources) directory, expand the layout directory, and open the XML file for the layout.

Once you've opened a layout, you can view its appearance in the graphical layout editor. Then, you can use the techniques described in the next few figures to add widgets to the layout, to align those widgets, and to set their properties. Then, if you want to review or edit the XML code that's generated for the layout, you can click on the Text tab. Whenever you want, you can click the Design tab to return to the graphical layout editor.

When you first open a layout, you may get an error message that indicates that the graphical editor is having problems rendering the layout. In that case, you can often solve the problem by clicking the Refresh button that's displayed in the toolbar above the graphical editor.

If you want to add layouts to a project, you can do that by right-clicking on the layout directory in the Project window and selecting the New→Layout Resource File item. Then, you can enter the name of the file and the type of layout. For now, you can accept the default layout.

Conversely, if you want to delete a layout from a project, you can do that by right-clicking on the layout in the Project window and selecting the Delete item.

The default layout for an activity in a new project

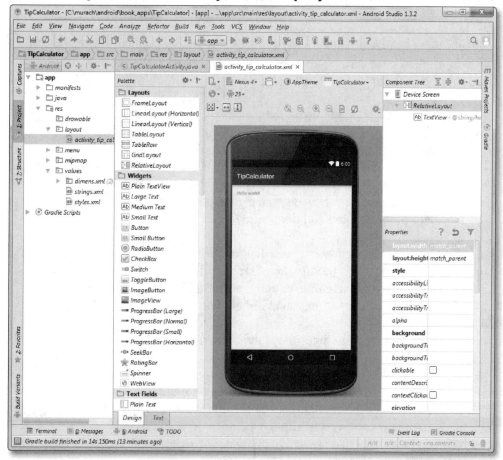

Description

- To open an existing layout, open the Project window, expand the res (resources) directory, expand the layout directory, and double-click the name of the XML file.

- To view the layout in the graphical editor, click the Design tab.

- To view the XML for the layout, click on the Text tab.

- If you get an error message that indicates that the graphical editor is having problems rendering the layout, click the Refresh button that's displayed in the toolbar above the graphical layout.

- To create a new layout, right-click on the layout directory in the Project window and select the New→Layout Resource File item. Then, enter the name of the file and the type of layout. For now, you can accept the default layout.

- To delete a layout, right-click on the layout in the Project window and select the Delete item.

Figure 2-3   How to work with a layout

How to add widgets to a layout

Figure 2-4 shows how to add widgets to a layout. To do that, you just drag a widget from the Palette onto the layout. This figure, for example, shows the ten widgets for the Tip Calculator after they have been added to the layout but before all of their properties have been set correctly.

As you add widgets to a layout, you often need to scroll through the categories in the Palette to help you find the widget you want to add. If necessary, you can click on the category for the widget to collapse or expand it. The Widgets category contains a variety of useful widgets including the TextView and Button widgets. The Text Fields category, on the other hand, only contains variations of the EditText widget. For this figure, I used the widget named Number (Decimal).

When you use the relative layout, it's a good practice to set the id property for the widget as soon as you add it to the layout to identify each widget. In this figure, for example, I set the id property of the last TextView widget to totalTextView. This ID clearly indicates that this widget is a TextView widget for the total amount. If you modify the id property for a widget after you have aligned other widgets relative to that widget, Android Studio needs to update all references to the id property that you modified. Fortunately, it can do that for you automatically.

By default, Android Studio displays the Properties window on the right side of the main window. If it isn't displayed there, you can display it by clicking on the Properties tab that's on the right side of the window.

In most cases, you can align a widget by dragging it to where you want it. This sets properties of the widget that begin with "layout:align" such as the layout:alignTop property, the layout:alignLeft property, and so on.

As you add widgets to a layout, you may need to set its *theme*, which is a group of styles that define the look of the layout. To set the theme, select it from the toolbar that's displayed above the layout when it's in the graphical editor. In this figure, the AppTheme has been selected, which is usually what you want.

A layout after some widgets have been added to it

Description

- To add a widget, drag the widget from the Palette onto the layout. TextView and Button widgets are in the Widgets category, and EditText widgets are in the Text Fields category.

- When you use the relative layout, it's a good practice to set the id property for the widget as soon as you add it to the layout. To do that, you can click on the widget and use the Properties window to set the id property.

- By default, the Properties window is displayed below the Component Tree window. If it isn't, you can display it by clicking on the Component Tree tab on the right side of the main window.

- To align a widget, you can drag it to where you want it to be aligned. This sets properties of the widget that begin with "layout:" such as the layout:alignTop and layout:alignLeft properties.

- To delete a widget, click on the widget in the graphical editor or the Component Tree window and press the Delete key.

- A *theme* is a group of styles that define the look of the layout. To set the theme for the layout, select the theme from the toolbar.

Figure 2-4   How to add widgets to a layout

How to set the display text

The Android development environment encourages separating the display text from the rest of the user interface. The advantage of this approach is that it's easier to create international apps that work in multiple languages. The disadvantage is that it takes a little extra work to set the display text for the user interface.

However, figure 2-5 shows a technique that makes it easy to set the display text, and this extra step shouldn't slow down your development much once you get used to it. In this figure, for example, I set the text property for the first TextView widget to a string named bill_amount_label that's stored in the strings.xml file. This is indicated by the value for the text property which is set to:

`@string/bill_amount_label`

This string contains a value of "Bill Amount", and this value is displayed in the layout when it's in the graphical editor.

A layout after the text has been set for the widgets

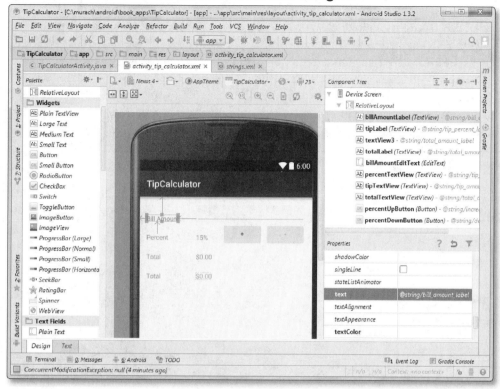

How to set the text property for a widget

1. In the graphical editor for the layout, select the widget.
2. In the Properties window, click the button (…) to the right of the text property.
3. In the Resources dialog box, click the New Resource button and select the New String Value item. Or, if the string value is already available, select the string and click the OK button.
4. In the New String Value Resource dialog box, enter a name and value for the string and click the OK button.

Description

- It's generally considered a best practice to store the display text for your app in a separate XML file. This makes it easy to create international apps that work in multiple languages.

Figure 2-5   How to set the display text for a widget (part 1 of 2)

After you click the button (…) to the right of the text property, Android Studio displays the Resources dialog box. If the string you want to display is available from this dialog box, you can select it and click the OK button. This sets the display text to the selected string.

However, if you need to enter a new string, you can click the New Resource button and select the New String Value item to display the New String Value Resource dialog box. Then, you can enter the string you want to display and a name for the string. In this figure, for example, I entered a name of bill_amount_label and a string of "Bill Amount".

The Resources dialog box

The New String Value Resource dialog box

Figure 2-5   How to set the display text for a widget (part 2 of 2)

How to work with resource files

When you use the technique shown in the previous figure to set the display text for your app, Android Studio stores the text in a strings.xml file like the one that's shown in figure 2-6. If you want, you can open the strings.xml file and edit it just as you would edit any other XML file.

This figure shows the strings.xml file for the Tip Calculator app. Here, the first string is set to "Tip Calculator". This string is displayed in the taskbar for the main activity of the app. In addition, it's used in the manifest file to identify the name of the app.

The last ten strings are used for the widgets. The strings that provide labels for widgets have names that end with "_label". Other strings provide default values for the bill amount, tip percent, tip amount, and total. These default values make it easier to use the graphical editor to set the properties for the widgets that display these values. In addition, they can make it easier to test an app. For example, when the app starts, it displays a value of 34.60 in the editable text view for the bill amount. As a result, you don't have to enter a bill amount to test the app.

However, before you put the app into production, you can modify the app so it sets these values to appropriate starting values for the app. For example, you probably want to supply an empty string for the bill amount. That way, when the app starts, the editable text view for the bill amount will be blank.

If you want to supply a strings.xml file for another country, you can create a values-*xx* directory within the res directory where *xx* is the ISO two-letter code for the country. For example, *fr* is the two-letter country code for France. Then, you can supply a strings.xml file within that directory that uses French instead of English. To find a list of these two-letter codes, you can search the Internet for "ISO two-letter country code".

When you use Android Studio to create a project, it stores some dimensions in a dimens.xml file like the one shown in this figure. If you want, you can open this file and edit it.

This figure shows the dimens.xml file for the Tip Calculator app. By default, Android Studio uses these dimensions to set the amount of space between the edge of the screen and the widgets contained by the layout.

Here, both dimensions are set to 16 *density-independent pixels* (*dp*). A density-independent pixel is a virtual pixel that you should use when defining a layout. Then, at runtime, Android scales these virtual pixels to make sure it displays your layout properly for the density of the screen of the device that your app is running on.

The location of the strings.xml file

```
res/values/strings.xml
```

The strings.xml file for the Tip Calculator app

```xml
<resources>
    <string name="app_name">Tip Calculator</string>

    <string name="bill_amount_label">Bill Amount</string>
    <string name="bill_amount">34.60</string>

    <string name="tip_percent_label">Percent</string>
    <string name="tip_percent">15%</string>
    <string name="increase">+</string>
    <string name="decrease">-</string>

    <string name="tip_amount_label">Tip</string>
    <string name="tip_amount">$0.00</string>

    <string name="total_amount_label">Total</string>
    <string name="total_amount">$0.00</string>
</resources>
```

The location of a strings.xml file for France

```
res/values-fr/strings.xml
```

The location of the dimens.xml file

```
res/values/dimens.xml
```

The dimens.xml file for the Tip Calculator app

```xml
<resources>
    <!-- Default screen margins, per the Android Design guidelines. -->
    <dimen name="activity_horizontal_margin">16dp</dimen>
    <dimen name="activity_vertical_margin">16dp</dimen>
</resources>
```

Description

- When you use the technique shown in the previous figure to set the display text for your app, Android Studio stores the text in the strings.xml file.

- If you want, you can open the strings.xml file and edit it just as you would edit any XML file.

- To provide a strings.xml file for a country that uses another language, you can create a values-*xx* directory within the res directory to store the strings.xml file. Here, you can replace *xx* with the ISO two-letter country code.

Figure 2-6   How to work with resource files

How to set properties

Figure 2-7 shows how to use the Properties window to set the properties of a widget. To do that, select the widget and use the Properties window to change the property. After you change a property from its default value, the Properties window displays the property in blue. In this figure, for example, the Properties window shows that the textStyle property has been set to bold.

A layout after some properties of the widgets have been set

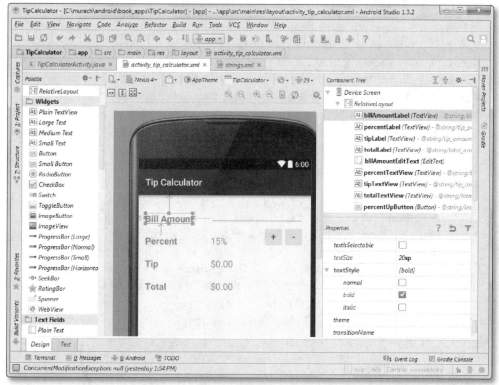

Description

- To set a property for a widget, select the widget and use the Properties window to change the property.

- The Properties window displays the properties that have been changed from their default in blue.

Figure 2-7   How to set properties

Common properties

Figure 2-8 shows some of the properties that are used by the Tip Calculator app. These properties are commonly used by most apps.

For a layout, the width and height properties are typically set to a pre-defined value of "match_parent". This expands the layout so it takes up the entire parent minus padding. In most cases, this causes the layout to take up the entire screen.

For the Tip Calculator app, the padding properties of the layout begin with a prefix of @dimen/. This indicates that you are specifying a dimension value that's stored in the dimens.xml file. The dimension named activity_vertical_margin is set to a value of "16dp". As a result, the layout includes 16 density-independent pixels (dp) of space between the edge of the screen and the widgets contained by the layout. Although Android supports other units of measurement such as inches, it's a best practice to use density-independent pixels whenever possible.

When you specify the id property for a widget, you begin the value of the id property with a prefix of @+id/. This indicates that you are specifying an ID value for the widget. Then, you can specify the ID for the widget. In this figure, the id property shows how to set the ID for a widget to billAmountLabel.

For a widget, the width and height properties are typically set to a predefined value of "wrap_content". This forces the width and height to be just big enough to contain the widget and its content. However, if you want, you can use density-independent pixels (dp) to specify the height or width.

For the text property, you begin with a prefix of @string/. This indicates that you are specifying a value that's stored in the strings.xml file. In this figure, the text property shows how to set the display text for the widget to the string in the strings.xml file that's named bill_amount_label.

For the textSize property, it's generally considered a best practice to use *scale-independent pixels* (*sp*) whenever possible. In this figure, the textSize property specifies a value of "20sp". Scale-independent pixels (sp) work much like density-independent pixels. However, scale-independent pixels provide a way for Android to adjust the size of the text so it scales correctly for different screen densities.

The last two properties are typically used with an EditText widget. Here, the ems property sets the width of the EditText widget so it's wide enough to contain the specified number of the letter *m*.

The inputType property sets the type of input that the EditText widget should accept. In this figure, the inputType property specifies a value of "numberDecimal". This indicates that the EditText widget should only accept numeric characters and the decimal point character. Of course, an EditText widget can accept many other types of input too such as plain text, email addresses, telephone numbers, passwords, and so on.

To specify a density-independent pixel, it's possible to use an abbreviation of dip instead of dp. For example, you could use "10dip" instead of "10dp". In this book, I use dp for two reasons. First, it's shorter. Second, it's more consistent with the sp abbreviation that's used for scale-independent pixels.

Common properties for layouts

Property	Example settings
width	match_parent
height	match_parent
paddingTop	@dimen/activity_vertical_margin
paddingLeft	@dimen/activity_horizontal_margin

Common properties for widgets

Property	Example settings
id	@+id/billAmountLabel
width	wrap_content
height	wrap_content
layout:alignParentLeft	true
layout:alignParentStart	true
layout:alignParentTop	true
layout:padding	10dp
text	@string/bill_amount_label
textSize	20sp
textStyle	bold
layout:marginLeft	130dp
ems	8
inputType	numberDecimal

Pre-defined values for setting height and width

Value	Description
wrap_content	Wraps the height or width so it is large enough to display the widget.
match_parent	Stretches the height or width to match the parent container.

Common units of measurement

Name	Abbreviation	Typical uses
Density-independent pixels	dp or dip	margins, padding, etc.
Scale-independent pixels	sp	font sizes

Description

- For the text property, you begin with a prefix of @string/ to indicate that you are specifying a value that's stored in the strings.xml file.
- Although Android supports other units of measurement, such as inches, it's a best practice to use density-independent and scale-independent pixels whenever possible.

Figure 2-8   Common properties

The XML for the user interface

Figure 2-9 shows the XML for the user interface. This code is stored in the file named activity_tip_calculator.xml file. Most of this code is generated when you use the graphical editor to work with the widgets of the user interface. However, you can edit the XML for the user interface whenever you want. For example, you may want to do this to make minor adjustments that are difficult to make in the graphical layout editor.

To start, the RelativeLayout element defines a relative layout that contains the widgets for the user interface. Here, the layout_width and layout_height attributes use a value of match_parent to specify that the layout should use the whole screen. And the padding attributes use the values stored in the dimens.xml file to specify the amount of space between the edge of the layout and its child elements.

The XML for this user interface includes comments to identify each row of widgets. These comments use the same syntax as an HTML comment (start with <!-- and end with -->).

The first TextView element sets the layout_width and layout_height attributes to a value of wrap_content to specify that the widget should be just wide enough and tall enough to fit its content. The text attribute sets the text that's displayed on the widget to the value that's stored in the bill_amount string of the strings.xml file. The id attribute defines an ID of billAmountLabel. The textSize attribute sets the size of the text to 20 scale-independent pixels (20sp). The textStyle attribute sets the style of the text to bold. And the paddingTop attribute sets the padding at the top of the attribute to 10 density-independent pixels (10dp).

The EditText element displays the widget that allows the user to enter a bill amount. By now, you should understand how most of the attributes of the EditText element work. Within a relative layout, the alignment attributes align the editable text view relative to other widgets. In this case, the alignment attributes align the editable text view to the right of the TextView element for the bill amount label with a margin of 16dp between the two elements. Also, the ems attribute specifies that the editable text view should be wide enough to fit 8 ems (8 of the letter *m*). And the inputType element specifies that the soft keyboard should only allow the user to enter decimal numbers.

The next two TextView elements define the widgets that display the tip percent. The Button elements define the buttons that allow the user to increase or decrease the tip percent.

The last four TextView elements define the widgets that display the tip and total amounts.

The XML for the user interface

Page 1

```
<RelativeLayout
    xmlns:android="http://schemas.android.com/apk/res/android"
    xmlns:tools="http://schemas.android.com/tools"
    android:layout_width="match_parent"
    android:layout_height="match_parent"
    android:paddingLeft="@dimen/activity_horizontal_margin"
    android:paddingRight="@dimen/activity_horizontal_margin"
    android:paddingTop="@dimen/activity_vertical_margin"
    android:paddingBottom="@dimen/activity_vertical_margin"
    tools:context=".TipCalculatorActivity">

    <!-- The bill amount -->

    <TextView
        android:layout_width="wrap_content"
        android:layout_height="wrap_content"
        android:text="@string/bill_amount_label"
        android:id="@+id/billAmountLabel"
        android:textSize="20sp"
        android:textStyle="bold"
        android:layout_alignParentTop="true"
        android:layout_alignParentLeft="true"
        android:layout_alignParentStart="true"
        android:paddingTop="10dp" />

    <EditText
        android:layout_width="wrap_content"
        android:layout_height="wrap_content"
        android:inputType="numberDecimal"
        android:ems="8"
        android:id="@+id/billAmountEditText"
        android:text="@string/bill_amount"
        android:textSize="20sp"
        android:layout_alignTop="@+id/billAmountLabel"
        android:layout_toRightOf="@+id/billAmountLabel"
        android:layout_toEndOf="@+id/billAmountLabel"
        android:layout_marginLeft="16dp" />
```

Figure 2-9   The XML for the user interface (part 1 of 3)

The XML for the user interface

```xml
<!-- The tip percent -->

<TextView
    android:layout_width="wrap_content"
    android:layout_height="wrap_content"
    android:text="@string/tip_percent_label"
    android:id="@+id/percentLabel"
    android:textSize="20sp"
    android:textStyle="bold"
    android:layout_alignBottom="@+id/percentDownButton"
    android:layout_alignParentLeft="true"
    android:layout_alignParentStart="true" />

<TextView
    android:layout_width="wrap_content"
    android:layout_height="wrap_content"
    android:text="@string/tip_percent"
    android:id="@+id/percentTextView"
    android:layout_alignTop="@+id/percentLabel"
    android:layout_alignLeft="@+id/billAmountEditText"
    android:layout_alignStart="@+id/billAmountEditText"
    android:textSize="20sp" />

<Button
    android:layout_width="45dp"
    android:layout_height="45dp"
    android:text="@string/decrease"
    android:id="@+id/percentDownButton"
    android:layout_below="@+id/billAmountEditText"
    android:layout_toLeftOf="@+id/percentUpButton"
    android:layout_toStartOf="@+id/percentUpButton"
    android:textSize="20sp" />

<Button
    android:layout_width="45dp"
    android:layout_height="45dp"
    android:text="@string/increase"
    android:id="@+id/percentUpButton"
    android:layout_alignTop="@+id/percentDownButton"
    android:layout_alignRight="@+id/billAmountEditText"
    android:layout_alignEnd="@+id/billAmountEditText"
    android:textSize="20sp" />
```

Figure 2-9   The XML for the user interface (part 2 of 3)

The XML for the user interface

```xml
<!-- The tip amount -->

<TextView
    android:layout_width="wrap_content"
    android:layout_height="wrap_content"
    android:text="@string/tip_amount_label"
    android:id="@+id/tipLabel"
    android:textSize="20sp"
    android:textStyle="bold"
    android:layout_marginTop="23dp"
    android:layout_below="@+id/percentDownButton"
    android:layout_alignParentLeft="true"
    android:layout_alignParentStart="true" />

<TextView
    android:layout_width="wrap_content"
    android:layout_height="wrap_content"
    android:text="@string/tip_amount"
    android:id="@+id/tipTextView"
    android:layout_alignTop="@+id/tipLabel"
    android:layout_alignLeft="@+id/percentTextView"
    android:layout_alignStart="@+id/percentTextView"
    android:textSize="20sp" />

<!-- The total amount -->

<TextView
    android:layout_width="wrap_content"
    android:layout_height="wrap_content"
    android:text="@string/total_amount_label"
    android:id="@+id/totalLabel"
    android:textSize="20sp"
    android:textStyle="bold"
    android:layout_below="@+id/tipLabel"
    android:layout_alignParentLeft="true"
    android:layout_alignParentStart="true"
    android:layout_marginTop="24dp" />

<TextView
    android:layout_width="wrap_content"
    android:layout_height="wrap_content"
    android:text="@string/total_amount"
    android:id="@+id/totalTextView"
    android:textSize="20sp"
    android:layout_alignTop="@+id/totalLabel"
    android:layout_alignLeft="@+id/tipTextView"
    android:layout_alignStart="@+id/tipTextView" />


</RelativeLayout>
```

Figure 2-9   The XML for the user interface (part 3 of 3)

Perspective

In this chapter, you learned how to use Android Studio to develop the user interface for an Android app. However, if you run this app, it displays the user interface, but it doesn't do anything else. That's because you still need to write the Java code that adds the functionality to the user interface. In the next chapter, you'll learn how to do that.

Terms

widgets	activity
controls	layout
text view	theme
label	density-independent pixels (dp)
editable text view	scale-independent pixels (sp)
text box	

Summary

- In Android development, the components that make up the user interface are known as *widgets*.
- In Android, an *activity* stores the code for a screen of an app.
- In Android, a *layout* stores the XML that defines the user interface of an app.
- In Android Studio, you can use the graphical editor to add widgets to a layout by dragging them from the Palette window onto the layout.
- It's generally considered a best practice to store the display text for your app in a separate XML file. This makes it easy to internationalize the text for your application.
- Although Android supports other units of measurement, such as inches, it's a best practice to use *density-independent* and *scale-independent pixels* whenever possible.

Exercise 2-1   Develop a Tester app

Create the app

1. Start Android Studio.
2. Create a project for an Android app named Tester and store it in this directory:

    `\murach\android\ex_starts`

    This project should be stored in a package named com.murach.tester, and it should be based on the Empty Activity template.
3. If you have an Android device, run the app on that device. Otherwise, run it in an emulator. This should display a message that says "Hello world!" in the center of the screen.

Modify the app

4. Open the layout for the activity. If necessary, click on the Design tab to display the layout in the graphical editor.

5. Drag the TextView widget to the top left corner of the screen.

6. Change the text size of the TextView widget to 40sp.

7. Change the text of the TextView widget from "Hello World!" to "Success!". To do that, you may want to modify the strings.xml file.

8. Click the Text tab to view the XML for the user interface.

9. Modify the XML for the layout so the text size of the TextView widget is 80sp.

10. Click the Design tab to switch back to the graphical editor. Note that the graphical editor shows the changes that were made to the XML for the layout.

11. If you have an Android device, run the app on your device. Otherwise, run the app in an emulator. This should display a message that says "Success!" at the top left corner of the screen.

Exercise 2-2   Create the user interface for the Invoice Total app

In this exercise, you'll create the Invoice Total app. When you're done, a test run should look something like this:

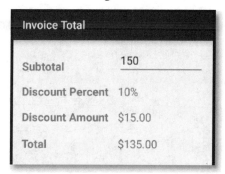

Create the project

1. Start Android Studio.

2. Create a new project for an Android app and store it in a project named Invoice in this directory:

   `\murach\android\ex_starts`

   This project should be stored in a package named com.murach.invoice, and it should be based on the Empty Activity template.

Create the user interface

3. Open the layout for the activity that's stored in the res\layout directory. If necessary, click the Design tab to display the graphical editor.

4. Delete the TextView widget that displays the "Hello world!" message.

5. Add the seven TextView widgets and one EditText widget to the layout. Set the id and text properties of each widget. When you're done, the user interface should have the widgets and text shown above. However, these widgets should look different since you haven't set their properties yet.

6. Set the textSize property for all eight widgets to 20sp.

7. Set the textStyle property for the four widgets in the left column to bold.

8. Click the Text tab to review the XML for this layout. Note how the XML attributes correspond with the properties that you have set with the graphical editor.

9. Open the strings.xml file that's in the res/values directory. Then, review the code.

10. Change the value of the string named app_name to "Invoice Total".

11. Test the user interface by running the app on a physical device. Or, if you don't have a physical device, run the app on an emulator. The app should allow you to enter a subtotal, but that shouldn't cause the discount percent, discount amount, or total values to change.

3

How to finish your first Android app

In the previous chapter, you learned how to develop the XML for the user interface of an app. Now, this chapter shows how to write the Java code that adds the functionality to this user interface. In addition, it shows some more skills that you can use to put the final touches on an app.

How to write the Java code

Once you have the user interface working the way you want, you can write the Java code for the app. This is the code that provides the functionality for the app.

How to work with an activity

When you create an Android project based on the Empty Activity template, Android Studio automatically creates a class for an activity like the one shown in figure 3-1. This class inherits the AppCompatActivity class, which in turn inherits the Activity class. Then, this class overrides the onCreate method of the AppCompatActivity class. Android runs this method when the activity is started.

Within the onCreate method, the first statement passes the Bundle parameter to the onCreate method of the superclass (the Activity class). This Bundle parameter contains the activity's previously saved state if it exists. Then, the second statement displays the user interface for the activity. To do that, it calls the setContentView method of the superclass. Within the parentheses for this method, the code specifies the resource for the layout. To do that, this code uses the R (resources) class.

The R class is created automatically when you build your project, and it provides a way to access the compiled resources for your project. In the onCreate method, for example, this code:

```
R.layout.activity_tip_calculator
```

refers to a compiled version of this resource:

```
res\layout\activity_tip_calculator.xml
```

By default, the activity class includes an onCreateOptionsMenu method that displays an options menu, and an onOptionsItemSelected method that provides some starting code for working with this menu. If your app doesn't need an options menu, you can delete both of these methods.

If you're creating an app like Tip Calculator that only uses a single activity, you can usually modify the generated class to get it to work the way you want. However, if you need to add an additional screen to your app, you need to add a new activity. To do that, you can right-click on a package directory in the Project window and select the New→Activity→Empty Activity item. Then, respond to the resulting dialog. This adds an XML file for the layout as well as a Java file that's similar to the one shown in this figure.

The default Java code for an activity

```
package com.murach.tipcalculator;

import android.support.v7.app.AppCompatActivity;
import android.os.Bundle;

public class TipCalculatorActivity extends AppCompatActivity {

    @Override
    protected void onCreate(Bundle savedInstanceState) {
        super.onCreate(savedInstanceState);
        setContentView(R.layout.activity_tip_calculator);
    }
}
```

Description

- When you create an Android project that's based on the Empty Activity template, it automatically creates a class for the first activity in your app. This class displays the user interface for the activity.

- To create a new activity, you can right-click a package in the Project window, select the New→Activity→Empty Activity item, and respond to the resulting dialog.

- The AppCompatActivity class is designed to make your app compatible with most versions of Android. This class inherits the Activity class, which is the base class for all activities.

- Android runs the onCreate method when the activity is started.

- The onCreate method usually passes the Bundle parameter to the onCreate method of the superclass and uses the setContentView method to specify the layout.

- The R class provides a way to access the compiled resources for your project. This class is created automatically when you build your project.

- The onCreateOptionsMenu and onOptionsItemSelect methods contain some starting code for the options menu.

- To remove the options menu from an app, delete these methods.

Figure 3-1   The default code for an activity

How to remove the v7 appcompat support library

For the code in the previous figure to work, your project must include the v7 appcompat library. This provides support for the action bar and the Material theme, which was introduced with Android 5.0 (API 21), all the way back to Android 2.1 (API 7). To provide this support, this library includes the AppCompatActivity class shown in the previous figure as well as the AppCompat theme, which supports most features of the Material theme.

However, there also are a couple drawbacks to using this support library. First, it makes your app larger and dependent on a support library. Second, it makes your code more complex. For example, to create a class for an activity, you inherit the AppCompatActivity class instead of just inheriting the Activity class.

Because of this, you can remove this library from your project whenever you decide that you don't need it. To do that, you can use a procedure like the one shown in figure 3-2. This removes a dependency from your project, and it allows you to use simpler code such as inheriting the Activity class and using the built-in Holo theme.

So, when would you want to remove the v7 appcompat support library? Well, if your project specifies a minimum API of 14 or higher and you want to use the built-in Holo theme, you don't need the v7 appcompat support library. Also, if you want to keep things simple while you're learning Android programming, it makes sense to remove this support library. As a result, we've removed this library from most of the apps described in this book.

And when would you want to keep the v7 appcompat support library? Well, if your project specifies a minimum API that's below API 14, or if you want to use the Material theme, you should keep the v7 appcompat support library. In that case, your app can support the action bar and use the Material theme on virtually all Android devices. For an example of an app that keeps this support library and uses the AppCompat (Material) theme, please see chapter 7.

The v7 appcompat support library

General description

- Provides support back to Android 2.1 (API 7).

Features

- Supports the action bar, which was introduced with Android 4.0 (API 14).
- Supports the Material theme, which was introduced with Android 5.0 (API 21).

The Gradle build script dependency

```
com.android.support:appcompat-v7:23.0.0
```

How to remove the v7 appcompat support library

1. Open the Gradle build script for the app. To do that, open the Project window, expand the Gradle Scripts directory, and double-click on the "build.gradle (Module: app)" item.

2. In the Gradle build script, remove the dependency for the v7 appcompat support library. For most projects, you can do that by deleting the following code from the end of the file:

```
dependencies {
    compile fileTree(dir: 'libs', include: ['*.jar'])
    compile 'com.android.support:appcompat-v7:23.0.0'
}
```

3. Open the class for the activity.

4. In the class for the activity, replace the import statement for the AppCompatActivity class with an import statement for the Activity class like this:

```
import android.app.Activity;
```

5. In the class for the activity, modify the class declaration so it extends the Activity class, not the AppCompatActivity class, like this:

```
public class TipCalculatorActivity extends Activity
```

6. Open the res/values/styles.xml file. Then, modify the style named AppTheme so it uses a Holo theme as the parent like this:

```
<style name="AppTheme" parent="android:Theme.Holo.Light.DarkActionBar">
    <!-- Customize your theme here. -->
</style>
```

Description

- If a project doesn't need the v7 appcompat support library, you can make the project smaller and simpler by removing it.

Figure 3-2   How to remove the v7 appcompat support library

How to get references to widgets

Figure 3-3 shows how to get references to the widgets on the user interface. This is the first step to adding functionality to a user interface.

To make it easy to work with widgets, you typically start by adding import statements that import the classes for the widgets. Most widgets are stored in the android.widgets package. In this figure, for example, the TextView and EditText classes are both stored in this package.

After you import the classes for the widgets, you typically declare the widgets as private instance variables of the activity. In this figure, for example, three widgets have been declared as private instance variables. Here, the first widget is an EditText widget, and the other two are TextView widgets.

In the onCreate method, you can use the findViewById method to get references to the widgets that are declared in the layout. To do that, you can use the R (resources) class to access compiled resources. In this figure, for example, the first findViewById method specifies an argument of:

```
R.id.billAmountEditText
```

This gets a reference to a compiled resource with an ID of billAmountEditText. If you look at the XML file for the user interface, you'll find that this ID corresponds with the ID of the EditText widget on the user interface. Similarly, the other two IDs specified in this figure correspond with the IDs of two of the TextView widgets.

The findViewById method returns a View object that can be cast to the correct type for each widget. In this figure, for instance, the first statement casts the View object for the bill amount to the EditText type. Then, the next two statements cast the View object that's returned to the TextView types. This works because the View class is the superclass for all widgets.

An activity that gets references to the widgets

```
package com.murach.tipcalculator;

import android.app.Activity;
import android.os.Bundle;
import android.widget.EditText;        // 1. Import the classes for the widgets
import android.widget.TextView;

public class TipCalculatorActivity extends Activity {

    // 2. Declare variables for the widgets
    private EditText billAmountEditText;
    private TextView tipTextView;
    private TextView totalTextView;

    @Override
    public void onCreate(Bundle savedInstanceState) {
        super.onCreate(savedInstanceState);
        setContentView(R.layout.activity_tip_calculator);

        // 3. Get references to the widgets
        billAmountEditText = (EditText) findViewById(R.id.billAmountEditText);
        tipTextView = (TextView) findViewById(R.id.tipTextView);
        totalTextView = (TextView) findViewById(R.id.totalTextView);
    }
}
```

Description

- You can add import statements to import the classes for the widgets from the android. widgets package.

- You can declare widgets as private instance variables of the activity.

- You can use the findViewById method to get references to the widgets that are declared in the layout.

- The findViewById method accepts an argument for the ID of the widget. The R class provides a way to get the ID for any of the compiled resources for your project.

- The findViewById method returns a View object for the widget with the specified ID. This View object can be cast to the correct type of the widget.

Figure 3-3   How to get references to the widgets

How to handle the EditorAction event

Now that you have references to some of the widgets on the user interface, you're ready to learn how to code an event handler. An *event handler* is a special type of method that's executed when an *event* occurs. Typically, an event occurs when a user interacts with widgets. In figure 3-4, for example, you'll learn how to handle the EditorAction event. This event typically occurs when a user interacts with the EditText widget to display a soft keyboard and presses an *action key* such as the Done key.

In general, there are three steps to handling an event. First, you need to import the interface that defines a *listener*, which is an object that listens for an event. In this figure, the sixth import statement imports the OnEditorActionListener interface.

Second, you need to create a listener. To do that, you can implement the listener interface. In this figure, the class declares that it implements the OnEditorActionListener interface, and then it implements the method that's defined by the interface: the onEditorAction method. This method contains the code that's executed when the listener detects that an *action event* occurred on the widget.

Third, you must connect, or *wire*, the listener to a widget. In this figure, the last statement in the onCreate method wires the EditText widget for the bill amount to the onEditorAction listener that's defined by the current class. To do that, this code calls the setOnEditorActionListener method from the EditText widget for the bill amount. Then, it uses the keyword named *this* to pass the object for the current class to that method.

When you're implementing interfaces, Android Studio can automatically generate the method or methods defined by the interface for you, which is usually what you want. To do that, you can code the implements keyword followed by the name of the interface. Then, you can click within the name of the interface, click the error icon that Android Studio displays to the left of this code, select the "Add methods" item, and respond to the resulting dialog. When you do, Android Studio generates the method or methods for you. If necessary, you can modify the names of the arguments. In this figure, for example, I changed the default names of the arguments to make them more readable.

Within the onEditorAction method, you can handle the events that occur on specific action keys such as the Done key by comparing the second parameter (the ID of the action key) with the constants of the EditorInfo class. In this figure, this method begins by checking whether the Done button was clicked or whether an unspecified action occurred such as an Enter key being pressed on an older device. If either condition is true, this code sets the tip to $10 and the amount to $110. If not, the tip and amount are left at their default values. Either way, this method returns a false value, which causes the soft keyboard to be hidden, which is usually what you want.

If you review the onEditorAction method, you'll notice that the first and third parameters aren't used in this code. That's because the first parameter is used to determine which widget triggered the event. For this code, there is

An activity that handles the EditorAction event

```
package com.murach.tipcalculator;

import android.app.Activity;
import android.os.Bundle;
import android.view.KeyEvent;
import android.view.inputmethod.EditorInfo;
import android.widget.EditText;
import android.widget.TextView;
import android.widget.TextView.OnEditorActionListener;   // 1. Import listener

public class TipCalculatorActivity extends Activity
implements OnEditorActionListener {   // 2a. Implement the listener

    private EditText billAmountEditText;
    private TextView tipTextView;
    private TextView totalTextView;

    @Override
    public void onCreate(Bundle savedInstanceState) {
        super.onCreate(savedInstanceState);
        setContentView(R.layout.activity_tip_calculator);

        billAmountEditText = (EditText) findViewById(R.id.billAmountEditText);
        tipTextView = (TextView) findViewById(R.id.tipTextView);
        totalTextView = (TextView) findViewById(R.id.totalTextView);

        // 3. Set the listener
        billAmountEditText.setOnEditorActionListener(this);
    }

    // 2b. Implement the listener
    @Override
    public boolean onEditorAction(TextView v, int actionId, KeyEvent event) {

        if (actionId == EditorInfo.IME_ACTION_DONE ||
            actionId == EditorInfo.IME_ACTION_UNSPECIFIED)
        {
            tipTextView.setText("$10.00");
            totalTextView.setText("$110.00");
        }
        return false;
    }
}
```

Description

- An *event handler* is a special type of method that's executed when an *event* occurs. Typically, an event occurs when a user interacts with widgets.

- A *listener* is an object that listens for an event. To create a listener, import the interface for the listener and implement that interface. After you create a listener, connect, or *wire*, the listener to an event that's available from a widget.

- The EditorAction event typically occurs when the user uses a soft keyboard to enter text into an editable text view.

Figure 3-4   How to handle the EditorAction event (part 1 of 2)

only one widget that has been wired to the event, so that widget is the only one that could possibly trigger the event. The third parameter, on the other hand, is used to determine whether the Enter key triggered the event. For now, this app assumes that the device is a touchscreen device that doesn't have a physical Enter key. As a result, this app doesn't handle the Enter key.

Part 2 of figure 3-4 shows a few of the constants that are available from the EditorInfo class. All of these constants begin with IME, which stands for input method editor. An *input method editor* (*IME*) is a user control such as a soft keyboard that enables users to input data, often text.

The first constant shown in this figure is used in part 1 of this figure to check whether the Done key was pressed. The next three constants can be used to check for the Go, Next, and Search keys. Although these keys aren't available from the soft keyboard for the Tip Calculator shown in this chapter, they are available from other types of soft keyboards. For example, the Go key is available from the soft keyboard for entering a URL. Finally, the last constant can be used to check for an unspecified action such as an Enter key that's available from an older device.

As you review this code, you should be aware that there are several possible ways to define a class that implements the interface for a listener. In this figure, I decided to use the class for the activity to implement this interface. That way, that class defines the activity, and it acts as the listener for the EditorAction event. However, in some cases, it makes sense to define a separate class for the listener. You'll learn more about how this works in chapter 6.

A few constants from the EditorInfo class

Constant	The action key performs...
IME_ACTION_DONE	A "done" operation, typically closing the soft keyboard.
IME_ACTION_GO	A "go" operation, typically taking the user to the target specified by the user.
IME_ACTION_NEXT	A "next" operation, typically taking the user to the next field that needs text input.
IME_ACTION_SEARCH	A "search" operation, typically searching for the text specified by the user.
IME_ACTION_UNSPECIFIED	No specified operation, typically an action key on an older device such as an "enter" key.

Description

- After you declare the interface for the listener, you can automatically generate the declaration for the method by clicking on the error icon and selecting the "Add unimplemented methods" item.

- An *input method editor* (*IME*) is a user control such as a soft keyboard that enables users to input data, often text.

- To determine which widget triggered the event, you can use the first argument of the onEditorAction method.

- To determine which *action key* triggered the event, you can use the second argument of the onEditorAction method. To do that, compare the second argument with the constants of the EditorInfo class.

- To determine whether the action was trigged by pressing the Enter key, you can use the third argument of the onEditorAction method.

- To hide the soft keyboard, you can return a false value from the onEditorAction method.

- To keep the soft keyboard displayed, you can return a true value from the onEditorAction method.

Figure 3-4   How to handle the EditorAction event (part 2 of 2)

How to get and set the text for widgets

Figure 3-5 shows how to get and set the text that's displayed on a widget. Typically, you need to get the text from widgets such as editable text views that allow the user to enter text. Conversely, you often need to set the text on widgets that display text such as labels. However, these techniques work for most types of widgets.

You can use the getText method to get the text that's displayed on a widget. To do that, you call the getText method from the widget. This returns an Editable object. In most cases, you want to call the toString method from the Editable object to convert it to a String object. In this figure, for instance, the first example gets the text from the editable text view for the bill amount and converts that text to a string.

You can use the setText method to set the text that's displayed on a widget. To do that, you call the setText method from the widget and you pass a string argument that contains the display text. In this figure, for instance, the second example sets the display text for the TextView widget to "Test 1".

The calculateAndDisplay method shown in this figure uses both the getText and setText methods. To start, the first statement gets the display text from the editable text view for the bill amount and converts that text to a string. Then, this code uses the parseFloat method of the Float class to convert that string to a float value. Since the editable text view for the bill amount only allows the user to enter an empty string or a string for a decimal number, this code doesn't include a try statement that handles the exception that occurs if the user enters non-numeric data. However, this code does use an if statement to convert an empty string to a zero.

After getting a float value for the bill amount, this code calculates the tip and total amount for the bill. Here, the code assumes that the tip is a float value of 15%, and it uses the float type for all of these variables. To specify the literal value for the tip percent, this code appends an f to the number to indicate that it is a float value.

Before this code can display the tip and total amounts, it must convert them to string values. To do that, this code uses the NumberFormat class. First, it calls the static getCurrencyInstance method of the NumberFormat class to return a NumberFormat object for currency formatting. Then, it uses the format method of the NumberFormat object to convert numeric values to strings that have currency formatting, and it displays these strings on the TextView widgets for the tip and total amounts.

Although the code in this figure shows how to work with float values, you can use a similar technique to work with double and integer values. For example, you can use the parseInt method of the Integer class to convert a string to an int value.

Although the code in this figure shows how to work with currency formatting, you can use a similar technique to work with percent formatting. In particular, you can use the static getPercentInstance method of the NumberFormat class to return a NumberFormat object that you can use to convert numbers to strings that have percent formatting.

Two methods for working with widgets

Method	Description
`getText()`	Returns an Editable object that represents the text that the widget is displaying. You can use the toString method to convert the Editable object to a string.
`setText(string)`	Sets the text that's displayed on the widget.

Examples

How to get text

```
String billAmountString = billAmountEditText.getText().toString();
```

How to set text

```
tipTextView.setText("Test 1");
```

A method that gets input, calculates amounts, and displays output

```
public void calculateAndDisplay() {

    // get the bill amount
    billAmountString = billAmountEditText.getText().toString();
    float billAmount;
    if (billAmountString.equals("")) {
        billAmount = 0;
    }
    else {
        billAmount = Float.parseFloat(billAmountString);
    }

    // calculate tip and total
    float tipPercent = .15f;
    float tipAmount = billAmount * tipPercent;
    float totalAmount = billAmount + tipAmount;

    // display the results with formatting
    NumberFormat currency = NumberFormat.getCurrencyInstance();
    tipTextView.setText(currency.format(tipAmount));
    totalTextView.setText(currency.format(totalAmount));
}
```

Description

- For most widgets, you can use the getText and setText methods to get and set the text that's displayed on the widget.

- If necessary, you can use the Integer, Double, and Float classes to convert the strings that are returned by widgets to int, double, and float values.

- If necessary, you can use the NumberFormat class to convert numeric values to string values that have currency or percent formatting so they can be displayed on widgets.

Figure 3-5   How to get and set the text for widgets

How to handle the Click event

Figure 3-6 shows how to handle the Click event. This event is available from most widgets, but it's almost always handled for buttons, which are designed to be clicked. The steps for handling the Click event are the same as the steps for handling the EditorAction event that you saw earlier in this chapter.

First, you import the interface for the listener. Often, Android Studio does this for you automatically when you declare the interface. As a result, you can often skip this step.

Second, you implement the interface by declaring it and by implementing any methods declared by the interface. In this figure, the class for the Tip Calculator activity declares that it implements the OnClickListener interface, and it includes the onClick method that's defined by this interface.

Third, you wire the listener to a widget. In this figure, the two statements wire the event handler shown in this figure to both of the buttons on the Tip Calculator app. To do that, this code calls the setOnClickListener method from each button and passes the object for the current class to that method.

Since the onClick method is wired to two buttons, the code in the onClick method includes a switch statement that checks which button was clicked and executes the appropriate code. To do that, the switch statement gets the ID of the widget by calling the getId method from the View object that's the first and only argument of the onClick method. Then, it compares this value to the IDs of the Button widgets. To get these IDs, this code uses the R class to access information about the compiled resources for the project. For instance, to get the ID for the button named percentDownButton, it uses this code:

```
R.id.percentDownButton
```

Earlier in this chapter, the onEditorAction method didn't need to include code like this because that event handler was only wired to a single widget.

Since a class can implement multiple interfaces, a class can handle multiple events. In this figure, for instance, step 2a shows that the class implements both the OnClickListener and OnEditorActionListener interfaces.

How to handle the Click event

Step 1: Import the interface for the listener

```
import android.view.View.OnClickListener;
```

Step 2a: Implement the interface for the listener

```
public class TipCalculatorActivity extends AppCompatActivity
implements OnEditorActionListener, OnClickListener {
```

Step 2b: Implement the interface for the listener

```
@Override
public void onClick(View v) {
    switch (v.getId()) {
        case R.id.percentDownButton:
        tipPercent = tipPercent - .01f;
        calculateAndDisplay();
        break;
        case R.id.percentUpButton:
        tipPercent = tipPercent + .01f;
        calculateAndDisplay();
        break;
    }
}
```

Step 3: Set the listeners

```
percentUpButton.setOnClickListener(this);
percentDownButton.setOnClickListener(this);
```

Description

- The steps for handling the Click event are the same as the steps for handling the EditorAction event. However, since the Click event is wired to two buttons, the code in the onClick method includes a switch statement that checks which button was clicked and executes the appropriate code.

- To specify a literal value for the float type, you can append an f or an F to the number.

Figure 3-6   How to handle the Click event

The lifecycle of an activity

The Activity class defines methods that are called by the Android operating system at different points in the lifecycle of an activity. This is shown by figure 3-7.

When you create an activity, you create a class that inherits the Activity class. Then, to get your app to work the way you want, you override the lifecycle methods to control the lifecycle of the activity. For example, you have already seen how you can override the onCreate method to display and prepare the user interface for an activity. However, you often need to override other methods to make sure that your app doesn't lose its progress if the user leaves the app and returns to it later. Similarly, you often need to make sure that your app doesn't lose its progress when the user switches between landscape and portrait orientation.

When an app starts for the first time, Android calls the onCreate, onStart, and onResume methods. As this happens, the activity quickly passes through the *created state* and the *started state*. Of these three methods, the onCreate method is typically used to create the user interface and prepare it for display. In the Tip Calculator activity, for example, this method displays the user interface, gets references to important widgets, and wires the event handlers.

After the onResume method executes, the app is in the *running state*, which is also known as the *active state* or *resumed state*. In this state, the app is visible and has the focus. Unlike the created and started states, an app may remain in the running state for a long time.

If the running activity loses the focus, becomes partially covered by another activity, or if the device goes to sleep, Android calls the onPause method. This causes the activity to go into the *paused state*. Android attempts to maintain the state of all activities in the paused state. As a result, if the activity regains the focus or the device wakes up, Android can call the onResume method and return to the running state without losing any of the user's progress.

If the user gets a phone call or navigates to another activity, Android calls the onPause method followed by the onStop method. This causes the activity to go into the *stopped state*. In the stopped state, Android attempts to maintain the state of the activity for as long as possible. However, an activity in the stopped state has a lower priority than an activity in the running or paused states. As a result, Android often kills activities that are in the stopped state.

Most lifecycle states have a corresponding method. If you implement one method, it often makes sense to implement the other. For example, if you use the onPause method to save data when the activity goes into the paused state, you typically want to use the onResume method to restore the data when the activity returns to the running state. Similarly, if you use the onStop method, you typically want to use the onStart method, and possibly the onRestart method.

Although this diagram shows the most commonly used lifecycle methods for an activity, it doesn't show them all. For example, in chapter 8, you'll learn more about working with the lifecycle methods for displaying menus. Or, you can learn more about other lifecycle methods by looking up the Activity class in the documentation for the Android API.

The lifecycle of an activity

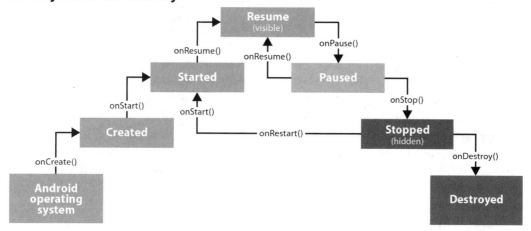

Three common states

State	Description
Resumed	This state is also known as the *active state* or the *running state*. A running activity is visible and has the focus. Android only destroys a resumed activity in extreme situations, such as if the activity tries to use more memory than is available on the device.
Paused	A paused activity has lost the focus and may be partially hidden by other activities, or the device has gone to sleep. A paused activity typically maintains all states. Android only destroys a paused activity if necessary to keep the activity that's in the running state stable and responsive.
Stopped	The activity has lost the focus and is completely hidden by another activity. Android tries to retain the state for a stopped activity for as long as possible. However, stopped activities have a lower priority than the other two states. As a result, Android destroys activities in the stopped state whenever it's necessary to allow activities with higher priorities to run.

Description

- When you create an activity, Android calls the methods of the Activity class at different points in its lifecycle. To get your app to work the way you want, you can override these methods.

- An activity passes through the *created state* and the *started state* quickly, but it can spend a long time in the *resumed state*, the *paused state*, and the *stopped state*.

- Although this diagram shows the most commonly used lifecycle methods for an activity, it doesn't show them all. To learn more about other lifecycle methods available to an activity, you can look up the Activity class in the documentation for the Android API.

Figure 3-7   The lifecycle of an activity

How to save and restore values

Android doesn't always save the data of an activity when the activity moves from the running state into the paused or stopped states. These states occur when the user navigates to another activity or when the user changes the orientation of the screen.

For the Tip Calculator app, this means that the activity won't always save the bill amount and the tip percent. As a result, if you only implement the onCreate method, the tip percent will be reset to its default value of 15% every time the user changes the orientation of the screen. Worse, the bill amount and the tip percent will be reset to their default values every time the user navigates to another activity and returns to the current activity. Since you usually want to save a user's progress, you typically want to save and restore these values as described in figure 3-8.

To start, you need a way to permanently save values. Fortunately, the SharedPreferences class provided by the Android API makes this easy to do. First, you import the SharedPreferences class and its Editor class. Then, you declare a SharedPreferences object as an instance variable, and you use the onCreate method to create an instance of this object. In this figure, for example, the onCreate method uses the getSharedPreferences method to get a SharedPreferences object named savedValues. This object saves values to a file named SavedValues, and this file is only available to this activity.

Once you have created a SharedPreferences object, you can override the onPause method and use the SharedPreferences object to save values. To do that, you can call the edit method of the SharedPreferences object to get an Editor object. Then, you can use the methods of the Editor object to store the bill amount string and the tip percent, which are both instance variables. Next, this code calls the commit method from the Editor object to save the values in the editor object to disk. Finally, the last statement calls the onPause method of the superclass, which is necessary for the superclass to work correctly.

To restore values that you have saved, you can override the onResume method. Within this method, the first statement calls the onResume method of the superclass, which is required for the superclass to work correctly. Then, the next two statements use the get methods of the SharedPreferences object to restore the instance variables for the bill amount string and tip percent.

When you work with the Editor and SharedPreferences objects, you can use them to put and get values for strings and most primitive types. In this figure, for example, the instance variable for the bill amount variable is a string value, and the instance variable for the tip percent is a float value. However, the Editor and SharedPreferences objects don't allow you to work directly with double values. So, if you want to use a double value, you must first cast it to a float value.

To test the onPause and onRestore methods in an emulator, you can run the Tip Calculator activity and navigate to an activity in another app. Then, you can navigate back to the Tip Calculator activity. Or, if you want, you can change the orientation of the Tip Calculator activity in your emulator by pressing Ctrl+F11 or Numpad 7.

How to import the SharedPreferences class and its Editor class

```
import android.content.SharedPreferences;
import android.content.SharedPreferences.Editor;
```

How to set up the instance variables

```
// define SharedPreferences object
private SharedPreferences savedValues;

@Override
public void onCreate(Bundle savedInstanceState) {
    // other code goes here

    // get SharedPreferences object
    savedValues = getSharedPreferences("SavedValues", MODE_PRIVATE);
}
```

How to use the onPause method to save values

```
@Override
public void onPause() {
    // save the instance variables
    Editor editor = savedValues.edit();
    editor.putString("billAmountString", billAmountString);
    editor.putFloat("tipPercent", tipPercent);
    editor.commit();

    super.onPause();
}
```

How to use the onResume method to restore values

```
@Override
public void onResume() {
    super.onResume();

    // get the instance variables
    billAmountString = savedValues.getString("billAmountString", "");
    tipPercent = savedValues.getFloat("tipPercent", 0.15f);
}
```

Description

- Android doesn't always save the data of an activity when you change orientation or when you navigate away from an activity to another activity or app.
- To save data for an activity, you can override the onPause method.
- To restore data for an activity, you can override the onResume method.
- You can use a SharedPreferences object and its Editor class to save and restore data.
- To change orientation in an emulator, you can press Ctrl+F11 or Numpad 7.

Figure 3-8   How to save and restore values

The Java code for the app

Figure 3-9 shows the Java code for the Tip Calculator app. By now, you should be able to understand how this code works.

To start, the package statement stores the class for the Tip Calculator activity in the package named com.murach.tipcalculator. Then, it imports all Java and Android classes and interfaces that are needed for this class. If you review these classes and interfaces, you should be familiar with all of them.

The TipCalculatorActivity class inherits the Activity class that's provided by Android. In addition, this class implements the listener interfaces for the EditorAction and Click events. As a result, this class must implement the onEditorAction and onClick methods.

Within the class, the first six statements define the instance variables for the widgets that the class needs to work with. Then, the seventh statement defines the SharedPreferences object that's used to save and restore values, and the next two statements define instance variables for the bill amount string and the tip percent.

The onCreate method begins by calling the onCreate method of the super-class, which is necessary for the superclass to work correctly. Then, it uses the setContentView method that's available from the superclass to display the user interface that's defined in the XML file for the activity.

After displaying the user interface, the onCreate method gets references to the six widgets that it declared earlier. To do that, it calls the findViewById method for each widget. Then, it casts the View object that's returned to the appropriate type for the widget.

After getting references to the widgets, this code sets the listeners. First, it sets the current class as the listener for the EditorAction event on the editable text view for the bill amount. Then, this code sets the current class as the listener for the Click event on both of the buttons.

The Java code

```java
package com.murach.tipcalculator;

import java.text.NumberFormat;

import android.app.Activity;
import android.os.Bundle;
import android.view.KeyEvent;
import android.view.View;
import android.view.View.OnClickListener;
import android.view.inputmethod.EditorInfo;
import android.widget.Button;
import android.widget.EditText;
import android.widget.TextView;
import android.widget.TextView.OnEditorActionListener;
import android.content.SharedPreferences;
import android.content.SharedPreferences.Editor;

public class TipCalculatorActivity extends Activity
implements OnEditorActionListener, OnClickListener {

    // define variables for the widgets
    private EditText billAmountEditText;
    private TextView percentTextView;
    private Button   percentUpButton;
    private Button   percentDownButton;
    private TextView tipTextView;
    private TextView totalTextView;

    // define SharedPreferences object
    private SharedPreferences savedValues;

    // define an instance variable for the tip percent
    private String billAmountString = "";
    private float tipPercent = 15f;

    @Override
    public void onCreate(Bundle savedInstanceState) {
        super.onCreate(savedInstanceState);
        setContentView(R.layout.activity_tip_calculator);

        // get references to the widgets
        billAmountEditText = (EditText) findViewById(R.id.billAmountEditText);
        percentTextView = (TextView) findViewById(R.id.percentTextView);
        percentUpButton = (Button) findViewById(R.id.percentUpButton);
        percentDownButton = (Button) findViewById(R.id.percentDownButton);
        tipTextView = (TextView) findViewById(R.id.tipTextView);
        totalTextView = (TextView) findViewById(R.id.totalTextView);

        // set the listeners
        billAmountEditText.setOnEditorActionListener(this);
        percentUpButton.setOnClickListener(this);
        percentDownButton.setOnClickListener(this);
```

Figure 3-9   The Java code (part 1 of 3)

The onPause method saves both of this activity's instance variables: the bill amount string and the tip percent. These instance variables need to be saved for the app to work correctly when orientation changes and when the user navigates away from and back to this activity.

The onResume method restores both of the activity's instance variables. Then, it sets the bill amount string as the display text on the bill amount EditText widget. This is necessary because Android sets the display text to its default value if the user navigates away from and back to this activity. Finally, this code calls the calculateAndDisplay method.

The calculateAndDisplay method calculates the tip and total amounts for the bill and displays all current data on the user interface. First, this method gets the bill amount from the EditText widget and converts this string to a float value. Then, it calculates the tip and total amounts and stores them as float values. Finally, it formats these float values and displays them on their corresponding widgets.

The Java code

```java
        // get SharedPreferences object
        savedValues = getSharedPreferences("SavedValues", MODE_PRIVATE);
    }

    @Override
    public void onPause() {
        // save the instance variables
        Editor editor = savedValues.edit();
        editor.putString("billAmountString", billAmountString);
        editor.putFloat("tipPercent", tipPercent);
        editor.commit();

        super.onPause();
    }

    @Override
    public void onResume() {
        super.onResume();

        // get the instance variables
        billAmountString = savedValues.getString("billAmountString", "");
        tipPercent = savedValues.getFloat("tipPercent", 0.15f);

        // set the bill amount on its widget
        billAmountEditText.setText(billAmountString);

        // calculate and display
        calculateAndDisplay();
    }

    public void calculateAndDisplay() {

        // get the bill amount
        billAmountString = billAmountEditText.getText().toString();
        float billAmount;
        if (billAmountString.equals("")) {
            billAmount = 0;
        }
        else {
            billAmount = Float.parseFloat(billAmountString);
        }

        // calculate tip and total
        float tipAmount = billAmount * tipPercent;
        float totalAmount = billAmount + tipAmount;

        // display the other results with formatting
        NumberFormat currency = NumberFormat.getCurrencyInstance();
        tipTextView.setText(currency.format(tipAmount));
        totalTextView.setText(currency.format(totalAmount));

        NumberFormat percent = NumberFormat.getPercentInstance();
        percentTextView.setText(percent.format(tipPercent));
    }
```

Figure 3-9   The Java code (part 2 of 3)

The onEditorAction method is executed whenever the user presses an action key on a soft keyboard such as the Done key. This method begins by using an if statement to check whether the action key is the Done key. If so, it calls the calculateAndDisplay method to perform the calculation and display the results on the user interface.

The onClick method is executed whenever the user clicks on either of the buttons. Within this method, a switch statement checks which button was clicked. Then, if the Decrease (-) button is clicked, this code decreases the tip percent by 1 percent and calls the calculateAndDisplay method to perform the calculation and display the results on the user interface. Conversely, if the Increase (+) button is clicked, this code increases the tip percent by 1 percent and calls the calculateAndDisplay method to perform the calculation and display the results on the user interface.

The Java code

```java
    @Override
    public boolean onEditorAction(TextView v, int actionId, KeyEvent event) {
        if (actionId == EditorInfo.IME_ACTION_DONE)
        {
            calculateAndDisplay();
        }
        return false;
    }

    @Override
    public void onClick(View v) {
        switch (v.getId()) {
            case R.id.percentDownButton:
            tipPercent = tipPercent - .01f;
            calculateAndDisplay();
            break;
            case R.id.percentUpButton:
            tipPercent = tipPercent + .01f;
            calculateAndDisplay();
            break;
        }
    }
}
```

Figure 3-9   The Java code (part 3 of 3)

More skills for finishing an app

Now that you understand the event handlers for the Tip Calculator app shown in this chapter, you may want to put a few final touches on it. To do that, you may need to modify the Gradle build script or the Android manifest file. Or, you may need to use the documentation for the Android API to learn more about the classes and methods that are available to you.

How to work with the Gradle build script

Android Studio uses an automated build tool known as Gradle to build its projects. This tool builds upon the concepts introduced by other build tools such as Apache Ant and Apache Maven.

When you use Android Studio to create a project, it automatically generates a *Gradle build script* like the one shown in figure 3-10. This generated script is usually adequate to get started. However, as you work on a project, you may want to view or modify the attributes in this script.

The minSdkVersion attribute specifies the minimum Android API for your app. A device running a lower API should refuse to install your app. And if you deploy your app to Google Play, Google Play won't display it for devices running a lower API. For most of the apps in this book, we've set the minimum API to 15 since that covered over 95% of all devices as this book went to press. As a result, devices running API 14 or lower should refuse to install the apps in this book.

The targetSdkVersion attribute specifies the version of Android that your app is designed to run on. For this book, we've set the target version to API 23 since that was the newest API available when this book went to press, and we were able to test all apps in this book against this API. A device running a higher version of Android should ignore any new features that change the appearance or behavior of the app. In other words, when a new API becomes available, the apps in the book should look and behave correctly on it, though they can't take advantage of its new features.

The compileSdkVersion attribute specifies the version of Android to use when building your app. This controls the features that are available to your app when you're developing it. For this book, we set the compile version for all apps to API 23. That way, we can use the latest API features that were available as this book went to press.

But what happens if you attempt to use a new feature from API 23 that isn't available from an older API that's supported by your app such as API 19? In that case, Android Studio should warn you. Then, you can decide how to handle this compatibility issue so that your app can run on all APIs from your minimum API to your target API. This is generally considered a best practice.

Another approach would be to set the compile and target APIs to the same level as the minimum API. However, this would prevent your app from using any features introduced by later APIs, which would prevent it from being as good as it could be. As a result, this is not considered a good practice.

The Gradle build script

```
apply plugin: 'com.android.application'

android {
    compileSdkVersion 23
    buildToolsVersion "23.0.1"

    defaultConfig {
        applicationId "com.murach.tipcalculator"
        minSdkVersion 15
        targetSdkVersion 23
        versionCode 1
        versionName "1.0"
    }
    buildTypes {
        release {
            minifyEnabled false
            proguardFiles
                getDefaultProguardFile('proguard-android.txt'),
                'proguard-rules.pro'
        }
    }
}
```

Description

- You can use the *Gradle build script* to control how Android Studio builds the project.

- To open the Gradle build script, open the Project window, expand the Gradle Scripts directory, and double-click on the "build.gradle (Module: app)" item.

- The minSdkVersion attribute specifies the minimum Android API for your app. A device running a lower API should refuse to install your app.

- The targetSdkVersion attribute specifies the Android API that your app is designed to run on. A device running a higher API should not use any new features that change the look or behavior of your app.

- The compileSdkVersion attribute specifies the Android API to use when building your app. This controls the Android features that are available to you as you develop your app.

- Whenever you change the Gradle build script, Android Studio prompts you to synchronize the build files with the new script. To do that, you can click on the Synch Now link that's displayed at the top of the text editor.

Figure 3-10   How to work with the gradle build script

Whenever you change the Gradle build script, Android Studio prompts you to synchronize the build files with the new script. To do that, you can click on the Synch Now link that's displayed at the top of the text editor. This rebuilds the project.

How to work with dependencies

A *dependency* specifies a library or file that your project depends on. The easiest way to add a dependency is to use Android Studio to display the Dependencies tab shown in figure 3-11. Once you display this Dependencies tab, you can use the Add button to add a dependency to your project. When you do this, Android Studio adds the corresponding compile attribute to the dependencies block at the end of the Gradle build script.

You can also use the Dependencies tab to remove a dependency. To do that, select the dependency and click the Remove button. However, it's also easy to remove a dependency by opening the Gradle build script and deleting the compile attribute for the dependency from the dependencies block at the end of the Gradle build script. Or, if you want to remove all dependencies, you can delete the entire dependencies block.

In this figure, the first compile attribute in the dependencies block specifies that each .jar file in the project's libs directory is a dependency. As a result, you can add a library as a dependency by copying its .jar file into this directory. This approach works well for any custom libraries that aren't maintained as part of the Android API.

The second compile attribute specifies that the v7 appcompat library is a dependency. This dependency specifies a library that hasn't been copied into your project folder. As a result, when you build this app, Gradle finds, downloads, and includes this dependency. This works well for standard libraries that are maintained as part of the Android API.

The dialogs for adding a dependency

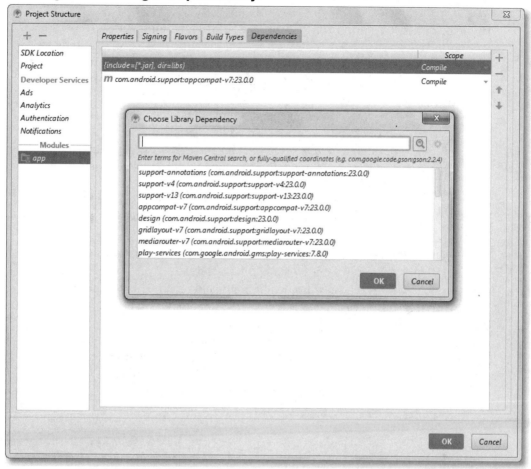

A dependencies block

```
dependencies {
    compile fileTree(dir: 'libs', include: ['*.jar'])
    compile 'com.android.support:appcompat-v7:23.0.0'
}
```

Description

- A *dependency* typically specifies a library or file that your project depends on.
- To add or remove a dependency, you can select the File→Project Structure item from the menu system. Then, you can click on the app module and the Dependencies tab.
- Once you display the Dependencies tab, you can use the Add and Remove buttons to add or remove a dependency.
- When you use the Dependencies tab to add or remove a dependency, Android Studio adds or removes the corresponding compile attribute from the dependencies block that's at the end of the Gradle build script.

Figure 3-11 How to work with dependencies

How to work with the Android manifest

When you use Android Studio to create a project, it automatically generates an *Android manifest* (the AndroidManifest.xml file) like the one shown in figure 3-12. This file specifies some essential information about an app that the Android system must have before it can run the app, including the first activity to launch when the app is started.

In most cases, Android Studio generates a manifest that's adequate for getting started. However, as you work on a project, you may want to modify the manifest. For example, if you don't want an app to respond to orientation changes, you can add a screenOrientation attribute to the activity element. Then, you can set the screenOrientation attribute to a value of "portrait" or "landscape". This may work better for your users who may not want the app to change orientation. Also, if you use this approach, you don't have to write the code that handles the orientation changes.

If you're working on an existing app, the Android manifest may include a uses-sdk element. This element typically includes minSdkVersion and target-SdkVersion attributes that specify the minimum and target APIs for the app. However, these attributes are overridden by the minSdkVersion and targetSdk-Version attributes in the Gradle build script. As a result, when you use Android Studio, you typically delete the uses-sdk element from the Android manifest and use the Gradle build script to set these attributes. That way, your project only specifies these settings in one location. This should reduce confusion and make the app easier for other programmers to understand and maintain.

The AndroidManifest.xml file

```
<?xml version="1.0" encoding="utf-8"?>
<manifest xmlns:android="http://schemas.android.com/apk/res/android"
    package="com.murach.tipcalculator" >

    <application
        android:allowBackup="true"
        android:icon="@mipmap/ic_launcher"
        android:label="@string/app_name"
        android:theme="@style/AppTheme" >
        <activity
            android:name=".TipCalculatorActivity"
            android:label="@string/app_name" >
            <intent-filter>
                <action android:name="android.intent.action.MAIN" />

                <category android:name="android.intent.category.LAUNCHER" />
            </intent-filter>
        </activity>
    </application>

</manifest>
```

An activity element that only allows portrait orientation

```
<activity
    android:name=".TipCalculatorActivity"
    android:label="@string/app_name"
    android:screenOrientation="portrait">
```

The element that sets the minimum and target SDKs

```
<uses-sdk
    android:minSdkVersion="8"
    android:targetSdkVersion="17" />
```

Description

- The *Android manifest* (the AndroidManifest.xml file) specifies some essential information about an app that the Android system must have before it can run the app, including the first activity to launch when the app is started.

- If you don't want an activity to respond to orientation changes, you can add a screenOrientation attribute to the activity element. Then, you can set the screenOrientation attribute to a value of "portrait" or "landscape".

- With some older apps, the Android manifest may include a uses-sdk element. This element typically includes minSdkVersion and targetSdkVersion attributes that specify the minimum and target APIs for the app. However, the corresponding attributes in the Gradle build script override these attributes. As a result, when you use Android Studio, you typically delete the uses-sdk element.

Figure 3-12   How to work with the Android manifest file

How to set the launcher icon for an app

When you use Android Studio to create a project, it automatically generates a *launcher icon* that displays the Android logo. Once you install an app on a device or emulator, you can use this icon to restart it. In addition, Android uses this icon in various places to identify the app. Although the Android logo is fine for getting started, you should set a custom launcher icon for your app as described in figure 3-13 before you distribute it.

To set the launcher icon, you begin by using a browser to find the web pages for the Android Asset Studio. Within these web pages, you navigate to the page for generating launcher icons. Then, after using that page to generate the icons, you can download a zip file that contains the icons and unzip them. The unzipped icon files should be stored in a directory structure that's similar to the one shown in this figure. Then, you can copy these directories and files into the res directory of your app.

In most cases, this will overwrite the default launcher icons generated by Android Studio. As a result, your app will use the new icons. However, if this isn't the case, you can open the Android manifest, and modify the application element so that its icon attribute points to the correct launcher icon. In this figure, for example, the icon attribute points to the launcher icons named ic_launcher that are stored in the mipmap directories as shown in this figure.

If you're working with older Android apps, you may find that they store the launcher icons in a series of directories that start with "layout" like this:

```
app/src/main/res/layout-hdpi/ic_launcher.png
                 layout-mdpi/ic_launcher.png
                 layout-xhdpi/ic_launcher.png
```

In that case, the Android manifest file points to the layout directories like this:

```
android:icon="@layout/ic_launcher"
```

For new development, it's preferable to use the mipmap directories for the launcher icons. However, the layout directories work fine for existing apps.

The Android Asset Studio web page for generating launcher icons

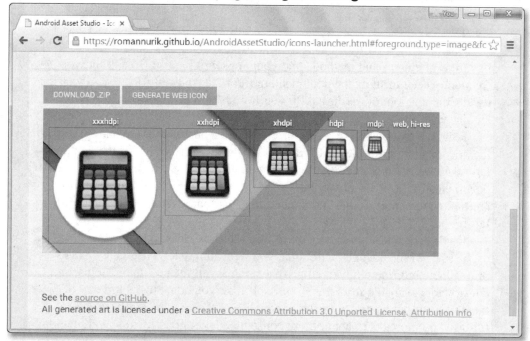

The directory structure for launcher icons

```
app/src/main/res/mipmap-hdpi/ic_launcher.png
                 mipmap-mdpi/ic_launcher.png
                 mipmap-xhdpi/ic_launcher.png
                 mipmap-xxhdpi/ic_launcher.png
                 mipmap-xxxhdpi/ic_launcher.png
```

A few attributes from the AndroidManifest.xml file

```
<application
    android:allowBackup="true"
    android:icon="@mipmap/ic_launcher"
    android:label="@string/app_name"
```

Procedure

1. Use a browser to find the Android Asset Studio by searching the Internet.
2. Navigate to the page for generating launcher icons.
3. Generate the launcher icons, download them, and unzip them.
4. Copy the generated mipmap directories and their files into the res directory of your app. This should overwrite any existing launcher icons.
5. If necessary, modify the Android manifest file so it points to the correct icon files.

Figure 3-13   How to set the launcher icon for an app

How to use the documentation for the Android API

One of the most difficult aspects of learning Android is mastering the hundreds of classes and methods that your apps will require. To do that, you frequently need to study the documentation for the Android API. If you've used the documentation for the Java API, you'll find that the Android API works similarly.

Figure 3-14 summarizes some of the basic techniques for navigating through the documentation for the Android API. This shows one screen of the documentation available from the Activity class, which has many screens. To view the documentation for the Activity class, you click on the package name (android. app) in the upper left frame and then on the class name (Activity) in the lower left frame.

If you scroll through the documentation for this class, you'll get an idea of the scale of the documentation that you're dealing with. After a few pages of overview information, you come to a summary of the constants, fields, and constructors available from the class. After that, you come to a summary of the dozens of public and protected methods that the class offers. That in turn is followed by more detail about these constants, fields, constructors, and methods.

One of the goals of this book is to introduce you to the dozens of classes and methods that you'll use in most of the Android apps that you develop. Once you've learned those, you'll have the background you need to understand the documentation for the Android API, and you'll be able to use that documentation to research classes and methods that aren't presented in this book.

It's never too early to start using the documentation, though. So by all means use the documentation to get more information about the methods that are presented in this book and to research the other methods that are offered by the classes that are presented in this book. After you learn how to use the Activity class, for example, take some time to do some research on that class.

The documentation for the Activity class

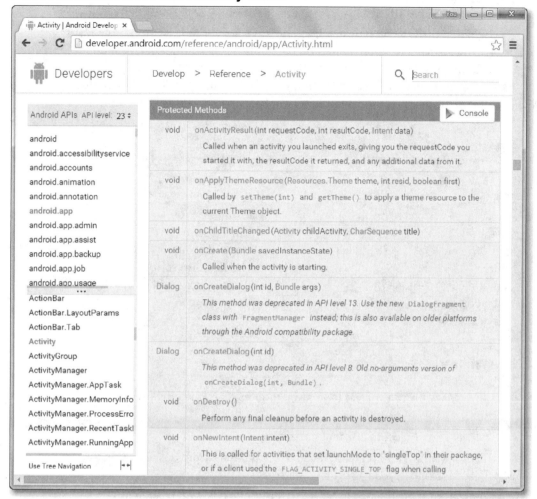

Description

- The Android API contains thousands of classes and methods that you can use in your apps.

- You can use a browser to view the documentation for the Android API by going to this address:

  `http://developer.android.com/reference/`

- You can select the name of the package in the top left frame to display information about the package and the classes it contains. Then, you can select a class in the lower left frame to display the documentation for that class in the right frame.

- Once you display the documentation for a class, you can scroll through it or click on a hyperlink to get more information.

Figure 3-14   How to use the documentation for the Android API

Perspective

We designed the previous chapter and this chapter to get you off to a fast start with Android development. Now, if you understand the Tip Calculator app, you should also be able to develop simple Android apps of your own. Keep in mind, though, that this chapter is just an introduction to the Android essentials described in section 2 of this book.

Terms

event handler	running state
event	active state
action key	resumed state
listener	paused state
action event	stopped state
wire	Gradle build script
input method editor	Android manifest
created state	launcher icon
started state	

Summary

- An *event handler* is a special type of method that's executed when an *event* occurs.
- A *listener* is an object that listens for an event.
- An EditorAction event typically occurs when the user presses an action key, such as the Done key, on a soft keyboard.
- A Click event typically occurs when the user clicks a widget such as a button.
- You can use the SharedPreferences class to permanently save values in your app.
- To save values, you can override the onPause method.
- To restore values you have saved, you can override the onResume method.
- The *Gradle build script* controls how Android Studio builds the project. This includes specifying any *dependencies* for the project.
- The *Android manifest* (the AndroidManifest.xml file) specifies some essential information about an app that the Android system must have before it can run the app, including the first activity to launch when the app is started.
- Once you install an app on a device or emulator, you can use its *launcher icon* to restart it.

Exercise 3-1   Create the Invoice Total app

In this exercise, you'll add functionality to an Invoice Total app. When you're done, a test run should look something like this:

Open the app and test it

1. Start Android Studio and open the project named ch03_ex1_Invoice that's in this directory:

    `\murach\android\ex_starts`

2. Run this project and test the app with a valid subtotal like 100. The app should accept this input, but it shouldn't perform any calculations or display any results.

Handle the EditorAction event

3. Open the InvoiceTotalActivity class that's in the java directory of the project.

4. Use the onCreate method to get references to the EditText widget and the three TextView widgets that display data.

5. Create a method named calculateAndDisplay. This method should get the subtotal value. Then, it should calculate the discount amount and total. It should give a 20% discount if the subtotal is greater than or equal to 200, a 10% discount if the subtotal is greater than or equal to 100, and no discount if the subtotal is less than 100.

6. Add code to the end of the calculateAndDisplay method that displays the results of the calculation on the widgets for the discount percent, discount amount, and total.

7. Handle the EditorAction event for the EditText widget so that it executes the calculateAndDisplay method when the Done key is pressed.

8. Test the app. It should display the starting values that are coded in the strings.xml file.

Set the starting values correctly

9. Modify the strings.xml file so it doesn't display a starting value for the subtotal.

10. Test the app again. This time, it shouldn't display a starting value for the subtotal. Enter some values for the subtotal and make sure it works correctly.

Handle orientation changes

11. Override the onResume method and use it to call the calculateAndDisplay method.

12. Test the app again. Make sure to change the orientation of the activity. The activity should retain all of its data.

Handle navigation

13. Press the Back key to navigate away from the app. Then, navigate back to the app. In an emulator, you can do this by clicking on the Apps icon and clicking on the Invoice Total app. The activity should lose all of its data.

14. Override the onPause method so it saves the string for the subtotal. Then, modify the onResume method so it gets the string for the subtotal. To get these methods to work correctly, you need to set up instance variables for the subtotal string and for a SharedPreferences object that you can use to save and get this string.

15. Test the app again. This time, the app should always remember the last subtotal that you entered even if you navigate away from the app and return to it. In addition, the app should always calculate and display the correct values for the discount percent, discount amount, and total.

Set a launcher icon (optional)

16. Set a launcher icon for the app. You should be able to download possible icons by searching the Internet. When you do that, make sure you have permission to use the image or that it is available under a license that allows you to use it legally.

17. Test the app again. Note that the launcher icon is available from your device or emulator, just like the launcher icon for any other app.

Exercise 3-2   Use the documentation for the Android API

This exercise steps you through the documentation for the Android API. This should give you an idea of how extensive the Android API is.

1. Open a browser and display the documentation for the Android API.

2. Click the android.app package in the upper left frame and the Activity class in the lower left frame to display the documentation for the Activity class. Then, scroll through this documentation to get an idea of its scope.

3. Skim through the overview information for the Activity class.

4. Scroll through the public methods. These methods include the findViewById and setContentView methods that you learned how to use in this chapter.

5. Scroll through the protected methods. These methods include the onCreate, onPause, and onResume methods that you learned how to override in this chapter.

6. Go to the documentation for the TextView class, which is in the android.widget package. Review the attributes for this class as well as the public methods for this class. The public methods include the getText and setText methods you learned about in this chapter.

7. Go to the documentation for the SharedPreferences interface, which is in the android.content package. Then, review the public methods. These methods provide a way to get strings as well as values of the boolean, int, long, and float types. However, they don't provide a way to get values of the double type.

4

How to test and debug an Android app

As you develop an Android app, you need to test it to make sure that it performs as expected. Then, if there are any problems, you need to debug your app to correct any problems. This chapter shows how to do both.

Basic skills for testing and debugging

When you *test* an app, you run it to make sure that it works correctly. As you test the app, you try every possible combination of input data and user actions to be certain that the app works in every case. In other words, the goal of testing is to find errors (*bugs*) and make an app fail.

When you *debug* an app, you fix the bugs that you discover during testing. Each time you fix a bug, you test again to make sure that the change you made didn't affect any other aspect of the app.

Typical test phases

When you test an app, you typically do so in phases, like the four that are summarized in figure 4-1.

In the first phase, you should test the user interface. To start, you can use the graphical layout editor to check the widgets to make sure they display properly. As you do this, you should test portrait and landscape orientation, all target screen sizes, and all target platforms. You'll learn how to do that in the next figure. Then, you should run the app on a device or emulator to make sure that all the widgets work correctly.

In the second phase, you should test the app with valid data. To start, you can run the app and enter data that you would expect a user to enter. Then, you should enter valid data that tests the limits of the app. You should change the orientation of the app to make sure it works correctly when the user switches between portrait and landscape orientation. And you should test other changes in the lifecycle of each activity. For instance, you should navigate away from an activity and return to it.

In the third phase, you should try to make the app fail by testing every combination of invalid data and user action that you can think of. For instance, if possible, you should try entering letters where numbers are expected.

In the first three phases, you can test the app on a single device or emulator. If possible, this device should be the primary target device. For example, if you primarily want your app to run on a current touchscreen smartphone, you can begin by testing on that device or an emulator that emulates that device.

In the fourth phase, the app should be working correctly on the primary target device or emulator. However, in this phase, you test the app on all other target devices or on emulators for those devices.

If possible, you should use physical devices since this is the only true way to test your app. However, whenever necessary, you can use an emulator instead of a physical device. As you test on various target devices, you will usually find some bugs that you'll need to fix. This is one of the most difficult aspects of Android programming.

Four test phases

1. Check the user interface to make sure that it works correctly.
   - Check in portrait and landscape orientation.
   - Check on all target screen sizes.
   - Check on all target platforms.

2. Test the app with valid input data to make sure the results are correct.
   - Test changing the orientation.
   - Test other changes in the lifecycle of the activity.

3. Test the app with invalid data or unexpected user actions. Try everything you can think of to make the application fail.

4. Test on all target devices. If possible, use a physical device. Otherwise, use an emulator.

Description

- The goal of *testing* is to find all errors (*bugs*) in the app.
- The goal of *debugging* is to fix all of the bugs that you find during testing.

Figure 4-1   An introduction to testing and debugging

How to check the layout

Figure 4-2 shows how to use the graphical editor to check the layout. To start, you can use the buttons on the toolbar to zoom in or out. These buttons can help you view the widgets on the user interface and make sure that they are being displayed correctly.

If you are going to allow the app to switch orientations, you need to make sure the layout works for both portrait and landscape orientation. To do that, you can click on the "Go to Next State" button in the toolbar to switch between portrait and landscape orientation. Then, you can check to make sure the layout displays correctly in both orientations. If it doesn't, you can tweak it until it does, or you can create a separate layout for each orientation as described in the next chapter.

Once the layout displays correctly in both orientations, you should test it for different screen sizes. To do that, you can use the drop-down list in the toolbar to select a device with a different screen size. In this figure, the Nexus S smartphone is selected. This device has a 4-inch screen. However, if you want you can select a tablet with a 7- or 10-inch screen.

Once you are satisfied with the layout in different screen sizes, you should test it on different platforms. To do that, you can select any platform that's installed on your system from the drop-down list in the toolbar. In this figure, for example, eleven platforms are installed on the user's system. When you do this, it may take your system a minute or so to render the user interface, so be patient. When the device is displayed on the new platform, the widgets may look different, but the layout should still display correctly.

A layout in the graphical editor displayed by Android 4.1.2 (API 16)

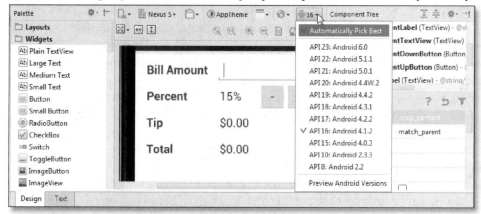

Description

- To zoom in or out, use the buttons on the toolbar to change the zoom level.
- To view on a different screen size, select a device with a different screen size from the drop-down list in the toolbar.
- To change the orientation, click on the "Go to next state" button in the toolbar.
- To change platforms, select a different platform from the drop-down list in the toolbar. This drop-down list only displays the platforms that are installed on your system.
- To change the theme, select a different theme from the drop-down list in the toolbar.

Figure 4-2   How to check the layout

How to handle runtime errors

When Android encounters a runtime error, the emulator or device displays an error message like the one in figure 4-3, and Android Studio prints some information about the error to its LogCat window. This information includes the name of the exception. In this figure, for example, a NumberFormatException caused the app to crash. If the LogCat window isn't visible in Android Studio, you can display it by clicking on the Android tab at the bottom left of the main window.

By examining the LogCat window closely, you can often figure out the cause of the bug, which is the first step in fixing the bug. In this figure, for example, the LogCat data indicates that there was a problem parsing a float value. As a result, you can begin by looking at the code that parses the float value to see if you can figure out what's causing the bug. If that doesn't work, you can use the debugging skills that are presented in the rest of this chapter to figure out what's causing the bug. Once you figure out what's causing the bug, fixing the bug is often the easy part.

The error that's displayed when an app crashes

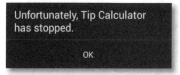

The LogCat window after a crash

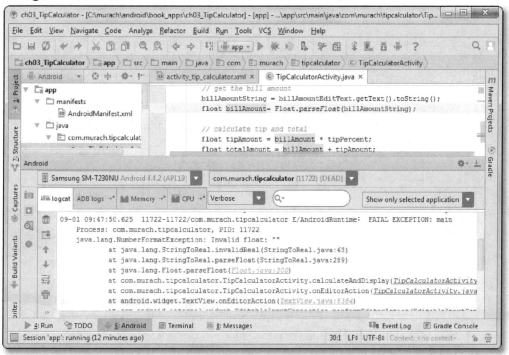

Description

- When an app encounters an error that causes it to end, the emulator or device displays an error message that indicates that the app has stopped, and Android Studio displays some information about the error in the LogCat window.

- If the LogCat window isn't displayed, you can display it by clicking the Android tab at the bottom of the main window. Then, if necessary, you can click the LogCat tab.

- When an error occurs, the LogCat window includes the name of the exception that caused the error.

Figure 4-3   How to handle runtime errors

How to trace code execution

When you *trace* the execution of an app, you add statements to your code that display messages or variable values at key points in the code. This is typically done to help you find the cause of a bug.

How to use LogCat

One way to trace code execution is to use the LogCat logging as shown in figure 4-4. To do that, you begin by importing the Log class that's in the android. util package. Then, you use one of the methods of the Log class to send data to the LogCat window.

When you use LogCat for debugging, you typically use the d method to send a debug message to the log. However, if necessary, you can use other methods to send other types of messages such as an error, warning, or informational message. Each type of message displays in a different color in the LogCat window, which makes it easy to distinguish between the types of errors.

All of the methods in this figure have two parameters. The first parameter is for a tag that's used to identify the message in the LogCat window. Often, the name of the activity is used as the tag for a message. As a result, it's a common practice to define a constant named TAG for each activity that includes the name of the activity. Then, you can use this constant as the first argument for all logging statements in the activity, and you can code any string as the second argument.

In this figure, the first Log statement prints a message that indicates that the calculateAndDisplay method is starting. Then, the second statement prints the value of the variable named billAmount. Printing the values of variables like this can often help you find a bug. Then, when you fix the problem, you can remove the logging statements.

When you use this technique, you usually start by adding just a few logging statements to the code. Then, if that doesn't help you find and fix the bug, you can add more. Once you find and fix the bug, you typically remove these logging statements from your code.

When you run the app, Android Studio displays the messages in the LogCat window. By default, Android Studio selects the "Show only selected application" option to make sure that the LogCat window is only showing information for the current application. You can also filter LogCat messages by different types of parameters from this drop down menu.

Android Studio with the LogCat window displayed

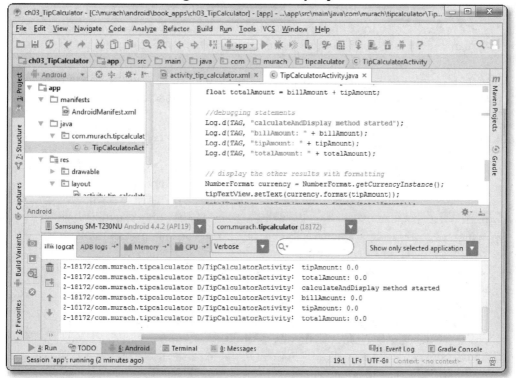

A few methods of the Log class

Method	Message type	Display color
d(tag, message)	Debug	Black
e(tag, message)	Error	Red
w(tag, message)	Warn	Blue
i(tag, message)	Info	Black

How to use the Log class

Import the Log class
```
import android.util.Log;
```

Declare a constant for the tag parameter
```
private static final String TAG = "TipCalculatorActivity";
```

Send messages to the log
```
Log.d(TAG, "calculateAndDisplay method started");
Log.d(TAG, "billAmount: " + billAmount);
```

Description

- A simple way to *trace* the execution of an app is to use methods of the Log class to send messages to the log at key points in the code. In Android Studio, these messages are shown in the LogCat window.

Figure 4-4   How to use LogCat to trace code execution

How to use toasts

Another way to trace code execution is to use *toasts*, which are messages that are briefly displayed on the user interface. In figure 4-5, for example, the bottom of the Tip Calculator app is displaying a toast that says, "onCreate method". Since toasts are only displayed briefly, they're only useful for certain situations. If you experiment with them, you'll quickly discover when they can be helpful.

The first step in displaying a toast is to import the Toast class. This class is stored in the android.widget package.

The second step is to create a Toast object. To do that, you call the static makeText method of the Toast class and pass this object three parameters. The first parameter is the context, and it determines where the toast is displayed. In most cases, you can use the this keyword to identify the current activity as the context. The second parameter is the message. You can specify any string you want for this parameter, including strings that display the value of a variable as shown in the previous figure. The third parameter is the length of time that the toast is displayed. You can use the LENGTH_SHORT and LENGTH_LONG constants of the Toast class for this parameter.

The third step is to call the show method from the Toast object that you created. There are two ways to do this. If you have created a variable for the object, you can call the method from that variable as shown in the second to last example. However, it's more common to use method chaining to call the show method directly from the makeText method as shown in the last example.

When you use toasts for debugging, you typically only use a toast or two. Then, when you're done debugging, you need to remove these statements.

Although it's common to use toasts for debugging, you can also use a toast in an app to display a message to the user. In an email app, for example, you may want to use a toast to display a message that says, "Message deleted" whenever a user deletes a message. As you progress through this book, you'll see some examples of this.

A toast displayed in an emulator

Two methods of the Toast class

Method	Description
makeText(context, message, length)	Returns a Toast object with a message that can be displayed in the specified context for the specified length.
show()	Displays the toast.

Two constants of the Toast class

Constant	Description
LENGTH_SHORT	A short length of time.
LENGTH_LONG	A long length of time.

How to display a toast

Import the Toast class

```
import android.widget.Toast;
```

Use two statements

```
Toast t = Toast.makeText(this, "onCreate method", Toast.LENGTH_SHORT);
t.show();
```

Use a single statement (method chaining)

```
Toast.makeText(this, "onCreate method ", Toast.LENGTH_SHORT).show();
```

Figure 4-5   How to use toasts to trace code execution

How to use the debugger

When you use LogCat logging or toasts for debugging, you have to add statements to your code. Then, when you're done debugging, you need to remove these statements. This creates extra work for you. Fortunately, Android Studio includes a powerful tool known as a *debugger* that you can use to debug an app without having to add or remove statements.

However, the debugger is a complex tool, and you may find it easier to use LogCat logging or toasts for some types of bugs. As a result, you'll have to choose the debugging technique that you prefer for any given situation.

How to set and remove breakpoints

The first step in debugging an app is to find the cause of the bug. To do that, it's often helpful to view the values of the variables at different points in the execution of the app.

The easiest way to view the variables while an app is running is to set a *breakpoint* as shown in figure 4-6. Then, when you run the app with the debugger, execution stops just prior to the statement at the breakpoint, and you can view the variables that are in scope at that point in the app.

To set a breakpoint, you click to the left of the line of code on the vertical bar that's on the left side of the code editor. Then, the breakpoint is marked by a red circle to the left of the line of code. Note, however, that you can only set a breakpoint on a line of code that can be executed, not on a declaration, comment, brace, or parenthesis.

When debugging, it's important to set the breakpoint before the line in the app that's causing the bug. Often, you can figure out where to set that breakpoint just by knowing which statement caused the crash. However, sometimes you need to experiment a little before finding a good location for a breakpoint.

After you set one or more breakpoints, you need to run the app with the debugger. To do that, you can use the Debug button that's available from the toolbar (just to the right of the Run button).

When you run an app with the debugger, Android Studio displays the Debug window. If it doesn't, you can display it by clicking the Debug tab at the bottom of the main window.

After you run an app with the debugger, the breakpoints remain where you set them. If you want to remove a breakpoint, you can do that by clicking on the red circle for the breakpoint.

Java code with a breakpoint

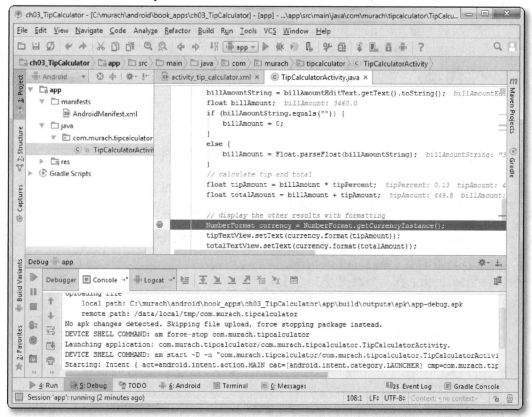

Description

- A *breakpoint* is indicated by a red circle icon that's placed to the left of the line of code. When an application is run in debug mode, it stops just before executing the statement at the breakpoint.

- To set a breakpoint, click on the vertical bar to the left of a line of code in the code editor. To remove a breakpoint, click on it.

- When the app arrives at the breakpoint, the red icon will turn into an orange circle icon with a checkmark.

- Once you set a breakpoint, you click the Debug button on the toolbar to begin debugging. This works much like the Run button described in chapter 1, except that it lets you debug the application. When the application encounters the breakpoint, Android Studio opens the Debug window as shown in the next figure.

Figure 4-6   How to set and remove breakpoints

How to step through code

When you run an app with the debugger and it encounters a breakpoint, execution stops before the statement at the breakpoint. In addition, Android Studio displays the Debug window, which contains several other windows such as the Variables and Frames windows shown in figure 4-7.

Along its top, the Debug window displays some toolbar buttons such as the Step Over button and Step Into buttons. You can use these buttons to *step through* the statements in the app, one statement at a time. This lets you observe exactly how and when the variable values change as the app executes, and that can help you determine the cause of a bug.

Along its left side, the Debug window displays some buttons that you can use when you're done stepping through code. For example, you can click the Resume Program button to continue execution until the next breakpoint. Or, you can click the Stop button to stop the app and end the debugging session.

How to inspect variables

When you set breakpoints and step through code, the Variables window automatically displays the variables that are in scope. In this figure, the execution point is in the calculateAndDisplay method of the TipCalculatorActivity class. Here, the billAmount variable is a local variable that's declared to store the amount of the bill. In addition, even though they aren't shown in this figure, the instance variables of this class are in scope. To view these variables, you can expand the variable named *this*.

For numeric variables and strings, the value of the variable is shown in the Variables window. However, you can also view the values for an object by expanding the variable that refers to the object. In this figure, for example, you can expand the variable named *this* by clicking on the arrow to its left. Then, you can view the values of its variables. Similarly, you can expand the variable named tipTextView by clicking on the arrow to its left.

How to inspect the stack trace

Within the Debug window, the Frames window shows the *stack trace*, which is a list of methods in the reverse order in which they were called. You can click on any of these methods to display the method and highlight the line of code that called the next method. This opens a class in a new code editor if necessary.

The Debug window

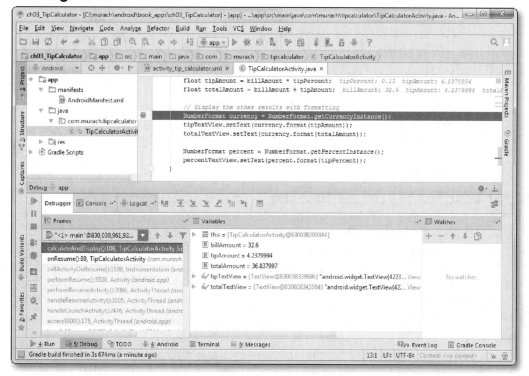

Description

- When program execution reaches a breakpoint, Android Studio typically displays the Debug window. This window has all the tools you need to debug your code.

- To display or hide the Debug window, click the Debug tab at the bottom of the main window.

- Within the Debug window, the Frames window shows the *stack trace*, which is a list of methods in the reverse order in which they were called. You can click on one of these methods to display it in the code editor.

- Within the Debug window, the Variables window shows the values of the variables that are in scope for the current method. This includes static variables, instance variables, and local variables. If a variable refers to an object, you can view the values for that object by expanding the object and drilling down through its variables.

- To step through code one statement at a time, click the Step Over button or the Step Into button.

- To execute the code until the next breakpoint, click the Resume Program button.

- To stop debugging the app, click the Stop button.

Figure 4-7   How to work with the Debug window

How to create emulators

Even after you've tested and debugged your app for your primary target device, you need to test and debug your app for the rest of your target devices. If possible, you should test and debug your app on physical devices. However, since it's not always possible to use physical devices to test all target devices, you'll probably need to use emulators to test some target devices.

In the appendix, you learned how to create an emulator for a Nexus S phone that's running Anroid 4.1 (API 16). Now, figure 4-8 shows how to create two more emulators.

How to create an emulator for a tablet

Part 1 of this figure shows how to create an emulator for a tablet. This provides a way to test your app on a device that has a large screen. In this figure, the emulator uses the built-in hardware profile for a Nexus 7 (2012) tablet. This device has a 7-inch screen with 800 by 1280 pixels, which provides for approximately the same number of dots per inch as a TV (tvdpi), which is somewhere between a medium density (mdpi) and high density (hdpi).

If the Android version you want isn't available from the dialog, you can download a system image for the Android version that you want. To do that, you can use this dialog or the SDK Manager described in the appendix. When you do that, you can select an image that runs efficiently on your CPU. For example, an x86 image runs more efficiently on a computer that has a CPU that uses the x86 architecture. As a result, you typically want to match the system image to your CPU as closely as possible.

Since emulators tend to run slowly on many systems, we recommend creating emulators that use fewer system resources when you're getting started. This helps the emulators to run a little faster. That's why we recommend selecting a hardware profile that has a relatively low density such as mdpi or hdpi. And that's why we recommend selecting a system image for an older version of Android that doesn't include extra Google APIs.

Of course, if your system is so powerful that it can run emulators without degrading performance to an unacceptable level, feel free to use whatever hardware profile and Android API that you like. Also, before you put an app into production, you will probably need to test on emulators that use extra high density (xhdpi) screens and the latest Android APIs. But at least you can wait to do that until you've finished basic testing on physical devices or emulators like the ones described in this figure.

A built-in hardware profile for a tablet

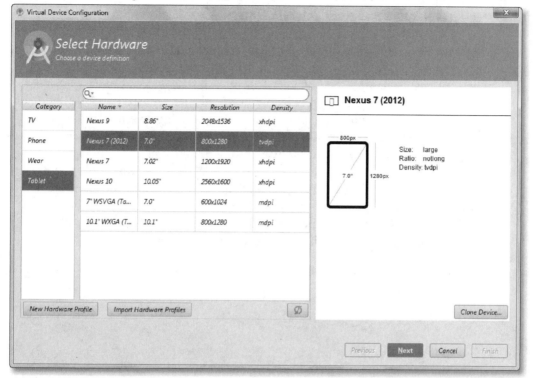

An emulator for a tablet

- Category: Tablet
- Hardware profile name: Nexus 7 (2012)
- System image: Jelly Bean (API 16) x86
- Emulator name: Nexus 7 (2012) API 16

How to create an emulator

1. Start Android Studio.
2. In the toolbar, click the button for the AVD Manager. This should start the Android Virtual Device Manager and show all available emulators.
3. Click the Create Virtual Device button.
4. Select a category such as phone or table.
5. Select one of the available hardware profiles. If necessary, you can create a new hardware profile.
6. Select one of the available system images. If necessary, you can download additional system images.
7. Enter a name for the emulator.

Figure 4-8   How to create emulators (part 1 of 2)

How to create an emulator for a phone with a hard keyboard and DPad

Part 2 of this figure shows how to create an emulator that has a hard keyboard and a DPad (directional pad), which provides up, down, left, right, and center buttons, much like a TV remote. This provides a way to test devices that aren't touchscreen devices and rely on hardware such a keyboard or a DPad.

The procedure for doing this is the same as for creating an emulator for a tablet as described in part 1 of this figure. However, at the Select Hardware screen, you need to create a new hardware profile for the emulator. To do that, you can click the New Hardware Profile button and specify a profile that includes a hard keyboard and a DPad. In this figure, for example, the hardware profile has a name of "Phone with keyboard and DPad". It specifies a 4-inch screen with 480x800 pixels (hdpi density). It specifies that the device has a hard keyboard. And it specifies that it uses a DPad for navigation.

A new hardware profile for a phone with keyboard and DPad

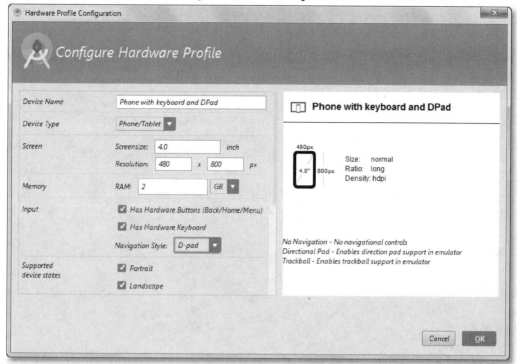

An emulator for a phone with a keyboard and a DPad

- Category: Phone
- Hardware profile name: Phone with keyboard and DPad
- System image: Jelly Bean (API 16) x86
- Emulator name: Phone with keyboard and DPad API 16

Description

- If the built-in device definitions aren't adequate for you, you can click the New Hardware Profile button to create your own hardware profile.
- If the Android version you want isn't available, you can download a system image for the Android version that you want. When you do that, you can select an image that runs efficiently on your CPU.
- If performance is your main consideration, you can select a hardware profile that has a relatively low density such as hdpi or mdpi. In addition, you can select a system image for an older version of Android that doesn't include extra Google APIs.

Figure 4-8   How to create emulators (part 2 of 2)

Perspective

Before you release an app, it should be thoroughly tested and debugged. Now that you've completed this chapter, you should be able to thoroughly test your apps. And if you find bugs during testing, you should be able to use the skills presented in this chapter to fix those bugs.

As your apps grow more complex, you may need testing and debugging skills that go beyond the skills covered in this chapter. For example, you may want to experiment with some of the other debugging windows and toolbar buttons. If you do, you can probably learn a few new debugging skills.

Terms

test	debugger
bugs	breakpoint
debug	step through
trace	stack trace
toasts	

Summary

- To *test* an app, you run it to make sure that it works properly no matter what combinations of valid or invalid data you enter.

- When you *debug* an app, you find and fix all of the errors (*bugs*) that you find when you test the app.

- A simple way to *trace* the execution of an app is to insert LogCat logging statements at key points in the code.

- Another way to trace code execution is to use *toasts*, which are messages that are briefly displayed on the user interface.

- Android Studio includes a powerful tool known as a *debugger* that can help you find and fix errors.

- You can set a *breakpoint* on a line of code to stop code execution just before that line of code. Then, you can *step through* the code and view the values of the variables as the code executes.

- A *stack trace* is a list of methods in the reverse order in which they were called.

Exercise 4-1   Test and debug the Tip Calculator app

This exercise guides you through the process of using Android Studio to test and debug an app.

Test the app with invalid data

1. Start Android Studio and open the project named ch04_ex1_TipCalculator that's in this directory:

   `\murach\android\ex_starts`

2. Run this project and test the app with a valid subtotal like 100. This should work correctly.

3. Test the app with an invalid subtotal by leaving the bill amount blank and pressing the Done key. This should cause the app to crash with a run-time error, and it should display an error message in the LogCat window.

4. Study the error message that's displayed in red. You can focus on the first few lines of this message. These lines give information about the exception that caused the app to crash. Based on this information, you should be able to figure out that the app crashed because an empty string isn't a valid float value.

5. Fix the bug by using a try/catch statement to handle the exception. The catch clause for the exception should set the billAmount variable to zero.

Use LogCat logging

6. At the end of the onCreate method, add a logging statement that prints "onCreate executed" to the LogCat window.

7. Add logging statements that print "onPause executed" and "onResume executed" to the ends of the onPause and onResume methods.

8. Run the app and view the logging statements as you change orientation. Note that this causes the onPause, onCreate, and onResume methods to be executed. This shows that the activity is destroyed and recreated when you change orientation.

9. Run the app and view the logging statements as you navigate away from and back to the app. Note that this only causes the onPause and onResume methods to be executed. This shows that the activity is paused when you navigate away from it and resumed when you navigate back to it.

Test on different emulators

10. Create an emulator for a tablet.

11. Run the app in that emulator. It should work as before.

12. Create an emulator for a phone with a hard keyboard and a DPad.

13. Run the app in that emulator. You should be able to use your computer's keyboard to enter a bill amount and press the Enter key to submit this entry.

Use toasts

14. Add a toast to the onEditorAction method that displays the value of the actionID parameter.

15. Run the app, use the soft keyboard to enter a bill amount, and press the Done key. Make a note of the value of the actionID parameter.

16. Run the app in an emulator that supports a hard keyboard. Then, use the hard keyboard to enter a bill amount and press the Enter key to submit this entry. Make a note of the value of the actionID parameter.

Use the debugger

17. In the calculateAndDisplay method, set a breakpoint on this line of code:
    ```
    billAmountString = billAmountEditText.getText().toString();
    ```

18. Click on the Debug button in the toolbar. This runs the project with the debugger.

19. When execution stops at the breakpoint, use the Variables window to examine the variables that are in scope. Then, expand the variable named *this* and view the value for the instance variable named billAmountString. Next, collapse the variable named *this* so you can easily view other variables.

20. Click the Step Into button in the toolbar 4 times to step through the app one statement at a time. After each step, review the values in the Variables window to see how they have changed. Note how the app steps through the try/catch statement based on the value of the bill amount string.

21. Click the Resume Program button in the toolbar to continue the execution of the app.

22. Switch to the emulator, enter a new bill amount, and click the Done button.

23. Switch back to Android Studio and use the Variables window to view the value of the instance variable named billAmountString.

24. When you're done inspecting the variables, click the Stop button to stop the app and end the debugging session. This should give you some idea of how useful the Android Studio debugging tools can be.

Section 2

Essential Android skills

Most of the chapters in this section expand upon the basic Android skills that you learned in section 1. In chapter 5, for example, you'll learn more about getting input from the user and displaying output. In chapter 6, you'll learn more about handling events. In chapter 7, you'll learn more about controlling the appearance of your app. And so on.

In addition, some of the chapters in this section present new skills. In chapter 8, for example, you'll learn how to work with menus and preferences. Then, in chapter 9, you'll learn how to work with fragments.

To illustrate these skills, this section adds new features to the Tip Calculator app that you learned about in section 1. This allows you to expand your skills while still working with an app that's familiar to you. When you're done with this section, you'll be ready to learn how to work with some more complicated apps such as the News Reader and Task List apps.

5

How to work with layouts and widgets

In chapter 3, you learned how to use a relative layout with three different types of widgets: text views, editable text views, and buttons. Now, this chapter shows you how to use some other types of layouts. In addition, this chapter shows you some new skills for working with widgets including how to use check boxes, radio buttons, spinners, and seek bars.

An introduction to layouts and widgets

This chapter begins by summarizing some of the layouts and widgets that are available from the Android API and by presenting the inheritance hierarchy for the classes for layouts and widgets.

A summary of layouts

Figure 5-1 begins by summarizing six layouts that are available from the Android API. Of these layouts, the relative layout described in chapter 3 is one of the most powerful since it provides a way to align widgets relative to one another. However, it is also one of the most difficult layouts to use.

That's why this chapter briefly presents three other layouts that can be useful. If you need to create a user interface that displays widgets in a vertical or horizontal line, you may want to use a linear layout. If you want to display widgets in rows and columns, you may want to use a table layout. Or, if you want to display widgets on top of each other in a stack, you may want to use the frame layout.

The grid layout was introduced with Android 4.0 (API level 14). As a result, it works for most modern Android devices without having to include a support library. However, you can usually get the same results with the relative layout or the table layout. That's why this book doesn't show how to use the grid layout. Still, if you're developing an app that needs to lay out components in a grid, you might want to learn more about the grid layout as it provides some nice improvements over the relative and table layouts.

This chapter doesn't present the absolute layout because this layout was deprecated in Android 1.5 (API level 3). As a result, you shouldn't use this layout for new development.

A summary of widgets

Figure 5-1 also summarizes some of the widgets that are available from the Android API. Keep in mind that this is just a summary, not a complete list. The Android API provides many other useful widgets.

You already learned about the TextView, EditText, and Button widgets in chapter 3. Now, you'll learn about the CheckBox, RadioButton, Spinner, SeekBar, ImageView, and ScrollView widgets.

Some of the widgets in this figure are often referred to using other terminology. For example, a TextView widget is often called a *label*. An EditText widget is often called an *editable text view*, a *text box*, or a *text field*. And a spinner is often called a *drop-down list*.

Once you understand how to use the widgets presented in this chapter, you should be able to learn about other widgets on your own. For example, once you learn how to use the Button and ImageView widgets, you should be able to use the ImageButton widget. Then, if you need help, you can look up the ImageButton class in the Android API documentation.

A summary of layouts

Layout	Description
RelativeLayout	Lays out widgets relative to one another. This layout is described in chapter 3.
LinearLayout	Lays out widgets in a vertical or horizontal line. This layout is described in figure 5-3.
TableLayout	Lays out widgets in a table. This layout is described in figure 5-4.
FrameLayout	Lays out widgets in a stack where one widget is displayed on top of the other. This layout is described in figure 5-5.
GridLayout	Lays out widgets in a grid. This layout was introduced with Android 4.0 (API level 14). This layout is not covered in this book.
AbsoluteLayout	This layout was deprecated in Android 1.5 (API level 3). This layout is not covered in this book.

A summary of widgets

Widget	Description
TextView	Also known as a *label*, this widget displays text.
EditText	Short for *editable text view*, this widget is also known as a *text box* or *text field*. This widget allows the user to enter text such as names, email addresses, passwords, phone numbers, dates, times, and numbers.
Button	Performs an action when the user clicks it.
CheckBox	Allows the user to check or uncheck an option.
RadioButton	Allows the user to select a single option from a group that's defined with a RadioGroup widget.
Spinner	Also known as a *drop-down list*, this widget allows the user to select an option from a list.
ProgressBar	Displays a visual indicator of the progress of an operation.
SeekBar	Allows the user to select a value by dragging a thumb to the left or right.
RatingBar	Allows the user to rate something by selecting one or more stars.
ImageView	Displays an image.
ImageButton	Works like a regular button but displays an image instead of text.
DatePicker	Allows the user to select a date.
TimePicker	Allows the user to select a time.
CalendarView	Allows the user to select a date.
ScrollView	Automatically displays a vertical scroll bar if the layout doesn't fit on the screen vertically.
WebView	Uses a built-in browser to display web content such as a web page that contains HTML, CSS, and JavaScript.

Figure 5-1   A summary of layouts and widgets

The View hierarchy

Figure 5-2 shows the inheritance hierarchy for the classes that define Android widgets and layouts. To start, the View class is the superclass for all widgets. As a result, any methods of this class are available to all widgets.

Under the View class, the ViewGroup class is the superclass for widgets that can contain other widgets. This includes layouts such as the RelativeLayout, LinearLayout, TableLayout, and FrameLayout classes. In addition, it includes the RadioGroup class that's used to group widgets created from the RadioButton class.

Under the View class, the TextView class is the superclass for widgets that display text such as the EditText and Button widgets. From the Button class, the CompoundButton class is the superclass for the CheckBox and RadioButton classes. As a result, any methods of the CompoundButton class are available to both CheckBox and RadioButton widgets.

The View class is also the superclass for other widgets that don't display text. For example, the View class is the superclass for the ProgressBar class, which is the superclass for the AbsSeekBar class, which is the superclass for the SeekBar and RatingBar classes. Similarly, the View class is the superclass for the ImageView class.

The View hierarchy

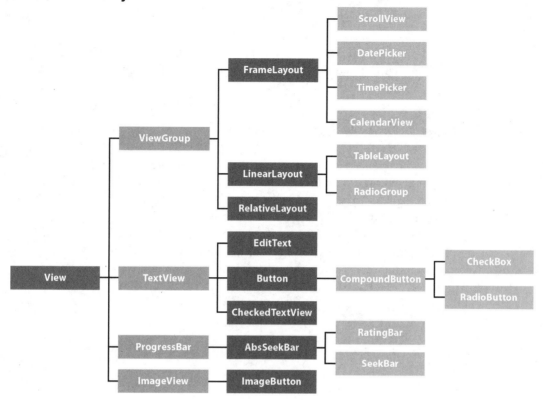

Description

- The View class is the superclass for all widgets.
- The ViewGroup class is the superclass for most widgets that can contain other widgets such as the RelativeLayout widget.
- The TextView class is the superclass for widgets that display text such as the EditText and Button widgets.

Figure 5-2   The View hierarchy

How to work with layouts

In chapter 3, you learned how to use a relative layout. Now, this chapter shows how to use three more layouts that are available from Android. In addition, it shows how to nest layouts and how to provide for separate landscape layouts.

How to use a linear layout

If you need to create a user interface that displays widgets in a vertical or horizontal line, you may want to use a linear layout as shown in figure 5-3. Here, the width and height attributes of the layout have been set to "match_parent". As a result, this layout stretches across the entire screen.

The orientation attribute of this layout has been set to "vertical". As a result, this layout displays the two buttons in a vertical line. In other words, this layout displays the two buttons in a column, one above the other.

A linear layout with vertical orientation and two buttons

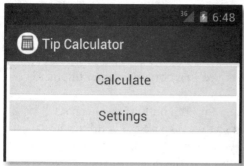

The XML for the linear layout

```xml
<?xml version="1.0" encoding="utf-8"?>
<LinearLayout xmlns:android="http://schemas.android.com/apk/res/android"
    android:layout_width="match_parent"
    android:layout_height="match_parent"
    android:orientation="vertical" >

    <Button
        android:id="@+id/calculateTipButton"
        android:layout_width="match_parent"
        android:layout_height="wrap_content"
        android:text="@string/calculate_tip" />

    <Button
        android:id="@+id/settingsButton"
        android:layout_width="match_parent"
        android:layout_height="wrap_content"
        android:text="@string/settings" />

</LinearLayout>
```

Description

- A linear layout displays a column or row of child widgets.

Figure 5-3   How to use a linear layout (part 1 of 2)

Part 2 of figure 5-3 shows some more examples for working with a linear layout. For all of these examples, the width of the buttons has been set to "wrap_content". That way, the buttons are only wide enough to display their text.

In the first example, the orientation attribute for the layout is set to a value of "horizontal". As a result, the buttons are displayed in a horizontal line. In other words, they are displayed in a row.

In the second example, the weight attribute for both buttons has been set to "1". As a result, both buttons are stretched so they take up the width of the layout, with each button taking half of the layout since both buttons have equal weight. If you want to have one button be wider than the other, you can experiment with the weight attributes of the buttons. For example, if you set the weight of the first button to 2 and the second button to 1, the first button takes 2/3 of the width, and the second button takes 1/3 of the width. Put another way, the first button expands at a rate twice that of the second button.

In the third example, the gravity attribute of both buttons has been set to "center". As a result, both buttons are centered horizontally in the layout, which has been set to vertical orientation.

Common attributes for working with linear layouts

Attribute	Description
orientation	Controls whether the linear layout uses vertical or horizontal orientation.
weight	Determines how much space Android allocates to a widget.
gravity	Aligns a widget with the top, bottom, center, left, or right of its layout.

A horizontal layout where the buttons have no weight

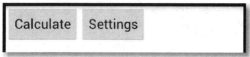

The orientation attribute for the layout
```
android:orientation="horizontal"
```

A horizontal layout where the buttons have equal weight

The weight attribute for both buttons
```
android:weight="1"
```

A vertical layout where the buttons are centered horizontally

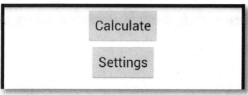

The gravity attribute for both buttons
```
android:gravity="center"
```

Description

- The attributes of a linear layout and its child widgets control the appearance of the user interface.

Figure 5-3   How to use a linear layout (part 2 of 2)

How to use a table layout

If you need to create a user interface that displays controls in a grid, you may want to use the table layout as shown in figure 5-4. Here, the table layout displays the first two rows of the Tip Calculator app. To do that, this layout uses four columns. In the first row, the TextView widget is displayed in the first column, and the EditText widget is displayed in the next three columns. In the second row, one widget is displayed in each column.

To create a table layout, you use the TableRow element to define each row. Within each row, you can put as many widgets as you'd like. Then, if you need a widget to span multiple columns, you can use the layout_span attribute to specify the number of columns to span. In this figure, for example, the layout_span attribute of the EditText widget has been set to "3". As a result, this widget spans 3 columns.

A table layout has a couple of advantages over a relative layout. To start, when you use a table layout, you only have to provide the id attribute for the widgets that you need to use with your Java code. In this figure, for example, the widgets that just display text don't have id attributes. In addition, you don't have to provide layout attributes for positioning the widget relative to other widgets. Instead, you just use the layout_span attribute wherever necessary. As a result, you don't need to set as many attributes when you use a table layout.

However, a relative layout has a couple advantages over a table layout. To start, it gives you more control over the positioning and alignment of widgets. Second, a relative layout typically runs faster than a table layout.

A table layout with two rows and four columns

Bill Amount	100		
Percent	15%	-	+

Common attribute for working with table layouts

Attribute	Description
`layout_span`	Specifies the number of columns that the widget should use.

The XML for the table layout

```xml
<?xml version="1.0" encoding="utf-8"?>
<TableLayout xmlns:android="http://schemas.android.com/apk/res/android"
    android:layout_width="match_parent"
    android:layout_height="match_parent" >

    <TableRow>
        <TextView
            android:text="@string/bill_amount_label"
            <!--    TextView widget attributes    -->    />
        <EditText
            android:id="@+id/billAmountEditText"
            android:text="@string/bill_amount"
            android:layout_span="3"
            <!--    EditText widget attributes    -->
            <requestFocus />
        </EditText>
    </TableRow>
    <TableRow>
        <TextView
            android:text="@string/tip_percent_label"
            <!--    TextView widget attributes    -->    />
        <TextView
            android:id="@+id/percentTextView"
            android:text="@string/tip_percent"
            <!--    TextView widget attributes    -->    />
        <Button
            android:id="@+id/percentDownButton"
            android:text="@string/decrease"
            <!--    Button widget attributes    -->    />
        <Button
            android:id="@+id/percentUpButton"
            android:text="@string/increase"
            <!--    Button widget attributes    -->    />
    </TableRow>

</TableLayout>
```

Description

- A table layout displays widgets in rows and columns.

Figure 5-4   How to use a table layout

How to use a frame layout

A frame layout is one of the simplest and most efficient types of layouts. You can use this type of layout to display one widget over another widget as shown in figure 5-5. Or, you can use a frame layout as a placeholder within another layout for displaying a single child layout or widget.

If a frame layout contains multiple child widgets, it stacks them on top of each other. This displays the first widget added to the layout on the bottom of the stack and the last widget added on the top. In this figure, for example, the first child widget is an ImageView widget that displays an image of a restaurant. Then, the second child widget is a TextView widget that displays white text. The result is that white text is displayed over an image.

The ImageView widget in this figure displays an image of a restaraunt. For now, don't worry about the details of how this widget works. These details are explained later in this chapter.

The TextView widget in this figure works like the TextView widgets presented in chapter 3. However, it uses the gravity attribute to center it horizontally. In addition, it uses the textColor attribute to set the text color to a hexadecimal color of "#ffffff", which is white.

If you're familiar with HTML, you may already be familiar with hexadecimal colors, which are a way of represeting RGB color values using the base 16 number system. In this case, "ff" is the hexadecimal representation of the value 255, so #ffffff, translates into red 255, green 255, blue 255, which produces a solid white color. A discussion of how to convert between hexadecimal notation and decimal notation is beyond the scope of this book. However, it's easy to find calculators online that can convert from RGB color notation to hexadecimal notation. Also, you can find charts and tables online that provide hexadimal notations for many of the most commonly used colors.

Interestingly, the DatePicker, TimePicker, and CalendarView classes inherit the FrameLayout class. That's because the DatePicker, TimePicker, and CalendarView widgets consist of different widgets stacked on each other.

A frame layout that displays an image behind some text

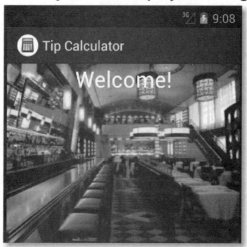

The XML for the frame layout

```xml
<?xml version="1.0" encoding="utf-8"?>
<FrameLayout xmlns:android="http://schemas.android.com/apk/res/android"
    android:layout_width="match_parent"
    android:layout_height="match_parent" >

    <ImageView
        android:layout_width="match_parent"
        android:layout_height="wrap_content"
        android:contentDescription="@string/photo"
        android:src="@drawable/restaurant" />

    <TextView
        android:layout_width="match_parent"
        android:layout_height="wrap_content"
        android:gravity="center"
        android:text="@string/welcome"
        android:textColor="#ffffff"
        android:textSize="30sp" />

</FrameLayout>
```

Description

- A frame layout is one of the simplest and most efficient types of layouts.
- A frame layout often displays only a single child layout or widget.
- If a frame layout contains multiple child widgets, it stacks them on top of each other displaying the first widget added to the layout on the bottom of the stack and the last widget added on the top.

Figure 5-5   How to use a frame layout

How to nest layouts

When working with linear layouts, it's common to *nest* one linear layout within another to align widgets in columns and rows as shown in figure 5-6. Here, two linear layouts with horizontal orientation are nested within a linear layout that has vertical orientation. This creates two rows, which can be filled with widgets. In this figure, the first row has a TextView and an EditText widget, and the second row has a Button widget.

If a layout only has a few levels of nested layouts, the nested layout probably won't degrade performance so much that it's noticeable to the user. However, nesting multiple levels can degrade performance significantly. As a result, if you are using multiple levels of nesting, and you notice that your app is taking too long to display, you should consider reducing the number of nested levels. One way to do that is to use a relative layout as described in chapter 3. When you use a relative layout, you typically don't need to nest any layouts, and that usually improves performance.

Interestingly, the table layout is a type of linear layout. If you look back to figure 5-2, you can see that the LinearLayout class is the superclass for the TableLayout class. As a result, the table layout uses nested layouts behind the scenes even though you don't explicitly specify nested layouts in the XML code for the layout.

Nested linear layouts

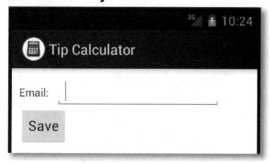

The XML for nested linear layouts

```xml
<?xml version="1.0" encoding="utf-8"?>
<LinearLayout xmlns:android="http://schemas.android.com/apk/res/android"
    android:layout_width="match_parent"
    android:layout_height="match_parent"
    android:orientation="vertical"
    android:padding="10dp" >

    <!-- the first row -->
    <LinearLayout
        android:layout_width="match_parent"
        android:layout_height="wrap_content"
        android:orientation="horizontal" >

        <!--  widgets go here -->

    </LinearLayout>

    <!-- the second row -->
    <LinearLayout
        android:layout_width="match_parent"
        android:layout_height="wrap_content"
        android:orientation="horizontal" >

        <!--  widgets go here -->

    </LinearLayout>

</LinearLayout>
```

Description
- You can *nest* one layout within another layout.
- Nesting multiple levels can degrade performance.

Figure 5-6   How to nest layouts

How to provide a landscape layout

By now, you should know that an Android activity needs to handle screen orientation changes. One way to do that is to prevent the activity from changing orientation. An easy way to do that is to edit the AndroidManifest.xml file for the app, and add a screenOrientation attribute to the activity that sets the screen orientation to portrait or landscape like this:

`android:screenOrientation="portrait"`

This forces the activity to display only in portrait orientation even if the user rotates the device to attempt to change to landscape orientation. This approach is shown in the Android manifest file from chapter 1.

Another way to handle screen orientation changes is to use the same layout for both portrait and landscape. This is the approach that's used for the Tip Calculator app in chapter 3.

However, there are times when you want to provide one layout for portrait orientation and another layout for landscape orientation. For example, you may want to rearrange the widgets on the activity to make better use of the space that's available in landscape orientation. To do that, you can create a directory named res/layout-land as shown in figure 5-7. Then, you can copy the layout for the portrait version of the layout into this directory. After that, you can modify the layout so it makes better use of the space that's available from landscape orientation. In this figure, for example, the two buttons from figure 5-3 have been rearranged to take advantage of landscape orientation.

The layout from figure 5-3 displayed in landscape orientation

The location of the XML files

Portrait
`res/layout/settings_activity.xml`

Landscape
`res/layout-land/settings_activity.xml`

The XML for landscape orientation

```
<?xml version="1.0" encoding="utf-8"?>
<LinearLayout xmlns:android="http://schemas.android.com/apk/res/android"
    android:layout_width="match_parent"
    android:layout_height="match_parent"
    android:orientation="horizontal" >

    <Button
        android:id="@+id/calculateTipButton"
        android:layout_width="match_parent"
        android:layout_height="wrap_content"
        android:layout_weight="1"
        android:text="@string/calculate_tip" />

    <Button
        android:id="@+id/settingsButton"
        android:layout_width="match_parent"
        android:layout_height="wrap_content"
        android:layout_weight="1"
        android:text="@string/settings" />

</LinearLayout>
```

Description

- To provide a separate XML file for landscape orientation, create the layout-land directory if necessary. Then, copy the XML file for the layout into this directory and modify it so it works correctly for landscape orientation.

Figure 5-7   How to provide a custom landscape layout

How to work with widgets

In chapter 3, you learned how to use three kinds of widgets: text views, editable text views, and buttons. Now, this chapter introduces several more widgets that are commonly used, and it presents some new skills for working with editable text views.

How to use editable text views

Figure 5-8 shows some new skills for working with editable text views. To start, this figure shows two editable text views. The first is for getting an email address from a user, and the second is for getting a password from a user. Here, the characters entered into the editable text view for the password are displayed as bullets, which is usually what you want for a password.

The XML for these two editable text views is mostly the same. However, for the first editable text view, the inputType attribute has been set to "textEmailAddress". For the second editable text view, the inputType attribute has been set to "textPassword". As a result, the first editable text view displays a soft keyboard like the one shown in this figure. This soft keyboard is optimized for entering an email address, and it only provides characters that are valid for an email address, which is usually what you want from an editable text view for an email address. Similarly, the second editable text view automatically displays bullets instead of characters, which is what you want from an editable text view for a password.

Two text views and two editable text views

Email:	joel@murach.com
Password:	●●●●●●●●

The soft keyboard for an editable text view for an email address

The XML for an editable text view for an email address

```
<EditText
    android:id="@+id/emailEditText"
    android:layout_width="wrap_content"
    android:layout_height="wrap_content"
    android:ems="10"
    android:inputType="textEmailAddress">
    <requestFocus />
</EditText>
```

The XML for an editable text view for a password

```
<EditText
    android:id="@+id/passwordEditText"
    android:layout_width="wrap_content"
    android:layout_height="wrap_content"
    android:ems="10"
    android:inputType="textPassword" />
```

Figure 5-8   How to use editable text views (part 1 of 2)

Part 2 of figure 5-8 begins by summarizing two commonly used attributes of an EditText widget. Of these, the inputType attribute is the most commonly used since you set this for almost every editable text view. However, you may occasionaly want to use the lines attribute to change the number of lines that are displayed by the editable text view. If you don't include this attribute, the editable text view only displays one line. The user can still use the Return key to enter more lines but will have to scroll up or down to view those lines. To allow the editable text view to display more than one line at a time, you can set the lines attribute to a value of 2 or higher. Then, the editable text view always displays the specified number of lines.

The common values of the inputType attribute show that Android provides for many types of editable text views. For example, you can change the input type so it's appropriate for a person's name, email address, password, number, decimal number, phone number, date, or time. In general, you should select an inputType attribute that's as specific as possible for the text that you want the user to enter. That way, Android can present a soft keyboard that's optimal. However, if you don't find an inputType attribute setting that's right for your app, you can always set this attribute to "text". This allows the user to enter any kind of text. Or, you can set this attribute to "textMultiline". This provides a soft keyboard that includes a Return key that you can use to start new lines.

When working with editable text views, the soft keyboard is sometimes displayed every time the user enters the activity. This typically happens because the activity gives the focus to the editable text view when the user enters the activity, which causes the soft keyboard to be displayed. In other words, every time you start an activity or change the orientation of an activity, Android displays the soft keyboard. Often, this isn't what you want. Instead, when the user enters the activity, you want the activity to display with no soft keyboard. That way, the user can view all of the widgets on the activity. Then, if the user moves the focus to the editable text view, you want to display the soft keyboard to the user.

To prevent the soft keyboard from being displayed automatically when the user enters an activity, you can edit the AndroidManifest.xml file. In this figure, for example, an attribute named windowSoftInputMode has been added to the activity element for the Tip Calculator activity. This attribute has been set to "stateUnchanged". As a result, Android only shows the keyboard if the keyboard was previously opened by the user. This is typically what you want. However, if you always want to hide the soft keyboard when entering an activity, you can set this attribute to "stateHidden".

In some rare cases, you may want to always show the soft keyboard when the user enters an activity. For example, if the widgets only fill the top half of the activity, you may always want to display the soft keyboard on the bottom half. In that case, you can change the windowSoftInputMode attribute to "stateVisible". That way, the user doesn't have to take any action to display the soft keyboard.

Two attributes of an EditText widget

Attribute	Description
inputType	The type of data for the editable text view. This typically determines the type of soft keyboard that's displayed for the user.
lines	The number of lines that are displayed by the editable text view. By default, this is set to 1.

Some common values for the inputType attribute

Value	Description
text	Any kind of text.
textPersonName	A person's name.
textEmailAddress	An email address.
textPassword	A password.
textMultiline	Multiple lines of text.
number	Any kind of number.
numberDecimal	A number that has a decimal.
phone	A phone number.
date	A date.
time	A time.

Some common values for the windowSoftInputMode attribute

Value	Description
stateHidden	Always hide the soft keyboard when entering the activity.
stateVisible	Always show the soft keyboard when entering the activity.
stateUnchanged	Only show the keyboard when entering the activity if it was previously opened by the user.

The activity element in the AndroidManifest.xml file

```
<activity
    android:name=".TipCalculatorActivity"
    android:label="@string/title_activity_tip_calculator"
    android:windowSoftInputMode="stateUnchanged" >
```

Description

- An *editable text view* lets the user enter text with a keyboard. A editable text view is also known as a *text field* or *text box*.

Figure 5-8   How to use editable text views (part 2 of 2)

How to use check boxes

Figure 5-9 shows show to use a *check box*, which is a widget that allows the user to check or uncheck an option. If you look at the XML for this widget, you should already be familiar with most of its attributes. The only new attribute is the checked attribute, which determines whether the box is checked or unchecked. In this figure, for example, the checked attribute has been set to "true", so the box is checked. However, you could set this attribute to "false" to uncheck the box.

When working with check boxes, you can use Java to work with a check box when the app runs. For example, it's common to need to set the checked attribute as shown in the first Java example. To do that, you can use the setChecked method to check or uncheck the box.

To determine whether the box is checked or not, you can use the isChecked method. In this figure, for example, the second Java example uses an if statement to determine whether the box is checked.

As you review this figure, note that the checked attribute and setChecked method both accomplish the same task. In Android, there's typically a corresponding set method for an attribute. As a result, you often have a choice as to whether you want to use XML or Java to work with a widget. Of course, you can only use XML to set attributes before the app is running. If you want to work with a widget dynamically, as the app is running, you need to use Java.

A check box

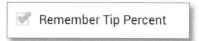

A common XML attribute for check boxes

Attribute	Description
checked	Checks or unchecks the box.

The XML code

```
<CheckBox
    android:id="@+id/rememberPercentCheckBox"
    android:layout_width="wrap_content"
    android:layout_height="wrap_content"
    android:checked="true"
    android:text="@string/remember_percent" />
```

Two common Java methods for check boxes

Method	Description
setChecked(boolean)	Checks or unchecks the box.
isChecked()	Returns a Boolean value that indicates whether the box is checked.

Java examples

Check or uncheck the box

```
rememberPercentCheckBox.setChecked(true);
```

Execute code if the box is checked

```
if (rememberPercentCheckBox.isChecked()) {
    // code to execute when the box is checked
}
else {
    // code to execute when the box is NOT checked
}
```

Description

- A *check box* allows the user to check or uncheck an option.

Figure 5-9   How to use check boxes

How to use radio buttons

Figure 5-10 shows how to use a *radio button*, which is a widget that lets the user select one option of several options. Selecting a radio button automatically deselects all other radio buttons in the same group.

When you add radio buttons, you typically begin by adding two or more radio buttons to a radio group. In this figure, for example, there are three radio buttons in a radio group.

When working with a radio group, you often need to set the orientation for the group. In this figure, for example, the first group of buttons has a vertical orientation, and the second group of buttons has a horizontal orientation.

The XML in this figure defines a radio group that displays three radio buttons in vertical orientation. Here, the orientation attribute of the RadioGroup element has been set to "vertical". Then, three RadioButton elements have been coded within the RadioGroup element. Here, the first radio button has a checked attribute that has been set to true. As a result, the first radio button is selected when the activity is first displayed.

The Java code for working with radio buttons is similar to the code for working with check boxes. For example, you can use the setChecked method to check a radio button, and you can use the isChecked method to determine which button is checked. The main difference is that only one radio button in a group can be checked. As a result, if you use the setChecked method to check one radio button, Android automatically unchecks all of the other radio buttons in that group.

Three radio buttons in a radio group with vertical orientation

- ◉ No Rounding
- ○ Round Tip
- ○ Round Total

Three radio buttons in a group with horizontal orientation

○ No Rounding ◉ Round Tip ○ Round Total

The XML code

```
<RadioGroup
    android:id="@+id/roundingRadioGroup"
    android:layout_width="wrap_content"
    android:layout_height="wrap_content"
    android:orientation="vertical" >

    <RadioButton
        android:id="@+id/noRoundingRadioButton"
        android:layout_width="wrap_content"
        android:layout_height="wrap_content"
        android:checked="true"
        android:text="@string/round_no" />

    <RadioButton
        android:id="@+id/roundTipRadioButton"
        android:layout_width="wrap_content"
        android:layout_height="wrap_content"
        android:text="@string/round_tip" />

    <RadioButton
        android:id="@+id/roundTotalRadioButton"
        android:layout_width="wrap_content"
        android:layout_height="wrap_content"
        android:text="@string/round_total" />

</RadioGroup>
```

Java examples

Check or uncheck the radio button
```
roundTipRadioButton.setChecked(true);
```

Execute code if a radio button is checked
```
if (roundTipRadioButton.isChecked()) {
    // code to execute when the button is checked
}
```

Description

- A *radio button* lets the user select one option of several options. Selecting a radio button automatically deselects all other radio buttons in the same group.

Figure 5-10   How to use radio buttons

How to use spinners

Figure 5-11 shows how to use a *spinner*, which is a widget that allows the user to select an item from a drop-down list. This type of widget is also known as a *drop-down list*. The top of this figure shows a spinner that allows the user to select how many ways to split a bill. The first item doesn't split the bill, the second item splits the bill 2 ways, the third item splits the bill 3 ways, and so on.

The XML for this spinner is simple. That's because most of the work for setting up a spinner is done with Java code as shown in part 2 of this figure.

An *array adapter* provides the list that a spinner should display. One easy way to provide a list is to store it in an array in the strings.xml file. In this figure, for example, the strings.xml file includes an element named split_array that provides the 4 items that are displayed in the spinner.

A spinner

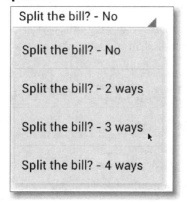

The XML code

```
<Spinner
    android:id="@+id/splitSpinner"
    android:layout_width="wrap_content"
    android:layout_height="wrap_content" />
```

The array in the strings.xml file

```
<string-array name="split_array">
    <item>Split the bill? - No</item>
    <item>Split the bill? - 2 ways</item>
    <item>Split the bill? - 3 ways</item>
    <item>Split the bill? - 4 ways</item>
</string-array>
```

Description

- A *spinner*, also known as a *drop-down list*, allows the user to select an item from a list.
- An *array adapter* provides the list that a spinner should display.

Figure 5-11   How to use spinners (part 1 of 2)

In part 2 of figure 5-11, the first Java example sets up the spinner. To start, it gets a reference to the Spinner object defined by the XML. Then, it creates an array adapter for the specified array and layout. To do that, it specifies the current activity as the context, it specifies an array from the strings.xml file as the array, and it specifies this layout for the spinner:

```
android.R.layout.simple_spinner_item
```

This is a built-in Android resource that's commonly used to specify the layout for a spinner.

After creating the array adapter, this code sets the layout for the drop-down list of the spinner. To do that, it specifies this layout:

```
android.R.layout.simple_spinner_dropdown_item
```

This is another built-in Android resource that's commonly used to specify the layout for the items in the drop-down list of a spinner.

The second Java example selects the item at the specified position. Here, 0 is the position of the first item, 1 is the second item, and so on. As a result, this code selects the first item.

The third Java example gets the position of the selected item from the spinner. Again, 0 is the position of the first item, 1 is position of the second item, and so on.

The fourth Java example gets the selected text from the selected item. To do that, it uses the getSelectedItem method to return a generic object. Then, it casts the generic object to a String object.

Common methods of the ArrayAdapter class

Method	Description
createFromResource(context, arrayID, layoutID)	Creates an array adapter for the specified array and layout for the widget.
setDropDownViewResource (resourceID)	Sets the layout for the items in the drop-down list.

Common methods for spinners

Method	Description
setAdapter(arrayAdapter)	Sets the array adapter for the spinner.
setSelection(index)	Selects an item from the spinner.
getSelectedItemPosition()	Returns an int value for the position of the selected item where 0 is the first item, 1 is the second item, and so on.
getSelectedItem()	Returns an Object type for the selected item.

Two Android resources for spinner layouts

```
android.R.layout.simple_spinner_item
android.R.layout.simple_spinner_dropdown_item
```

Java examples

Code that sets up the spinner

```java
// get a reference to the spinner
splitSpinner = (Spinner) findViewById(R.id.splitSpinner);

// create array adapter for specified array and layout
ArrayAdapter<CharSequence> adapter = ArrayAdapter.createFromResource(
    this, R.array.split_array, android.R.layout.simple_spinner_item);

// set the layout for the drop-down list
adapter.setDropDownViewResource(
    android.R.layout.simple_spinner_dropdown_item);

// set the adapter for the spinner
splitSpinner.setAdapter(adapter);
```

Code that selects an item

```java
splitSpinner.setSelection(0);  // select the first item
```

Code that gets the position of the selected item

```java
int position = splitSpinner.getSelectedItemPosition();
```

Code that gets the selected text from the selected item

```java
String selectedText = (String) splitSpinner.getSelectedItem();
```

Description

- You can use built-in Android resources to specify the layouts for the spinner and the items in the drop-down list of a spinner.

Figure 5-11   How to use spinners (part 2 of 2)

How to use seek bars

Figure 5-12 shows a *seek bar*, which is a widget that lets the user specify a value by dragging a *thumb* to the right or left. In this figure, the thumb of the seek bar is identified by a round circle. This thumb is set at a progress of 15 out of a total of 30, and the value for this progress is displayed in a label to the right of the seek bar.

By default, the max attribute of a seek bar is set to 100. As a result, a seek bar typically provides progress values from 0 to 100, which is usually what you want.

In this figure, however, the XML sets the max attribute for the seek bar to 30. As a result, the seek bar provides progress values from 0 to 30, which are more appropriate for a tip percent. Then, this XML sets the progress attribute to 15. As a result, when the seek bar is first displayed, the thumb is half way across the seek bar.

The first Java example sets the progress of the seek bar to a value of 20. To do that, it passes an int value of 20 to the setProgress method.

The second Java example gets the progress of the seek bar. To do that, it uses the getProgress method to return an int value for the progress.

A seek bar and a text view

Two common XML attributes for seek bars

Attribute	Description
`max`	Sets the maximum value to the specified int value.
`progress`	Sets the value of the seek bar to the specified int value.

The XML code

```
<SeekBar
    android:id="@+id/percentSeekBar"
    android:layout_width="200dp"
    android:layout_height="wrap_content"
    android:max="30"
    android:progress="15" />

<TextView
    android:id="@+id/percentTextView"
    android:layout_width="0dp"
    android:layout_height="wrap_content"
    android:text="@string/percent" />
```

Two common Java methods for seek bars

Method	Description
`setProgress(int)`	Sets the value of the seek bar to the specified int value.
`getProgress()`	Gets the current value of the seek bar.

Java examples

How to set progress
```
percentSeekBar.setProgress(20);
```

How to get progress
```
int percent = percentSeekBar.getProgress();
```

Description

- A *seek bar* lets the user specify a value by dragging a *thumb* to the right or left.

Figure 5-12   How to use seek bars

How to display images

The ImageView widget in figure 5-13 displays an image of a restaraunt. Here, the contentDescription attribute points to a string in the strings.xml file. Android displays this text instead of the image if there's a problem displaying the image. Then, the src attribute points to an image in one of the drawable directories of the project.

An image that can be drawn on the screen is known as a *drawable resource*. Drawable resources are stored in one of the res/drawable directories of the project. When you work with image files, you typically store them in a directory that corresponds with the density of the screen. For example, you store images for extra high-density screens in this directory:

`res/drawable-xhdpi`

Extra high-density screens use approximately 320 dots per inch (dpi).

For this example, I put a JPG file for the image in the medium-density (mdpi) directory because medium-density is the baseline size for Android. Then, if there's no corresponding image for screens of other densities, Android does its best to scale the medium-density image for screens of other densities.

If you need more control over the size of your image on various screens, you can create image files for screens of different densities. For example, let's say you have a 300x300 pixel image that displays correctly on a high-density screen. Then, you can put the file for that image in this directory:

`res/drawable-hdpi`

For a medium-density screen, you can resize that image so its 200x200 pixels and put it in this directory:

`res/drawable-mdpi`

This works because, to be displayed at the same size, an image for a high-density (240dpi) screen should be 50% larger than the same image for a medium density (160dpi) screen.

To help automate the resizing of images, you can go to the Android Asset Studio web page and use it to generate images of the correct size for the different density screens. Then, you can put those images in the correct directories. Although this web page isn't an official Google tool, it's widely used by Android developers, and you should be able to find it by searching the Internet.

Android supports PNG, JPG, and GIF images. It's generally considered a best practice to use PNG images for illustrations or drawings and JPG images for photographs. In general, it's considered a best practice to avoid using GIF images.

An image

The location of the image file

```
res/drawable-mdpi/restaurant.jpg
```

Two attributes of an ImageView widget

Value	Description
contentDescription	A string that describes the image. This string is only displayed if Android isn't able to display the image.
src	The location of the *drawable resource*, which is an image that can be drawn on the screen.

Four qualifiers for the drawable folder

Qualifier	Description
xhdpi	Extra high-density screen (approximately 320dpi).
hdpi	High-density screen (approximately 240dpi).
mdpi	Medium-density screen (approximately 160dpi).
ldpi	Low-density screen (approximately 120dpi).

The XML code

```
<ImageView
    android:layout_width="match_parent"
    android:layout_height="wrap_content"
    android:contentDescription="@string/photo"
    android:src="@drawable/restaurant" />
```

Description

- Android supports PNG, JPG, and GIF images. However, its API documentation discourages using GIF images.

Figure 5-13   How to display an image

How to show and hide widgets

Figure 5-14 shows how to use Java code to show or hide widgets. To start, this figure shows a layout with three rows. Here, the second row is the "Per Person" label and amount. The code in this figure shows this row when the bill is being split between multiple people, but hides this row when the bill is not being split.

In the Java code example, the first statement gets the position of the selected item from the Spinner widget. Then, the second statement adds an int value of 1 to the selected position and stores the result in a variable named split. That way, the split variable is equal to 1 if the bill is not being split, 2 if the bill is being split by two people, and so on. After that, an if statement checks whether the bill is being split. If not, this code uses the setVisible method to hide both widgets in the "Per Person" row. Otherwise, this code calculates the amount per person and uses the setVisible method to show both widgets in the "Per Person" row.

A layout with three rows

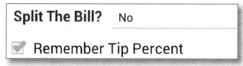

The same layout with the per person amount hidden

A method of the View class

Method	Description
`setVisibility(visibility)`	Hides the widget when set to the GONE constant of the View class.
	Shows the widget when set to the VISIBLE constant of the View class.

Code that shows and hides the per person amount

```
int splitPosition = splitSpinner.getSelectedItemPosition();
int split = splitPosition + 1;
float perPersonAmount = 0;
if (split == 1) {  // no split - hide widgets
    perPersonLabel.setVisibility(View.GONE);
    perPersonTextView.setVisibility(View.GONE);
}
else { // split - show widgets
    perPersonAmount = totalAmount / split;
    perPersonLabel.setVisibility(View.VISIBLE);
    perPersonTextView.setVisibility(View.VISIBLE);
}
```

Description

- You can use Java code to dynamically show or hide widgets.

Figure 5-14   How to show and hide widgets

How to add scroll bars

Figure 5-15 shows how to add scroll bars to a layout. To do that, you typically add a layout within a ScrollView element. Then, if the layout doesn't fit on the screen, Android displays a *vertical scroll bar* so you can scroll through it.

In this figure, the XML begins by declaring a ScrollView element as the root element of a layout file. This ScrollView element contains a single TableLayout element. In turn, this layout can contain multiple TableRow elements, which can contain elements for other widgets. Then, if these widgets become too tall for the screen to display, the user can scroll up or down through these widgets.

In most cases, you only need to provide a vertical scroll bar for your layouts. However, you may occasionally need to provide a *horizontal scroll bar* that lets the user scroll right or left. To do that, you can use a HorizontalScrollView element. Since this element works similarly to the ScrollView element, you shouldn't have much trouble figuring out how to use it.

A scrollable layout

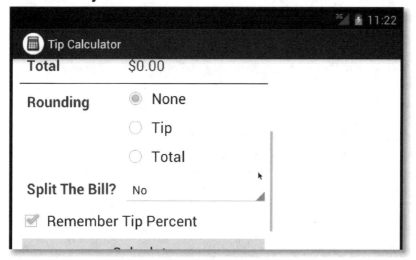

The XML

```xml
<ScrollView xmlns:android="http://schemas.android.com/apk/res/android"
    android:layout_width="wrap_content"
    android:layout_height="wrap_content">

  <TableLayout
    android:layout_width="match_parent"
    android:layout_height="wrap_content"
    android:padding="10dp" >

    <!-- All table rows and widgets go here -->

  </TableLayout>
</ScrollView>
```

Description

- A ScrollView widget can only have one child element, typically a layout that contains other elements.

- A ScrollView widget displays a *vertical scroll bar* that lets the user scroll up or down the child element.

- To display a *horizontal scroll bar* that lets the user scroll right or left across the child element, use a HorizontalScrollView element.

Figure 5-15   How to display scroll bars

How to display web content

Figure 5-16 shows a *web view*, which is a widget that you can use to display web content. To do that, you typically add a WebView widget to an activity. This widget uses a built-in browser to display web content. As a result, it can display any type of content that a web browser can display.

One common use of a web view is to create an app whose only purpose is to display a website that has been designed to work with mobile devices. This creates an app that works much like using the browser app on your Android device to open a website. However, it includes a title bar for the app, and it makes it possible for you to deploy the app to the Google Play store. When you create such an app, you typically want the web view to stretch across the entire screen. In this figure, for example, the app uses a web view to display the website for murach.com.

It's also possible to use a web view to display web content on part of the screen. For example, you might want to display widgets on the top of the screen that let you determine what web content to display on the bottom part of the screen.

When you use a web view to load a URL from the web, you often use a ProgressBar widget to display a *progress bar* while the WebView widget loads the URL. In this figure, for example, the app displays a circular progress bar on top of the WebView widget while content is loading. This occurs when the app starts and when the user clicks on a link that loads another URL within the website.

Before you can use a web view to load content from the web, you must add the INTERNET permission to the Android manifest. To do that, you must make sure to add a uses-permission element like the one shown in this figure after the manifest element but before the application element. If you add this element anywhere else, such as just before the activity element, Android doesn't give the WebView widget permission to access the Internet.

When an activity only displays a web view and a progress bar as shown in this figure, you typically want to add a configChanges attribute like the one shown in this figure to the activity element. If you don't, the web view has to completely reload the page each time the user changes orientation, which can slow down your app significantly. However, if you include this attribute, the activity automatically handles the change when the user opens or closes a physical keyboard, changes orientation, or changes the screen size (which happens when the user changes orientation). This is usually what you want for an activity like the one shown in this figure, but it isn't usually what you want for other types of activities such as the Tip Calculator activity described earlier in this book.

An activity that displays web content

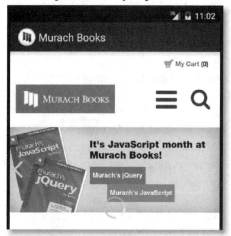

The Android manifest

An element that must be added just before the application element

```
<uses-permission android:name="android.permission.INTERNET" />
```

An attribute that you often want to add to the activity element

```
<activity
    android:name="com.murach.books.MainActivity"
    android:label="@string/title_activity_main"
    android:configChanges="keyboardHidden|orientation|screenSize">
```

The layout of the activity

```
<FrameLayout xmlns:android="http://schemas.android.com/apk/res/android"
    xmlns:tools="http://schemas.android.com/tools"
    android:layout_width="match_parent"
    android:layout_height="match_parent"
    tools:context="com.murach.books.MainActivity">

    <WebView
        android:layout_width="match_parent"
        android:layout_height="match_parent"
        android:id="@+id/webView" />

    <ProgressBar
        android:layout_width="wrap_content"
        android:layout_height="wrap_content"
        android:id="@+id/progressBar"
        android:layout_gravity="center" />

</FrameLayout>
```

Figure 5-16   How to display web content (part 1 of 2)

Part 2 of this figure shows the Java code that works with the web view and progress bar shown in part 1. To start, the top of this figure shows the package that stores the classes for working with a web view, the android.webkit package. This differs from the package for working with most other widgets, the android.widget package.

Most of the code for working with the web view and progress bar is stored in the onCreate method of the activity. To start, this code gets references to these widgets. Then, this code enables JavaScript for the web view. By default, JavaScript is disabled. As a result, if you want to allow JavaScript to run, you need to add the line of code shown in this figure. This line calls the getSettings method to get the settings for the web view. Then, it calls the setJavaScriptEnabled method to enable JavaScript.

After enabling JavaScript, this code specifies what to do if the user clicks on a link within the web view that loads another URL. By default, this would start a browser app on the device and load the URL in that browser. However, for a web view like the one in this figure, you want the web view to load the URL. To do that, this code creates a new WebViewClient object and overrides the method that controls URL loading so that it returns a false value. Then, it uses the setWebViewClient method to set that WebViewClient object in the web view.

After setting the URL loading option, this code specifies when to display the progress bar. To do that, this code creates a new WebChromeClient object and overrides the method that's executed when the progress of loading a URL changes. Within this method, an if statement begins by checking the progress parameter. If this parameter is equal to 100, the URL has finished loading. As a result, the code hides the progress bar. Otherwise, the code displays the progress bar. Finally, this code uses the setWebChromeClient method to set the WebChromeClient object in the web view.

The WebViewClient and WebChromeClient classes provide other methods that you can override to control other aspects of how the web view works. However, you only need to override the methods shown in this figure to create an app for a website that's designed to work with a mobile device.

After specifying when to display the progress bar, this code finishes by using the loadURL method to load the URL. Here, the URL is for our company's website (www.murach.com). For this statement to work correctly, it must be coded after the statements that configure the web view. In other words, it must be coded last.

If you aren't familiar with anonymous inner classes or event handling, you might have trouble understanding some of this code. However, most of this code is boilerplate code that you can copy into your apps to get started. Then, you can learn more about anonymous inner classes and events in the next chapter.

The package that stores the classes for working with web views

```
android.webkit
```

Code from the onCreate method of the activity

Get references to the web view and progress bar

```
WebView webView =
    (WebView) findViewById(R.id.webView);
final ProgressBar progressBar =
    (ProgressBar) findViewById(R.id.progressBar);
```

Enable JavaScript for the web view

```
webView.getSettings().setJavaScriptEnabled(true);
```

Load URLs in the web view instead of in a separate browser app

```
webView.setWebViewClient(new WebViewClient() {
    @Override
    public boolean shouldOverrideUrlLoading(WebView view, String url) {
        return false;
    }
});
```

Display the progress bar until the page is 100% loaded

```
webView.setWebChromeClient(new WebChromeClient() {
    public void onProgressChanged(WebView view, int progress) {
        if (progress == 100) {
            progressBar.setVisibility(View.GONE);
        }
        else {
            progressBar.setVisibility(View.VISIBLE);
        }
    }
});
```

Load the content from the specified URL into the web view

```
webView.loadUrl("http://www.murach.com/");
```

Description

- A *web view* uses a built-in browser to display web content. As a result, it can display any type of content that a web browser can display such as a web page that includes HTML, CSS, and JavaScript.

- A *progress bar* displays a visual indicator of the progress of an operation.

Figure 5-16   How to display web content (part 2 of 2)

Perspective

Now that you've finished this chapter, you should understand how to work with widgets on a layout. Although this chapter presented some of the most commonly used widgets, you'll learn how to use more widgets as you progress through this book. For example, in chapter 10, you'll learn how to work with the ListView widget.

In addition, once you learn how to use the widgets presented in this chapter, it should become easier to learn how to use other widgets. In general, you can use the same process to work with most widgets. To start, you can drag a widget from the Palette onto a layout. Then, you use the graphical layout editor to set the properties of the widget. Finally, you write Java code that uses the methods of the widget to work with it. To learn more about a widget, you can use the Android API documentation, or you can search the Internet for other information such as tutorials.

To get started with the widgets presented in this chapter, you can handle events like the Click event described in chapter 3. Then, you can use the methods shown in this chapter to work with the widgets when a Click event occurs. However, to provide a more responsive user interface, you often need to handle other events such as the ones described in the next chapter. That's why the next chapter starts by describing how to handle high-level events that occur when the user interacts with some of the widgets presented in this chapter.

Terms

label	array adapter
editable text view	seek bar
text box	thumb
text field	drawable resource
drop-down list	vertical scroll bar
nest	horizontal scroll bar
check box	web view
radio button	progress bar
spinner	

Summary

- The View class is the superclass for all widgets.
- A linear layout displays a column or row of child widgets.
- The attributes of a linear layout and its child widgets control the appearance of the user interface.
- A table layout displays widgets in rows and columns.
- A frame layout is one of the simplest and most efficient types of layouts, and often displays only a single child layout or widget.

- You can *nest* one layout within another layout, but it can degrade performance.

- An *editable text view* lets the user enter text with a keyboard. An editable text view is also known as a *text box* or *text field*.

- A *check box* allows the user to check or uncheck an option.

- A *radio button* lets the user select one option of several options.

- A *spinner*, also known as a *drop-down list*, allows the user to select an item from a list.

- An *array adapter* provides the list that a spinner should display.

- A *seek bar* lets the user specify a value by dragging a *thumb* to the right or left.

- Android supports PNG, JPG, and GIF images.

- You can use Java code to dynamically show or hide widgets.

- A ScrollView widget displays a *vertical scroll bar* that lets the user scroll up or down the child element.

- To display a *horizontal scroll bar* that lets the user scroll right or left across the child element, use a HorizontalScrollView element.

- A *web view* uses a built-in browser to display web content such as a web page that's designed for mobile devices.

- A *progress bar* displays a visual indicator of the progress of an operation.

Exercise 5-1 Modify the layout for the Tip Calculator app

In this exercise, you'll modify the Tip Calculator app so it uses a table layout instead of a relative layout.

Test the relative layout version of the app

1. Open the project named ch05_ex1_TipCalculator that's in the ex_starts directory.

2. Test the Tip Calculator app. It should work as it did in previous chapters.

3. Open the layout for the activity and view it in the graphical editor.

Use a table layout

4. Click on the Text tab to view the XML for the layout.

5. Change the RelativeLayout element to a TableLayout element.

6. Add TableRow elements to identify the rows of the table.

7. Review the XML for each widget and delete any old attributes that were needed for a relative layout but aren't needed for a table layout. Many of these will be marked as invalid by the XML editor.

8. Switch to the graphical editor to see if the widgets are being displayed correctly. Note that the buttons for setting the tip percent aren't displayed correctly.

9. Add the layout_span attribute to any widgets that should span multiple columns. For example, the EditText widget should span 3 columns. This should result in a grid that has four rows and four columns.

10. Switch to the graphical layout editor to view the widgets. They should be displayed correctly.

11. Test the new layout by running the app. The app should work correctly.

Exercise 5-2   Add radio buttons and a spinner

In this exercise, you'll modify the Tip Calculator app so that it uses radio buttons and a spinner. When you're done, the bottom of the layout should look like this:

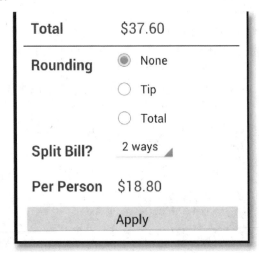

Test the app

1. Open the project named ch05_ex2_TipCalculator that's in the ex_starts directory.

2. Test this app to see how it works. Note that the Apply button doesn't do anything yet.

Add radio buttons

3. Open the layout and add a radio group that contains the three radio buttons shown above.

4. Open the class for the activity and modify the calculateAndDisplay method so it uses the selected radio button to determine whether to round the tip, total, or nothing. To round the tip, you can use code like this:

    ```
    tipAmount = StrictMath.round(billAmount * tipPercent);
    totalAmount = billAmount + tipAmount;
    ```

    To round the total, you can use code like this:

    ```
    float tipNotRounded = billAmount * tipPercent;
    totalAmount = StrictMath.round(billAmount + tipNotRounded);
    tipAmount = totalAmount - billAmount;
    ```

5. Test this change to make sure it works correctly. Note that the rounding is only displayed when you click the Apply button.

6. Modify the code so it automatically selects the None radio button when the user clicks on the increase (+) or decrease (-) buttons.

Add a spinner

7. Add a TextView widget and Spinner widget to the layout so the user can select the number of ways to split the bill as shown above.

8. Add TextView widgets to the layout that can display the amount per person when the bill is split.

9. Open the strings.xml file and add the array for the spinner.

10. Open the class for the activity and add the code that loads the spinner with the array.

11. Test this change to make sure it works correctly.

12. Modify the calculateAndDisplay method so it calculates and displays the correct amount per person when the bill is split. However, when the bill isn't split, this code should hide the TextView widgets that display the amount per person.

13. Test this change to make sure it works correctly.

6

How to handle events

In chapter 3, you learned one technique for handling events that occur on EditText and Button widgets. Now, this chapter expands on that knowledge to show you several techniques for handling events. In addition, it shows how to work with different types of events including events that occur on the CheckBox, RadioButton, RadioGroup, Spinner, and SeekBar widgets described in the previous chapter.

A summary of listeners

A *listener* is an object that listens for *events* that can occur on a widget. When an event occurs on a widget, the appropriate method of the listener is executed. Figure 6-1 summarizes some common listeners and divides them into two groups. Although this summary is far from complete, it shows how listeners work.

High-level events

High-level events only occur on specific types of widgets. For example, the EditorAction event only occurs on certain types of widgets such as an EditText widget. However, this event can't occur on other types of widgets such as a Button widget.

An interface that defines a listener for a high-level event is typically nested in the class for its corresponding widget. For example, the OnEditorActionListener interface is nested within the TextView class. As a result, this listener can work with TextView widgets or other widgets that inherit the TextView class such as the EditText widget.

Most listener interfaces define a single method for the event handler. For example, the OnEditorActionListener interface defines a single method named onEditorAction. However, some listener interfaces define multiple methods. For example, the OnItemSelectedListener interface defines two methods.

Low-level events

Low-level events occur on all types of widgets. The listeners for low-level events are nested within the View class. Since this class is the superclass for all widgets, these listeners can be wired to any type of widget. For example, the OnClickListener can be wired to any widget. This is necessary because a user can click on any type of widget.

Besides clicking on a widget and quickly releasing, it's common to click on a widget and hold the click for more than a second. This is known as a *long click*. You can use the OnLongClickListener to handle this type of event.

For phones that have hardware such as a keyboard or a DPad, it's common to use that hardware to enter text or to navigate through an app by moving the *focus* from one widget to another. Typically, Android handles these low-level events the way you want. However, if you need to modify Android's default behavior, you can use the OnKeyListener or OnFocusChangedListener to control how the hardware works with your app.

For some apps, such as games, you need more control over what happens when the user touches the screen. To get this control, you can implement the OnTouchListener.

Listeners for high-level events in the android.widget package

Class	Nested Interface	Methods
EditText	OnEditorActionListener	onEditorAction
CompoundButton	OnCheckedChangedListener	onCheckedChanged
RadioGroup	OnCheckedChangedListener	onCheckedChanged
AdapterView	OnItemSelectedListener	onItemSelected
		onNothingSelected
SeekBar	OnSeekBarChangeListener	onProgressChanged
		onStartTrackingTouch
		onStopTrackingTouch

Listeners for low-level events in the android.view package

Class	Nested Interface	Methods
View	OnClickListener	onClick
	OnLongClickListener	onLongClick
	OnKeyListener	onKey
	OnFocusChangeListener	onFocusChange
	OnTouchListener	onTouch

Description

- A *listener* listens for *events* that can occur on a widget. When an event occurs on a widget, the appropriate method of the listener is executed.

- *High-level events* are events that occur on a specific type of widget. An interface that defines a listener for a high-level event is typically nested within the class that defines the widget. These listeners can only be wired to an appropriate widget.

- *Low-level events* are events that occur on all types of widgets. An interface that defines a listener for a low-level event is typically nested within the android.view.View class. These listeners can be wired to any type of widget.

Figure 6-1   A summary of listeners

Four techniques for handling events

In chapter 3, you learned one technique for handling an event. Now, figure 6-2 reviews that technique. Then, it shows three other techniques for handling the same event.

Regardless of the technique you choose, you usually start by importing the class for the listener. That's why this is shown as step 1 for all four techniques.

How to use the current class as the listener

The first example shows how to handle the Click event for two buttons using the current class as the listener. By now, you should be familiar with this code since it's the same code that was presented in chapter 3. This code uses the current class (the TipCalculatorActivity class) to implement the onClick method defined by the OnClick listener interface. Then, it wires the listener to the two buttons. To do that, it uses the keyword named *this* to identify the current class as the listener.

How to use a named class as the listener

The second example shows how to handle the Click event for two buttons by creating a separate class named ButtonListener that handles the event. To do that, the ButtonListener class implements the listener interface. This class is stored in the same file as the class for the activity. Then, the code in the activity class creates a listener object from the listener class and assigns it to a variable named buttonListener. Finally, the activity class wires the listener to the two buttons.

Step 1: Import the interface for the listener

```
import android.view.View.OnClickListener;
```

Use the current class as the listener

Step 2a: Implement the interface for the listener

```
public class TipCalculatorActivity extends Activity
implements OnClickListener {
```

Step 2b: Implement the interface for the listener

```
@Override
public void onClick(View v) {
    switch (v.getId()) {
        case R.id.percentDownButton:
            tipPercent = tipPercent - .01f;
            calculateAndDisplay();
            break;
        case R.id.percentUpButton:
            tipPercent = tipPercent + .01f;
            calculateAndDisplay();
            break;
    }
}
```

Step 3: Set the listeners

```
percentUpButton.setOnClickListener(this);
percentDownButton.setOnClickListener(this);
```

Use a separate named class as the listener

Step 2: Code a separate class that implements the listener

```
class ButtonListener implements OnClickListener {
    @Override
    public void onClick(View v) {
        switch (v.getId()) {
            case R.id.percentDownButton:
                tipPercent = tipPercent - .01f;
                calculateAndDisplay();
                break;
            case R.id.percentUpButton:
                tipPercent = tipPercent + .01f;
                calculateAndDisplay();
                break;
        }
    }
}
```

Step 3: Create an instance of the listener

```
ButtonListener buttonListener = new ButtonListener();
```

Step 4: Set the listeners

```
percentUpButton.setOnClickListener(buttonListener);
percentDownButton.setOnClickListener(buttonListener);
```

Figure 6-2   Four techniques for handling events (part 1 of 2)

How to use an anonymous class as the listener

The third example shows how to handle the Click event for two buttons without specifying a name for the class that implements the listener. In other words, this code uses an *anonymous class* as the listener. Here, step 2 defines a private instance variable for an OnClickListener object named buttonListener. Then, it uses the new keyword to create the listener object, and it supplies all code for the class that defines the object within the braces ({ }). Next, step 3 uses the name of the instance variable to wire the listener to the two buttons.

How to use an anonymous inner class as the listener

The fourth example shows how to handle the Click event for two buttons without assigning the listener object to an instance variable. To do that, you can create the object for the listener within the parentheses for the method that wires the widget to the listener. This is known as an *anonymous inner clas*.

An anonymous inner class is useful when you want to use a different event handler for each widget. In this figure, for example, the onClick method must be coded for both buttons. Here, the first onClick method is for the button that increases the tip percent, and the second onClick method is for the button that decreases the tip percent. When you use this technique, you know that the code for the onClick method will only be executed for a specific widget. As a result, you don't need to begin this method by using a switch statement to check which widget was clicked.

Conversely, the first three techniques described in this figure are useful when you want to use one event handler to handle an event for multiple widgets. In this figure, for example, the code for the first three examples uses a single onClick method to handle the Click event for two buttons.

When to use each technique

In most cases, you can use any of these four techniques for your app. So, which technique should you use? In most cases, that's largely a matter of personal preference. However, if you need to wire the exact same event handler to multiple widgets, you should reduce code duplication by using one of the first three techniques.

In general, I recommend using the technique that results in the most readable and maintainable code. I also recommend using the same technique throughout an app whenever possible for the sake of consistency. Throughout this book, I have used the first technique because I think it's the easiest to understand when you're getting started with Android. However, all four techniques are commonly used by professional Android programmers.

Use an anonymous class as the listener

Step 2: Create an instance variable for the listener

```
private OnClickListener buttonListener = new OnClickListener() {
    @Override
    public void onClick(View v) {
        switch (v.getId()) {
            case R.id.percentDownButton:
                tipPercent = tipPercent - .01f;
                calculateAndDisplay();
                break;
            case R.id.percentUpButton:
                tipPercent = tipPercent + .01f;
                calculateAndDisplay();
                break;
        }
    }
};
```

Step 3: Set the listeners

```
percentUpButton.setOnClickListener(buttonListener);
percentDownButton.setOnClickListener(buttonListener);
```

Use an anonymous inner class as the listener

Step 2: Set the listeners and implement the interfaces for the listeners

```
percentUpButton.setOnClickListener(new OnClickListener() {
    @Override
    public void onClick(View v) {
        tipPercent = tipPercent + .01f;
        calculateAndDisplay();
    }
});

percentDownButton.setOnClickListener(new OnClickListener() {
    @Override
    public void onClick(View v) {
        tipPercent = tipPercent - .01f;
        calculateAndDisplay();
    }
});
```

Description

- You can create an instance variable that creates an object from a class that implements the listener interface. Since this class doesn't have a name, it's known as an *anonymous class*.

- You can create an instance of a listener interface without assigning it to an instance variable. This is known as an *anonymous inner class*.

Figure 6-2   Four techniques for handling events (part 2 of 2)

How to handle high-level events

In chapter 3, you learned how to handle the EditorAction event, which is a high-level event. Now, you'll learn how to handle high-level events for a few other widgets including check boxes, radio buttons, radio groups, spinners, and seek bars.

How to handle events for check boxes and radio buttons

Figure 6-3 shows how to handle the CheckedChanged event that occurs when a check box or radio button is checked or unchecked. To do that, you implement the OnCheckedChangedListener that's nested in the CompoundButton class and wire it to a check box or radio button. In this figure, for example, the onCheckedChanged method is executed when the user checks or unchecks the Remember Tip Percent check box.

The first code example shows how to use both parameters of the onCheckedChanged method. Here, the first parameter is a CompoundButton object for the check box or radio button that was clicked. As a result, you can call the getId method from this object to get its ID. Then, you can use a switch statement to execute the appropriate code for each widget. In this figure, there's only one check box, so this switch statement isn't necessary. However, if you have multiple check boxes or radio buttons, this statement is usually necessary.

The second parameter of the onCheckedChanged method is a Boolean value that indicates whether the widget is checked. As a result, you can use an if statement to determine whether the widget is checked. Then, you can execute the appropriate code. In this figure, the code just sets the value of a Boolean variable named rememberTipPercent.

The second code example shows simplified code for the event handler. This event handler assumes that it is only wired to a single check box. As a result, it doesn't use a switch statement to determine which widget was checked. Similarly, this code doesn't use an if statement to execute different code depending on whether the check box is checked. Instead, it uses the second parameter to set the value of the variable named rememberTipPercent.

Although the code in this figure is for check boxes, you can use similar skills for working with radio buttons whenever that's necessary. However, when you work with radio buttons, you often handle the CheckedChanged event that occurs on a radio group as described in the next figure instead of handling this event for individual buttons.

A check box

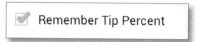

Remember Tip Percent

An event handler for a check box

```
@Override
public void onCheckedChanged(CompoundButton widget, boolean isChecked) {
    switch (widget.getId()) {
        case R.id.rememberPercentCheckBox:
            if (isChecked) {
                rememberTipPercent = true;
            }
            else {
                rememberTipPercent = false;
            }
            break;
    }
}
```

Another event handler for a check box

```
@Override
public void onCheckedChanged(CompoundButton widget, boolean isChecked) {
    rememberTipPercent = isChecked;
}
```

A method of the View class

Method	Description
getId()	Gets the ID of the widget on which the event occurred.

Description

- The CheckedChanged event of a check box or radio button occurs when a check box or radio button is checked or unchecked.
- The first parameter of the onCheckedChanged method is a CompoundButton object. This object can be cast to a CheckBox or RadioButton object.
- The second parameter of the onCheckedChanged method is a Boolean value that indicates whether the check box or radio button is checked.

Figure 6-3   How to handle events for check boxes and radio buttons

How to handle events for radio groups

Figure 6-4 shows how to handle the CheckedChanged event that occurs when a new radio button within a radio group is checked. To do that, you implement the OnCheckedChangedListener that's nested in the RadioGroup class and wire it to a radio group. In this figure, for example, the onChecked-Changed method is executed when the user selects a new radio button from the radio group.

The first code example shows how to use the second parameter of this onCheckedChanged method, which provides the ID of the new button that is checked. This code example uses the second parameter in a switch statement to execute the appropriate code depending on which radio button was clicked. Here, the code in the switch statement sets the value of a variable named rounding to a constant. After the switch statement, this code calls the calculateAndDisplay method. Although this method isn't shown here, it performs a calculation using the round variable and displays the results of the calculation on the user interface. As a result, the calculation is made whenever the user selects a new radio button from the group.

Three radio buttons in a group

| ○ No Rounding ◉ Round Tip ○ Round Total |

An event handler for a radio group

```
@Override
public void onCheckedChanged(RadioGroup group, int checkedId) {
    switch (checkedId) {
        case R.id.noRoundingRadioButton:
            rounding = ROUND_NONE;
            break;
        case R.id.roundTipRadioButton:
            rounding = ROUND_TIP;
            break;
        case R.id.roundTotalRadioButton:
            rounding = ROUND_TOTAL;
            break;
    }
    calculateAndDisplay();
}
```

Another event handler for a radio group

```
@Override
public void onCheckedChanged(RadioGroup group, int checkedId) {
    calculateAndDisplay();
}
```

Description

- The CheckedChanged event of a radio group occurs when a new button within that group is checked.

- The first parameter of the onCheckedChanged method is the RadioGroup object for the radio group.

- The second parameter of the onCheckedChanged method is the ID of the radio button within the group that is checked.

Figure 6-4   How to handle events for radio groups

How to handle events for spinners

Figure 6-5 shows how to handle the events that occur when a spinner is first displayed and when a new item is selected from a spinner. To do that, you implement the OnItemSelectedListener that's nested in the AdapterView class and wire it to a spinner. In this figure, for example, the onItemSelected method is executed when the user selects a new item from the spinner.

Within the onItemSelected method, the third parameter provides the position of the selected item. In most cases, this is the only parameter that you need. However, if necessary, the first parameter provides the AdapterView object for the list of items, the second parameter provides a View object for the selected item, and the fourth parameter provides the ID of the selected item.

Within the onItemSelected method, the first statement adds 1 to the position of the selected item and stores this value in a variable named split. That way, the split variable stores a value of 1 for the first item, 2 for the second item, and so on. This is necessary because the first item in a spinner has a position of 0. Then, the second statement calls the calculateAndDisplay method to perform a calculation that uses the split variable and to display the results to the user.

Since the onNothingSelected method is defined by the interface for the listener, you must implement this method. However, the onNothingSelected method is only executed if the selection disappears. For example, the selection can disappear when the adapter for the spinner becomes empty. Since this rarely happens, you typically don't need to include code for this method.

A spinner

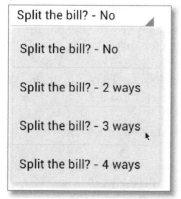

An event handler for a spinner

```java
@Override
public void onItemSelected(AdapterView<?> parent, View v, int position,
        long id) {
    split = position + 1;
    calculateAndDisplay();
}

@Override
public void onNothingSelected(AdapterView<?> parent) {
    // You typically don't need to include any code here
}
```

Description

- The onItemSelected method is executed when the spinner is first displayed and whenever a new item is selected. However, it isn't executed when the user selects an item that was already selected.

- Within the onItemSelected method, the third parameter provides the position of the selected item.

- The onNothingSelected method is only executed when the selection disappears. Since this rarely happens, you typically don't need to include code for this method.

Figure 6-5   How to handle events for spinners

How to handle events for seek bars

Figure 6-6 shows how to handle the events that occur on a seek bar. To do that, you implement the OnSeekBarChangeListener that's nested in the SeekBar class and wire it to a seek bar.

The interface for this listener specifies three methods: onStartTrackingTouch, onProgressChanged, and onStopTrackingTouch. The onStartTrackingTouch method is executed when the user begins to change the progress of the seek bar, the onProgressChanged method is executed as the user changes the progress of the seek bar, and the onStopTrackingTouch method is executed when the user finishes changing the progress of the seek bar. Since all three methods are specified by the interface, you must implement all three of these methods. However, you only need to provide code for the methods that you intend to use. In this figure, for example, the onStartTrackingTouch method doesn't contain any code.

The onProgressChanged method contains a single statement. This statement uses the second parameter to get the progress value for the seek bar. Then, it creates a string by appending a percent sign (%) to the progress value, and it uses the setText method of a TextView widget named percentTextView to display this string. As a result, users get immediate feedback on the progress value as they change the value of the seek bar. This is necessary to provide a responsive user interface.

The seek bar can also be used as a progress tracking bar that's updated by the app instead of by the user. For example, the app could use a seek bar to display the progress of a download. As a result, you might need to determine whether the user updated the seek bar by moving the slider, or whether the seek bar was updated programmatically. To do this, you can check the value of the fromUser parameter. If the user updated the seek bar, this parameter is true. Otherwise, it's false.

The onStopTrackingTouch method contains three statements. The first statement uses the getProgress method of the SeekBar parameter to get the final progress value for the seek bar. The second statement converts that int value to a float value by dividing it by 100 and uses the result to set the new tip percent. And the third statement calls the calculateAndDisplay method. As a result, the calculation is made and displayed when the user is done changing the value of the seek bar.

A seek bar and a label

An event handler for a seek bar

```java
@Override
public void onProgressChanged(SeekBar seekBar, int progress,
        boolean fromUser) {
    percentTextView.setText(progress + "%");
}

@Override
public void onStartTrackingTouch(SeekBar seekBar) {
    // TODO Auto-generated method stub
}

@Override
public void onStopTrackingTouch(SeekBar seekBar) {
    int progress = seekBar.getProgress();
    tipPercent = (float) progress / 100;
    calculateAndDisplay();
}
```

Description

- The onStartTrackingTouch method is executed when the user begins to change the value of the seek bar.

- The onProgressChanged method is executed as the user changes the value of the seek bar.

- The onStopTrackingTouch method is executed when the user finishes changing the value of the seek bar.

- Within the onProgressChanged method, the second parameter provides the progress value for the seek bar.

- To determine whether the user updated the progress value, or whether it was updated programmatically, you can check the fromUser parameter of the onProgressChanged method.

Figure 6-6   How to handle events for seek bars

How to handle low-level events

In chapter 3, you learned how to handle the Click event, which is a low-level event. Now, you'll learn how to handle some other low-level events including the Key and Touch events.

How to handle Key events

Most Android devices are touchscreen devices that the user can interact with by touching the screen. However, some Android devices include hardware components. This hardware may be a keyboard, a DPad, or a trackball. Or, it may be other buttons such as Power, Home, or Back buttons. When a user presses a hardware button, a Key event occurs.

Figure 6-7 shows how to handle a Key event. To do that, you implement the OnKeyListener that's nested in the View class and wire that listener to any widget. For example, the listener in this figure is wired to an EditText widget. As a result, when the EditText widget has the focus and the user presses a hardware key, the onKey method in this figure is executed.

Within this method, the code begins by using a switch statement to determine which hardware key was pressed by the user. To do that, the switch statement checks the key code value that's stored in the method's second parameter and compares that key code to the constant values that are stored in the KeyEvent class.

If the user presses the Enter key or the Center key on a DPad, this code calls the calculateAndDisplay method to perform a calculation and display the results to the user. Then, the next two statements hide the soft keyboard if it is displayed. To do that, the first statement uses the getSystemService method to get an InputMethodManager object. Then, the second statement uses that object to hide the soft keyboard. For now, don't worry if you don't understand this code. You can learn more about working with system services in chapter 11.

After hiding the soft keyboard, this code returns a true value for the onKey method. This indicates that this method has consumed the event. As a result, the event is not passed on to the parents of the current widget. In other words, if the user presses the Center or Enter keys, the processing of this event stops and isn't passed on to the EditText widget or any of its parent widgets.

If the user presses the Left or Right keys on a DPad, this code begins by checking which widget the Key event occurred on. If the event occurred on the SeekBar widget, this code calls the calculateAndDisplay method. Then, the break statement exits the switch statement.

Outside the switch statement, this code returns a false value for the onKey method. This indicates that this method has not consumed the event. As a result, the event is passed on to the parents of the current widget. In other words, when the user presses most keys, such as the number or letter keys, the Key event is passed on to the EditText widget and its parents so it can be processed normally, which is usually what you want.

An event handler for the Key event

```
@Override
public boolean onKey(View view, int keyCode, KeyEvent event) {
    switch (keyCode) {
        case KeyEvent.KEYCODE_ENTER:
        case KeyEvent.KEYCODE_DPAD_CENTER:

            calculateAndDisplay();

            // hide the soft keyboard
            InputMethodManager imm = (InputMethodManager)
                    getSystemService(Context.INPUT_METHOD_SERVICE);
            imm.hideSoftInputFromWindow(
                    billAmountEditText.getWindowToken(), 0);

            // consume the event
            return true;
        case KeyEvent.KEYCODE_DPAD_RIGHT:
        case KeyEvent.KEYCODE_DPAD_LEFT:
            if (view.getId() == R.id.percentSeekBar) {
                calculateAndDisplay();
            }
            break;
    }
    // don't consume the event
    return false;
}
```

Some constants from the KeyEvent class

Constant	Description
KEYCODE_ENTER	The Enter key on a hard keyboard.
KEYCODE_DPAD_CENTER	The Center key on the DPad.
KEYCODE_DPAD_LEFT	The Left key on the DPad.
KEYCODE_DPAD_RIGHT	The Right key on the DPad.
KEYCODE_SPACE	The Space key on a hard keyboard.

Description

- The KeyEvent class contains constants for almost every possible hardware key on a device including the keys on a hard keyboard or a DPad.

- The onKey method returns a Boolean value that indicates whether this method has consumed the event. If this method returns a true value, the event is not passed on to the parents of the current widget.

- If you don't want to wire an event handler for a Key event to multiple child widgets, you can wire the Key event to a parent widget such as the root layout. Then, the event handler handles any Key events that aren't consumed by child widgets.

- You can use the getSystemService method to get an InputMethodManager object. Then, you can use that object to hide the soft keyboard. For more information about how system services work, see chapter 11.

Figure 6-7   How to handle Key events

If you don't want to wire an event handler for a Key event to multiple child widgets, you can wire the Key event to a parent widget such as the root layout. Then, the event handler handles any Key events that aren't consumed by child widgets.

Figure 6-7 only shows five constants from the KeyEvent class. However, the KeyEvent class contains constants for almost every possible hardware key on a device.

How to handle Touch events

When a user touches a widget on a touchscreen device, a Touch event occurs on the widget. Most of the time, you don't need to know how to handle this event. Instead, you can use another event such as a Click event to get your app to work the way you want.

Sometimes, though, you may need to handle a Touch event. To do that, you implement the OnTouchListener that's nested in the View class and wire that listener to any widget. In figure 6-8, for example, the onTouch method includes code that sends data to the LogCat view.

Within the onTouch method, the first statement declares four float variables for the X and Y values when the user presses down and lifts up. Then, the second statement gets an int value for the type of MotionEvent action that has occurred. Next, an if statement compares this int value to some constants of the MotionEvent class to determine what type of event occurred.

If the user presses down, the if statement sets the values of the downX and downY variables. Then, it sends these variables to the LogCat view. These X and Y values indicate the location of the down touch on the horizontal and vertical axis of the widget.

If the user lifts up, the if statement sets the values of the upX and upY variables. Then, it sends these variables to the LogCat view. Although this type of logging doesn't perform a useful task, it's often helpful when you're getting started with Touch events.

Like the onKey method, the onTouch method returns a Boolean value that indicates whether this method has consumed the event. In this figure, the onTouch method returns a value of true if the user presses down or lifts up to indicate that it has consumed the event. However, for all other touch actions, this method returns a false value to indicate that it has not consumed the event and that other elements can continue processing this event.

When the user touches the screen and moves the touch, it's common that the device can't process events quickly enough. In that case, Android stores the Touch events in a batch. Then, if necessary, you can use the "history" methods of the MotionEvent object to access these Touch events. To start, you can use the getHistorySize method to get the number of Touch events in the batch. Then, you can create a loop that loops through these events, and you can use the getHistoricalX and getHistoricalY methods to get the X and Y values for each event.

An event handler for a Touch event

```
@Override
public boolean onTouch(View v, MotionEvent event) {
    float downX, downY, upX, upY;
    int action = event.getAction();
    if (action == MotionEvent.ACTION_DOWN) {
        Log.d("MotionEvent", "ACTION_DOWN");
        downX = event.getX();
        downY = event.getY();
        Log.d("MotionEvent", "downX = " + downX);
        Log.d("MotionEvent", "downY = " + downY);
        return true;
    }
    else if (action == MotionEvent.ACTION_UP){
        Log.d("MotionEvent", "ACTION_UP");
        upX = event.getX();
        upY = event.getY();
        Log.d("MotionEvent", "upX = " + upX);
        Log.d("MotionEvent", "upY = " + upY);
        return true;
    }
    else {
        return false;
    }
}
```

Some constants of the MotionEvent class

Constant	Description
ACTION_DOWN	The start of the touch.
ACTION_MOVE	The touch has moved between the down and up actions.
ACTION_UP	The end of the touch.

Some methods of the MotionEvent class

Method	Description
getAction()	Gets an int value for the type of action.
getX()	Gets a float value for the location of the X axis of the touch.
getY()	Gets a float value for the location of the Y axis of the touch.
getHistorySize()	The number of historical events.
getHistoricalX(int i)	Gets a float value for the X axis at the specified position of the stored historical values.
getHistoricalY(int i)	Gets a float value for the Y axis at the specified position of the stored historical values.

Description

- The onTouch method returns a Boolean value that indicates whether this method has consumed the event.

Figure 6-8   How to handle Touch events

The Tip Calculator app

This chapter finishes by showing a new and improved Tip Calculator app that uses one seek bar, three radio buttons, a radio button group, and a spinner. The code for this app handles the events that occur on these widgets.

The user interface

Figure 6-9 shows the user interface for the new and improved Tip Calculator app. When you use this app, you can set a new tip percent by dragging the thumb on the seek bar to the right or left.

If you want, you can round the tip or total to the nearest dollar by selecting the appropriate radio button. Similarly, you can split the bill between multiple people by selecting the item for the number of people from the spinner. When you do that, the user interface displays the Per Person label and amount. Otherwise, it hides this label and amount.

The user interface

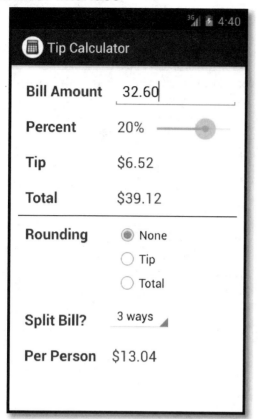

Description

- This version of the Tip Calculator includes a seek bar, a radio group, three radio buttons, and a spinner.

- To set a new tip percent, you can drag the thumb on the seek bar to the right or left.

- To round the tip or the total to the nearest dollar, you can select the appropriate radio button.

- To split the bill, you can select the appropriate item from the spinner.

Figure 6-9   The user interface

The Java code for the activity

Figure 6-10 shows the Java code for the new and improved Tip Calculator app. Since this code expands upon the Tip Calculator app described in chapter 3, I'll focus on the code that handles the events for the new widgets.

To start, this code imports the classes for all of the new widgets. In addition, it imports the five listeners needed to handle the events for this app. Then, the class declares that it implements all five of these listeners.

Within the class, the code defines instance variables for the widgets. Then, the onCreate method gets references to all necessary widgets and wires the five listeners to the appropriate widgets. Here, the listener for the Key event is wired to multiple widgets. In particular, it's wired to the EditText, SeekBar, and RadioGroup widget.

The onPause and onResume methods work much like they did in chapter 3. However, the onPause method contains new code that saves the instance variables that can be used to set the state of the widgets. Similarly, the onResume method contains new code that gets these variables and uses them to set the state of the widgets. Here, the event handlers for the RadioGroup and Spinner widgets are triggered by the code that sets the selected radio button and the position of the spinner. These event handlers both execute the calculateAndDisplay method.

The Java code for the activity

```java
package com.murach.tipcalculator;

import java.text.NumberFormat;

import android.os.Bundle;
import android.view.KeyEvent;
import android.view.View;
import android.view.View.OnKeyListener;
import android.view.inputmethod.EditorInfo;
import android.view.inputmethod.InputMethodManager;
import android.widget.AdapterView;
import android.widget.AdapterView.OnItemSelectedListener;
import android.widget.ArrayAdapter;
import android.widget.EditText;
import android.widget.RadioButton;
import android.widget.RadioGroup;
import android.widget.RadioGroup.OnCheckedChangeListener;
import android.widget.SeekBar;
import android.widget.SeekBar.OnSeekBarChangeListener;
import android.widget.Spinner;
import android.widget.TextView;
import android.widget.TextView.OnEditorActionListener;
import android.app.Activity;
import android.content.Context;
import android.content.SharedPreferences;
import android.content.SharedPreferences.Editor;

public class TipCalculatorActivity extends Activity
implements OnEditorActionListener, OnSeekBarChangeListener,
OnCheckedChangeListener, OnItemSelectedListener, OnKeyListener {

    // define variables for the widgets
    private EditText billAmountEditText;
    private TextView percentTextView;
    private SeekBar percentSeekBar;
    private TextView tipTextView;
    private TextView totalTextView;
    private RadioGroup roundingRadioGroup;
    private RadioButton roundNoneRadioButton;
    private RadioButton roundTipRadioButton;
    private RadioButton roundTotalRadioButton;
    private Spinner splitSpinner;
    private TextView perPersonLabel;
    private TextView perPersonTextView;

    // define the SharedPreferences object
    private SharedPreferences savedValues;

    // define rounding constants
    private final int ROUND_NONE = 0;
    private final int ROUND_TIP = 1;
    private final int ROUND_TOTAL = 2;
```

Figure 6-10   The Java code for the activity (part 1 of 6)

The Java code for the activity

```java
// define instance variables
private String billAmountString = "";
private float tipPercent = .15f;
private int rounding = ROUND_NONE;
private int split = 1;

@Override
public void onCreate(Bundle savedInstanceState) {
    super.onCreate(savedInstanceState);
    setContentView(R.layout.activity_tip_calculator);

    // get references to the widgets
    billAmountEditText = (EditText) findViewById(R.id.billAmountEditText);
    percentTextView = (TextView) findViewById(R.id.percentTextView);
    percentSeekBar = (SeekBar) findViewById(R.id.percentSeekBar);
    tipTextView = (TextView) findViewById(R.id.tipTextView);
    totalTextView = (TextView) findViewById(R.id.totalTextView);
    roundingRadioGroup = (RadioGroup)
            findViewById(R.id.roundingRadioGroup);
    roundNoneRadioButton = (RadioButton)
            findViewById(R.id.roundNoneRadioButton);
    roundTipRadioButton = (RadioButton)
            findViewById(R.id.roundTipRadioButton);
    roundTotalRadioButton = (RadioButton)
            findViewById(R.id.roundTotalRadioButton);
    splitSpinner = (Spinner) findViewById(R.id.splitSpinner);
    perPersonLabel = (TextView) findViewById(R.id.perPersonLabel);
    perPersonTextView = (TextView) findViewById(R.id.perPersonTextView);

    // set array adapter for spinner
    ArrayAdapter<CharSequence> adapter = ArrayAdapter.createFromResource(
        this, R.array.split_array, android.R.layout.simple_spinner_item);
    adapter.setDropDownViewResource(
        android.R.layout.simple_spinner_dropdown_item);
    splitSpinner.setAdapter(adapter);

    // set the listeners
    billAmountEditText.setOnEditorActionListener(this);
    billAmountEditText.setOnKeyListener(this);
    percentSeekBar.setOnSeekBarChangeListener(this);
    percentSeekBar.setOnKeyListener(this);
    roundingRadioGroup.setOnCheckedChangeListener(this);
    roundingRadioGroup.setOnKeyListener(this);
    splitSpinner.setOnItemSelectedListener(this);

    // get SharedPreferences object
    savedValues = getSharedPreferences("SavedValues", MODE_PRIVATE);
}
```

Figure 6-10   The Java code for the activity (part 2 of 6)

The Java code for the activity

```
@Override
public void onPause() {
    // save the instance variables
    Editor editor = savedValues.edit();
    editor.putString("billAmountString", billAmountString);
    editor.putFloat("tipPercent", tipPercent);
    editor.putInt("rounding", rounding);
    editor.putInt("split", split);
    editor.commit();

    super.onPause();
}


@Override
public void onResume() {
    super.onResume();

    // get the instance variables
    billAmountString = savedValues.getString("billAmountString", "");
    tipPercent = savedValues.getFloat("tipPercent", 0.15f);
    rounding = savedValues.getInt("rounding", ROUND_NONE);
    split = savedValues.getInt("split", 1);

    // set the bill amount on its widget
    billAmountEditText.setText(billAmountString);

    // set the tip percent on its widget
    int progress = Math.round(tipPercent * 100);
    percentSeekBar.setProgress(progress);

    // set rounding on radio buttons
    // NOTE: this executes the onCheckedChanged method,
    // which executes the calculateAndDisplay method
    if (rounding == ROUND_NONE) {
        roundNoneRadioButton.setChecked(true);
    }
    else if (rounding == ROUND_TIP) {
        roundTipRadioButton.setChecked(true);
    }
    else if (rounding == ROUND_TOTAL) {
        roundTotalRadioButton.setChecked(true);
    }

    // set split on spinner
    // NOTE: this executes the onItemSelected method,
    // which executes the calculateAndDisplay method
    int position = split - 1;
    splitSpinner.setSelection(position);
}
```

Figure 6-10   The Java code for the activity (part 3 of 6)

The calculateAndDisplay method works much like it did in chapter 3. However, it contains new code that rounds the tip and total if necessary. In addition, it contains new code that calculates the split amount, shows or hides the Per Person label and amount when necessary, and formats the Per Person amount.

The event handler for the EditText widget is executed when the user clicks the Done button on the soft keyboard that's displayed for the EditText widget. This works as described in chapter 3.

The event handler for the SeekBar widget is executed when the user moves the thumb on the seek bar. Here, the onProgressChanged method displays immediate feedback to the user by updating the tip percent whenever the user changes the progress value on the seek bar. Then, when the user finishes changing the progress value on the seek bar, the onStopTrackingTouch method calculates a new tip percent and calls the calculateAndDisplay method. This method uses the new tip percent to calculate and display new tip and total amounts.

The event handler for the RadioGroup widget is executed when the user selects a new radio button. Here, the onCheckedChanged method begins by using the second parameter to check which radio button is selected. Then, it sets the rounding variable to the appropriate constant value. For example, if the No Rounding radio button is selected, this code sets the rounding variable to the ROUND_NONE constant. After it sets the rounding variable, this code calls the calculateAndDisplay method. This method uses the rounding variable to determine what type of rounding the app uses for the calculation.

The event handler for the Spinner widget is executed when the user selects a new item. Here, the onItemSelected method begins by using the second parameter of the method to set the value of the split variable. Since the position of the first item is 0, this code adds a value of 1 to get the value for the split variable. After that, this event handler calls the calculateAndDisplay method. This method uses the split variable to determine how many ways to split the bill.

The event handler for the Key event is executed when the user presses a hardware key and the focus is on the EditText, SeekBar, or RadioGroup widgets. Here, the onKey method begins by using the second parameter of the method to execute code depending on the hardware key. If the user pressed the Enter key on a hard keyboard or the Center key on a DPad, this code hides the soft keyboard, calls the calculateAndDisplay method, and consumes the event.

If the user presses the Right or Left keys on a DPad and the focus is on the SeekBar widget, this code calls the calculateAndDisplay method. However, in this case, it does not consume the event. As a result, Android can use this event to update the user interface accordingly.

The Java code for the activity

```java
public void calculateAndDisplay() {
    // get the bill amount
    billAmountString = billAmountEditText.getText().toString();
    float billAmount;
    if (billAmountString.equals("")) {
        billAmount = 0;
    }
    else {
        billAmount = Float.parseFloat(billAmountString);
    }

    // get tip percent
    int progress = percentSeekBar.getProgress();
    tipPercent = (float) progress / 100;

    // calculate tip and total
    float tipAmount = 0;
    float totalAmount = 0;
    if (rounding == ROUND_NONE) {
        tipAmount = billAmount * tipPercent;
        totalAmount = billAmount + tipAmount;
    }
    else if (rounding == ROUND_TIP) {
        tipAmount = StrictMath.round(billAmount * tipPercent);
        totalAmount = billAmount + tipAmount;
    }
    else if (rounding == ROUND_TOTAL) {
        float tipNotRounded = billAmount * tipPercent;
        totalAmount = StrictMath.round(billAmount + tipNotRounded);
        tipAmount = totalAmount - billAmount;
    }

    // calculate split amount and show/hide split amount widgets
    float splitAmount = 0;
    if (split == 1) {   // no split - hide widgets
        perPersonLabel.setVisibility(View.GONE);
        perPersonTextView.setVisibility(View.GONE);
    }
    else {              // split - calculate amount and show widgets
        splitAmount = totalAmount / split;
        perPersonLabel.setVisibility(View.VISIBLE);
        perPersonTextView.setVisibility(View.VISIBLE);
    }

    // display the results with formatting
    NumberFormat currency = NumberFormat.getCurrencyInstance();
    tipTextView.setText(currency.format(tipAmount));
    totalTextView.setText(currency.format(totalAmount));
    perPersonTextView.setText(currency.format(splitAmount));

    NumberFormat percent = NumberFormat.getPercentInstance();
    percentTextView.setText(percent.format(tipPercent));
}
```

Figure 6-10   The Java code for the activity (part 4 of 6)

The Java code for the activity

```java
//****************************************************
// Event handler for the EditText
//****************************************************
@Override
public boolean onEditorAction(TextView v, int actionId, KeyEvent event) {
    if (actionId == EditorInfo.IME_ACTION_DONE ||
        actionId == EditorInfo.IME_ACTION_UNSPECIFIED) {
        calculateAndDisplay();
    }
    return false;
}


//****************************************************
// Event handler for the SeekBar
//****************************************************
@Override
public void onStartTrackingTouch(SeekBar seekBar) {
    // TODO Auto-generated method stub
}


@Override
public void onProgressChanged(SeekBar seekBar, int progress,
        boolean fromUser) {
    percentTextView.setText(progress + "%");
}


@Override
public void onStopTrackingTouch(SeekBar seekBar) {
    calculateAndDisplay();
}


//****************************************************
// Event handler for the RadioGroup
//****************************************************
@Override
public void onCheckedChanged(RadioGroup group, int checkedId) {
    switch (checkedId) {
        case R.id.roundNoneRadioButton:
            rounding = ROUND_NONE;
            break;
        case R.id.roundTipRadioButton:
            rounding = ROUND_TIP;
            break;
        case R.id.roundTotalRadioButton:
            rounding = ROUND_TOTAL;
            break;
    }
    calculateAndDisplay();
}
```

Figure 6-10   The Java code for the activity (part 5 of 6)

The Java code for the activity   **Page 6**

```java
//*****************************************************
// Event handler for the Spinner
//*****************************************************
@Override
public void onItemSelected(AdapterView<?> parent, View v, int position,
        long id) {
    split = position + 1;
    calculateAndDisplay();
}

@Override
public void onNothingSelected(AdapterView<?> parent) {
    // Do nothing
}


//*****************************************************
// Event handler for the keyboard and DPad
//*****************************************************
@Override
public boolean onKey(View view, int keyCode, KeyEvent event) {
    switch (keyCode) {
        case KeyEvent.KEYCODE_ENTER:
        case KeyEvent.KEYCODE_DPAD_CENTER:

            calculateAndDisplay();

            // hide the soft keyboard
            InputMethodManager imm = (InputMethodManager)
                    getSystemService(Context.INPUT_METHOD_SERVICE);
            imm.hideSoftInputFromWindow(
                    billAmountEditText.getWindowToken(), 0);

            // consume the event
            return true;
        case KeyEvent.KEYCODE_DPAD_RIGHT:
        case KeyEvent.KEYCODE_DPAD_LEFT:
            if (view.getId() == R.id.percentSeekBar) {
                calculateAndDisplay();
            }
            break;
    }
    // don't consume the event
    return false;
}
}
```

Figure 6-10   The Java code for the activity (part 6 of 6)

Perspective

Now that you've finished this chapter, you should understand how high-level and low-level events work. In addition, you should be able to handle events that occur on the widgets described in this chapter. More importantly, you should have all the skills you need to figure out how to handle events that occur on other kinds of widgets. To do that, you can search the Android API or the Internet to find the event listener that you need. Then, you can use the techniques described in this chapter to handle the event by implementing that listener.

Terms

listener	long click
events	focus
high-level events	anonymous class
low-level events	anonymous inner class

Summary

- A *listener* listens for *events* that can occur on a widget. When an event occurs on a widget, the appropriate method of the listener is executed.

- *High-level events* are events that occur on a specific type of widget. An interface that defines a listener for a high-level event is typically nested within the class that defines the widget. These listeners can only be wired to an appropriate widget.

- *Low level events* are events that occur on all types of widgets. An interface that defines a listener for a low-level event is typically nested within the android.view.View class. These listeners can be wired to any type of widget.

- You can create an instance variable that creates an object from a class that implements the listener interface. Since this class doesn't have a name, it's known as an *anonymous class*.

- You can create an instance of a listener interface without assigning it to an instance variable. This is known as an *anonymous inner class*

- The CheckedChanged event of a check box or radio button occurs when a check box or radio button is checked or unchecked.

- For a spinner, the onItemSelected method is executed when the spinner is first displayed and whenever a new item is selected. The onNothingSelected method is only executed when the selection disappears.

- For a seek bar, the onStartTrackingTouch method is executed when the user begins to change the value of the seek bar. The onProgressChanged method is executed as the user changes the value of the seek bar. The onStopTrackingTouch method is executed when the user finishes changing the value of the seek bar.

- The KeyEvent class contains constants for almost every possible hardware key on a device including the keys on a hard keyboard or a DPad.

- The onKey and onTouch methods both return a Boolean value that indicates whether the method has consumed the event. If this method returns a true value, the event is not passed on to the parents of the current widget.

Exercise 6-1   Use anonymous classes for event listeners

In this exercise, you'll modify the Tip Calculator app that's presented in this chapter so it uses anonymous classes for event listeners.

1. Open the project named ch06_ex1_TipCalculator that's in the ex_starts directory.

2. Open the class for the activity and review its code. Note that the activity class implements five listener interfaces.

3. Test the app to make sure the EditText, SeekBar, RadioGroup, and Spinner widgets all work correctly.

4. Modify the code so it uses an anonymous class (not an anonymous inner class) for the OnEditorActionListener. To do that, modify the declaration for the class so it doesn't implement this interface, create an instance variable for an object that implements this interface, and set that object as the listener.

5. Test the app to make sure the EditText widget still works correctly.

6. Repeat steps 4 and 5 for the listeners for the SeekBar, RadioGroup, and Spinner widgets.

Exercise 6-2   Improve the listener for the Key events

In this exercise, you'll modify the Tip Calculator app that's presented in this chapter so it uses the OnKeyListener interface to handle Key events instead of using the OnEditorActionListener interface to handle Key events.

To test this exercise, you'll need to be able to run an emulator on your computer.

Test the Enter key and DPad keys

1. Open the project named ch06_ex2_TipCalculator that's in the ex_starts directory.

2. Run the app on an emulator that has a hard keyboard and a DPad like the one shown at the end of chapter 4.

3. Enter a bill amount and use the Enter key on the keyboard and the Center key on the DPad to finish the entry. This should work correctly.

4. Use the DPad to move the focus to the seek bar for the tip percent. Then, use the Left and Right keys on the DPad to increase or decrease the tip amount. This should work correctly.

Improve the listener for the Key events

5.   Enter a new bill amount and use the Down key on the DPad to move the focus to the seek bar. Note that this does not update the tip and total amounts or hide the soft keyboard.

6.   Modify the code for the onKey method so the Down key on the DPad executes the calculateAndDisplay method, hides the soft keyboard, and moves the focus to the next widget. To get this to work correctly, the Down key should not consume the event, but the Center key should continue to consume the event.

7

How to work with themes and styles

In chapter 3, you learned how to manually set the properties of the widgets in your app to control how they look. For example, you set the textSize property for every widget in the app so that each widget uses the same text size. Now, this chapter shows how to use themes and styles to automatically set the properties for widgets.

Themes and styles allow you to separate the design from the content, which is generally considered a best practice. In addition, they can help you keep the appearance of your app consistent, they can help you reduce code duplication, and they can make your app easier to develop and maintain.

An introduction to themes and styles

By default, an Android app uses themes and styles that are built-in to the operating system or available from a support library. This topic begins by examining a few of these themes. But first, you should know that a *style* is a collection of properties that apply to a widget, and a *theme* is a collection of styles that apply to an entire activity or app. In other words, the main difference between a style and a theme is where it is applied: a style is applied to a widget, and a theme is applied to an app or activity.

Three common themes

Figure 7-1 shows three themes that are commonly used. Each theme provides the look for the title bar and widgets of the app. The first is the Holo theme. More specifically, it's the Holo.Light.DarkActionBar variation of the Holo theme. Here, the Light modifier specifies that the screen should use a white background. Then, the DarkActionBar modifier specifies that the screen should include a dark *action bar*, which is a title bar that can display buttons that allow the user to perform actions. Although the action bar shown in this figure doesn't include buttons, you'll learn how to add buttons to the action bar in the next chapter. Note that this theme is available with Android 4.0 (API 14) and later. As a result, it's available on most Android devices.

The second theme is the Material theme. More specifically, it's the Material.Light.DarkActionBar variation of the Material theme. This theme works much like the Holo theme shown in this figure. However, the Material theme is only available on Android 5.0 (API 21) and later. As a result, it's not directly available on most Android devices. The Material theme provides a new appearance that includes informative touch animations for widgets and transitions between activities that can make an app easier and more enjoyable to use.

The third theme is the AppCompat theme. This theme is available from the v7 appcompat support library. It looks like the Material theme, and it provides a way to port the appearance of the Material theme all the way back to Android 2.1 (API 7). As a result, if you use the v7 appcompat support library, you can make the appearance of the Material theme available to virtually all Android devices. However, these devices won't be able to support the animations and transitions of the Material theme.

Until now, most of the applications in this book have used the Holo.Light.DarkActionBar theme for two reasons. First, this theme is available for all devices supported by the apps in this book, which specify a minimum API of 15 or 16. Second, this theme doesn't depend on any support libraries. This keeps the apps in this book clean and simple.

However, if this isn't what you want, you can use another approach. One approach is to use the Material theme for API 21 and later and fall back on the Holo theme for older devices as described in the next figure. Another approach is to use the AppCompat theme as described in figure 7-3.

The Holo theme (light with dark action bar)

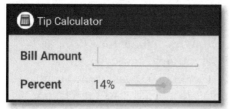

Description

- Is available from Android 4.0 (API 14) and later.

The Material theme (light with dark action bar)

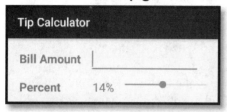

Description

- Is available from Android 5 (API 21) and later.
- Provides a new appearance that includes animations and transitions.

The AppCompat theme (light with dark action bar)

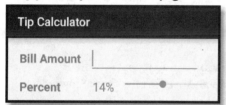

Description

- Is available with Android 2.1 (API 7) and later if you use the v7 appcompat support library.
- Provides the same appearance as the Material theme but without animations and transitions.

Description

- A *style* is a collection of properties that are applied to a widget.

- A *theme* is a collection of styles that are applied to an entire activity or app. Several themes are available from Android and its support libraries. When a theme is applied, every widget applies each style of the theme that it supports.

- With Android 4.0 (API 14) and later, an activity can include an *action bar* that displays the activity's title and, optionally, buttons that perform actions.

Figure 7-1   Three common themes

How to supply different themes for different APIs

If you want to use the Material theme for API 21 and later and fall back on the Holo theme for older devices, you can use one styles.xml file for API 21 and later and another for earlier APIs. In figure 7-2, for example, the styles.xml file in the res\values-v21 directory applies to API 21 (Android 5.0) and later. Then, the styles.xml file in the res\values directory applies to all earlier API levels. For example, if the minimum API for the app is set to 15, this styles.xml file would apply to APIs 15 through 20.

Developers often used this technique in the early days of Android programming. As a result, you may see this technique if you're working with an older Android app. If so, this technique may specify styles.xml files for API levels that you no longer need to support. In that case, you can delete any styles.xml files that are below the minimum API for your app.

Within the styles.xml files, a style element can specify a theme or a style. Within a style element, the name attribute specifies the name for the style or theme, and the parent attribute specifies the style or theme that the current style element inherits. In this figure, both of the style elements define themes.

In the styles.xml file in the res\values directory, the style element defines a theme named AppTheme that inherits the theme named Holo.Light. DarkActionBar. As a result, this is the default theme for all APIs.

In the styles.xml file in the res\values-v21 directory, the first style element defines a theme named AppTheme that inherits the theme named Material.Light. DarkActionBar. Since this theme has the same name (AppTheme) as the theme in the previous styles.xml file, this theme overrides the theme in the previous styles.xml file. As a result, API 21 and later use the Material theme, not the Holo theme.

In the AndroidManifest.xml file, the application element specifies the theme for the app. In this figure, the theme attribute specifies the theme named AppTheme. This applies an appropriate built-in theme for each API level.

For this to work, the activities in your app need to extend the Activity class, not the AppCompatActivity class described in the next figure. If your activity inherits the AppCompatActivity class, it will crash when it tries to load a Holo or Material theme.

The application element of the AndroidManifest.xml file

```
<application
    android:icon="@drawable/ic_launcher"
    android:label="@string/app_name"
    android:theme="@style/AppTheme"
    android:allowBackup="true">
```

Two styles.xml files

In the res\values directory (prior to API 21)

```
<resources>
    <style name="AppTheme"
            parent="android:Theme.Holo.Light.DarkActionBar" />
</resources>
```

In the res\values-v21 directory (API 21 and later)

```
<resources>
    <style name="AppTheme"
            parent="android:Theme.Material.Light.DarkActionBar" />
</resources>
```

Description

- If necessary, you can use multiple styles.xml files to set appropriate themes for the different API levels.

- The AndroidManifest.xml file uses its application element to specify the theme for the application.

- The styles.xml file in the res\values directory is applied to all API levels unless it is overridden by a styles.xml file in a res\values-vXX directory.

- A style element can specify a theme or a style.

- The name attribute of a style element specifies the name for the style or theme.

- The parent attribute of a style element specifies the style or theme that the current style element inherits.

- The style element named AppTheme inherits a theme that's appropriate for the API level.

- To add a styles.xml file and create a directory for it, right-click on the values directory, select the New→Values Resource File item, and use the resulting dialog to add a new file named styles.xml to the directory with the specified name.

Figure 7-2   How to supply different themes for different APIs

How to convert an existing project to the AppCompat theme

When you use Android Studio to create a project based on the Blank Activity template, it automatically sets the theme of the app to the AppCompat theme. However, if you're working with an older app and you want to convert it so it uses the AppCompat theme, you can use the procedure shown in figure 7-3 to do that.

To start, you can add the v7 appcompat library to your app. To do that, right-click on the app node of your project and select the Open Module Settings item. This should display the Project Structure dialog. Then, click the Dependencies tab, click the Add (+) button on the right side of the main window, and use the resulting dialog to add the appcompat-v7 library.

After you add the library, Gradle should automatically rebuild your project. However, if the version of the v7 appcompat support library isn't set correctly, you may get dozens of errors about missing widget styles when Gradle attempts to rebuild your project. To fix this, you may need to modify the Gradle build script, so the dependency for the v7 appcompat library matches the version that's available on your system. To check which version of the support library is available on your system, you can use the SDK Manager as described in the appendixes. Once the support library version is set correctly, Gradle should build the project with no errors.

After you import the v7 appcompat library, you can modify the classes for the activities so they extend the AppCompatActivity class instead of the Activity class. Then, you can change the theme for your app so it uses an AppCompat theme such as the AppCompat.Light.DarkActionBar theme shown in the figure.

Since the AppCompat theme is available from the support library, not from the Android operating system, you don't prefix the name of the theme with the android: namespace. In other words, you specify a theme like this:

```
parent="Theme.AppCompat.Light.DarkActionBar"
```

However, for a built-in theme such as Holo, you must prefix the name of the theme with the android: namespace like this:

```
parent="android:Theme.Holo.Light.DarkActionBar"
```

This indicates that the theme is available from the Android operating system.

If an activity extends the AppCompatActivity class, you can only use AppCompat themes. Otherwise, your app will crash when the activity tries to display a built-in theme such as the Material or Holo theme. As a result, when you convert an app to use the AppCompat theme, you should make sure you remove any styles that use built-in themes. Or, you should update them so they use an AppCompat theme instead.

How to add a dependency for the v7 appcompat library

1. In Android Studio, right-click on the app node of your project and select the Open Module Settings item.

2. Click the Dependencies tab, click the Add (+) button, and use the resulting dialog to add the v7 appcompat library as shown here:

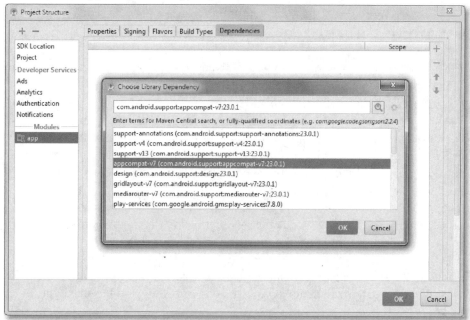

This should add a dependency to the end of the Gradle build script like this:

```
dependencies {
    compile 'com.android.support:appcompat-v7:23.0.1'
}
```

3. Open the class for the activity. Then, replace the import statement for the Activity class with an import statement for the AppCompatActivity class like this:

```
import android.support.v7.app.AppCompatActivity;
```

4. In the class for the activity, modify the class declaration so it extends the AppCompatActivity class, not the Activity class, like this:

```
public class TipCalculatorActivity extends AppCompatActivity
```

5. Open the res/values/styles.xml file. Then, modify the style named AppTheme so it inherits an AppCompat theme like this:

```
<style name="AppTheme" parent="Theme.AppCompat.Light.DarkActionBar">
    <!-- Customize your theme here. -->
</style>
```

Description

- If you are working with a project that uses an older theme such as the Holo theme, you can convert that project so it uses the AppCompat theme to port some features of the Material theme back to Android 2.1 (API 7).

Figure 7-3   How to convert an existing project to the AppCompat theme

How to work with styles

Now that you have a general idea of how three common themes work, you're ready to learn how to work with styles. To start, you can define a style. Then, you can apply that style to a widget.

How to define a style

Figure 7-4 shows how to define a style. To start, you typically add a style element to the XML file named styles.xml in the res/values/ directory of your project. If necessary, you can create this file.

The first example uses a style element to create a style named TextView. This name is appropriate for a style that's going to be applied to most TextView widgets in the app. Within the style element, this style uses a single item element to specify a property of the style. More specifically, this item element sets the textSize property of the style to 20sp.

The second example uses a style element to create a style named EditText. This style inherits the TextView style that was defined in the first example. As a result, the EditText style also specifies a text size of 20sp.

The third example uses a style element to create a style named Label. This style inherits the style named TextView that was defined in the first example. In addition, it uses two item elements to specify 10dp of padding and to make the text bold.

The fourth example uses a style element to create a style named TextView. Label. Since this name begins with TextView followed by a period, it automatically inherits the TextView style defined in the first example. When you use this syntax, you should be aware that you can only use it with user-defined styles. In other words, you can't use it to inherit built-in Android styles.

The fifth example shows how to use the syntax in the fourth example to inherit multiple styles. This style inherits the TextView and Label styles and adds another style named Indent that indents the left margin by 10dp.

When you work with styles, you can set the same properties that are available from the Properties window in the graphical layout editor. As a result, if you have already used the Properties window to set the properties for your app, you can look in the XML file for the layout to find the names of the properties and their settings. Then, if you want, you can create styles to set those properties more consistently. In this figure, for instance, the examples show how to use a style to set the textSize, textStyle, padding, and layout_marginLeft properties. However, you can use a style to set any of the properties that are available from the Properties window.

When a style inherits another style, its properties override the properties of the inherited style. Let's assume, for example, that I create a style that inherits the Label style. And let's assume that this new style has only a single item element setting the textSize property to 15sp. This would override the textSize property of the Label style, but it wouldn't override the textStyle and padding properties of the Label style.

A style that overrides one property

```
<style name="TextView">
    <item name="android:textSize">20sp</item>
</style>
```

A style that inherits a user-defined style

```
<style name="EditText" parent="@style/TextView" />
```

A style that inherits a user-defined style and overrides two properties

```
<style name="Label" parent="@style/TextView">
    <item name="android:textStyle">bold</item>
    <item name="android:padding">10dp</item>
</style>
```

Another way to code the previous style

```
<style name="TextView.Label">
    <item name="android:textStyle">bold</item>
    <item name="android:padding">10dp</item>
</style>
```

How to inherit multiple user-defined styles

```
<style name="TextView.Label.Indent">
    <item name="android:layout_marginLeft">10dp</item>
</style>
```

Description

- To define a style, you typically add a style element to the XML file named styles.xml in the res/values/ directory of your project. If necessary, you can create this file.

- The name attribute of a style element is required and must specify the name of the style.

- The parent attribute of a style element is optional. However, it can be used to specify a style that the current style inherits. The inherited style can be a built-in Android style or a user-defined style.

- A style element can contain one or more item elements. Each item element specifies a *property* of the style. If a style inherits another style, its properties override the properties of the inherited style.

- A style can specify any properties that are available from the graphical layout editor, such as textSize, textStyle, padding, margin, textColor, and background.

- When working with user-defined styles, you can also inherit other styles by coding the name of the inherited style, followed by a period, followed by the name of your new style.

Figure 7-4   How to define a style

How to apply a style

Figure 7-5 shows how to apply styles like the ones that were defined in the previous figure. One way to do that is to display a layout in the graphical editor, click the widget, and click the button to the right of the style property in the Properties window. When you do, Android Studio displays a dialog box like the one shown in this figure. Then, you can select the style you want to apply. In this figure, for example, I have selected the style named TextView.Label.

Another way to apply a style is to edit the XML for the layout. To do that, you just add a style attribute to the widget and use it to specify the name of the style that's in the styles.xml file. In this figure, for instance, the first XML example applies the style named TextView.Label to a TextView widget. Then, the second XML example applies the style named TextView to the TextView widget.

Unlike most attributes, the style attribute does not use the android: namespace prefix. In this figure, for example, the id and text attributes use that prefix, but the style attribute doesn't.

The Resources dialog for a style

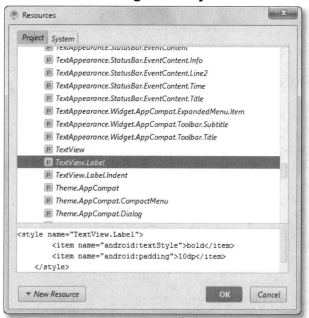

A TextView widget with the TextView.Label style

```
<TextView
    android:id="@+id/percentLabel"
    android:layout_width="wrap_content"
    android:layout_height="wrap_content"
    android:layout_alignLeft="@+id/billAmountLabel"
    android:layout_below="@+id/billAmountLabel"
    android:text="@string/tip_percent_label"
    style="@style/TextView.Label" />
```

A TextView widget with the TextView style

```
<TextView
    android:id="@+id/percentTextView"
    android:layout_width="wrap_content"
    android:layout_height="wrap_content"
    android:layout_alignBaseline="@+id/percentLabel"
    android:layout_alignLeft="@+id/billAmountEditText"
    android:padding="5dp"
    android:text="@string/tip_percent"
    style="@style/TextView" />
```

Description

- From the graphical layout editor, you can specify a style for a widget by using the Properties window to select a style.

- In XML, you can apply a style by using the style attribute to specify the name of the style in the styles.xml file. This attribute doesn't use the android: namespace prefix.

Figure 7-5   How to apply a style

How to create a style sheet

A *style sheet* is a collection of styles that can be applied throughout an application. To create a style sheet, you store the styles for your application in a styles.xml file like the one shown in figure 7-6.

The first element defines the custom theme named AppTheme. This theme inherits the AppCompat theme described earlier in this chapter.

The next four style elements define the custom styles that are used to apply formatting to all of the widgets of the Tip Calculator app. To start, the styles named TextView, EditText, and Button are designed to be applied to their corresponding widgets. Then, the style named TextView.Label is designed to be applied to the TextView widgets that label other widgets.

A styles.xml file in the res\values directory with four user-defined styles

```xml
<resources>
    <!-- The theme -->
    <style name="AppTheme"
            parent="Theme.AppCompat.Light.DarkActionBar" />

    <!-- Four custom styles -->
    <style name="TextView">
        <item name="android:textSize">20sp</item>
    </style>

    <style name="EditText" parent="@style/TextView" />

    <style name="Button" parent="@style/TextView">
        <item name="android:textStyle">bold</item>
    </style>

    <style name="TextView.Label">
        <item name="android:textStyle">bold</item>
        <item name="android:padding">10dp</item>
    </style>
</resources>
```

Description

- A *style sheet* is a collection of styles that can be applied throughout an application.

Figure 7-6   How to create a style sheet

How to work with themes

Now that you know how to work with styles, you're ready to take another look at themes. For example, you're ready to learn how to modify a theme.

How to modify a built-in theme

Figure 7-7 shows how to modify a built-in theme such as the Holo theme. Here, the theme named AppTheme is a custom theme that contains three item elements. Each of these item elements uses its name attribute to specify an attribute that corresponds with a style from the inherited theme, the Holo.Light. DarkActionBar theme.

After specifying the style that you want to override, each of these item elements uses its body to specify the custom style that overrides the inherited style. In this figure, the first item element overrides the TextView style in the Holo theme with the custom TextView style defined in this styles.xml file.

This custom TextView style inherits the built-in style named Widget. TextView. As a result, you only need to override the properties that you want to change. In this case, the custom TextView style only overrides the textSize property. Similarly, the custom styles named EditText and Button inherit their corresponding built-in styles.

Since the first three custom styles defined in this figure override the corresponding styles in the Holo theme, they're applied automatically to any TextView, EditText, or Button widgets in the app. In other words, you don't need to manually apply these styles. The fourth custom style, on the other hand, needs to be applied manually.

If you want to modify a theme, you often need to do some research and experimentation to figure out which attributes and styles to modify. When working in Android Studio, you can often use the autocomplete feature to view lists of available attributes, styles, and properties. Often, this is all you need.

However, if you need more information, you can view the standard attributes for a theme by viewing the API documentation for the Theme array that's defined in the R.styleable class. Then, you can find the styles that you want to inherit by viewing the styles.xml file that's available from the Android SDKs on your computer. This file is available from the data\res\values directory that corresponds with each Android API. Alternately, you can view the API documentation for the R.styles class. However, since styles aren't well documented, it's often helpful to view the styles.xml file that contains the source code for the styles.

A styles.xml file that customizes the built-in Holo theme

```xml
<resources>
    <style name="AppTheme"
            parent="android:Theme.Holo.Light.DarkActionBar">

        <!-- Set new styles for three widgets in the theme -->
        <item name="android:textViewStyle">@style/TextView</item>
        <item name="android:editTextStyle">@style/EditText</item>
        <item name="android:buttonStyle">@style/Button</item>
    </style>


    <!-- These styles are applied automatically -->
    <style name="TextView" parent="@android:style/Widget.TextView">
        <item name="android:textSize">20sp</item>
    </style>

    <style name="EditText" parent="@android:style/Widget.EditText">
        <item name="android:textSize">20sp</item>
    </style>

    <style name="Button" parent="@android:style/Widget.Button">
        <item name="android:textSize">20sp</item>
        <item name="android:textStyle">bold</item>
    </style>

    <!-- This style needs to be applied manually -->
    <style name="TextView.Label">
        <item name="android:textStyle">bold</item>
        <item name="android:padding">10dp</item>
    </style>
</resources>
```

The API documentation for the standard attributes in a theme

```
https://developer.android.com/reference/android/R.styleable.html#Theme
```

The styles.xml file for Android 6.0 (API 23)

```
\android\sdk\platforms\android-23\data\res\values\styles.xml
```

Description

- Within a theme, the name attribute of an item element can specify an attribute that corresponds to a style in the theme. Then, the body of the item element can specify a custom style that overrides the inherited style.

- When you create a custom style, you typically begin your style by inheriting another style. Then, you can use the custom style to override only the properties that you want to change.

- When you work with a built-in theme or style, such as a Holo theme or style, you typically prefix the name of the theme or style with android: namespace. This shows that the theme is built-in to the Android operating system.

Figure 7-7   How to modify a built-in theme

How to modify the AppCompat theme

Figure 7-8 shows how to modify the AppCompat theme. This works much like modifying a built-in theme such as the Holo theme. However, it's a little trickier since you need to specify whether to override a style from the appcompat support library or a style from Android operating system. Often, this requires some trial and error.

In this figure, the theme named AppTheme contains three item elements. Each of these item elements uses its name attribute to specify an attribute that corresponds with a style. Here, the first attribute overrides a style that's available from the Android operating system. To do that, it prefixes the name of the style with the android: namespace. Then, the second and third styles override a style that's available form the support library. Note that these styles don't include the android: namespace.

In this figure, the first style inherits the built-in style named Widget. TextView. However, the custom styles named EditText and Button inherit a style from the support library. More specifically, the second style inherits the Widget. AppCompat.EditText style, and the third style inherits the Widget.AppCompat. Button style.

If you want to modify the AppCompat theme, you can use the techniques described in the previous figure to research and experiment with the attributes and styles that you want to modify. However, when you view the attributes of the AppCompat theme, you'll see that this theme doesn't include attributes for all widgets.

For example, the AppCompat theme doesn't include the textViewStyle attribute. This indicates that the AppCompat theme is using the attribute that's built-in to the Android operating system. As a result, you need to override the attribute that's built-in to the Android system.

However, the AppCompat theme does provide editTextStyle and buttonTextStyle attributes. As a result, you can override the attributes that are stored in the appcompat support library.

A styles.xml file that customizes the AppCompat theme

```xml
<resources>
    <style name="AppTheme"
            parent="Theme.AppCompat.Light.DarkActionBar">

        <!-- Set new styles for three widgets in the theme -->
        <item name="android:textViewStyle">@style/TextView</item>
        <item name="editTextStyle">@style/EditText</item>
        <item name="buttonStyle">@style/Button</item>
    </style>


    <!-- These styles are applied automatically -->
    <style name="TextView" parent="@android:style/Widget.TextView">
        <item name="android:textSize">20sp</item>
    </style>

    <style name="EditText" parent="Widget.AppCompat.EditText">
        <item name="android:textSize">20sp</item>
    </style>

    <style name="Button" parent="Widget.AppCompat.Button">
        <item name="android:textSize">20sp</item>
        <item name="android:textStyle">bold</item>
    </style>

    <!-- This style needs to be applied manually -->
    <style name="TextView.Label">
        <item name="android:textStyle">bold</item>
        <item name="android:padding">10dp</item>
    </style>
</resources>
```

The API documentation for the attributes from the AppCompat theme

```
https://developer.android.com/reference/android/support/v7/appcompat
/R.attr.html
```

The styles.xml file for the v7 appcompat support library

```
\android\sdk\extras\android\support\v7\appcompat\res\values\styles.xml
```

Description

- When you work with a theme or style from a support library, such as an AppCompat theme or style, you don't prefix the name of the theme or style. This shows that the theme or style is available from a support library, not from the Android operating system.

Figure 7-8   How to modify the AppCompat theme

How to modify a theme depending on the API

If you use the AppCompat theme, you can modify it as shown in the previous figure. However, if you want to use the Material theme for API 21 and later and fall back on the Holo theme for older APIs, you can modify the themes by using a similar technique like the one shown in figure 7-9.

In the styles.xml file in the res\values directory, the first style element defines a theme named AppTheme that inherits the theme named AppBaseTheme. Within this style element, you can put any customizations that are not specific to an API. For example, this style element contains three item elements that specify the custom styles that override the built-in styles for the TextView, EditText, and Button widgets.

The second style element defines a theme named AppBaseTheme that inherits the Holo.Light.DarkActionBar theme. As a result, this is the default theme for all APIs.

In the styles.xml file in the res\values-v21 directory, the first style element defines a theme named AppBaseTheme that inherits the Material.Light. DarkActionBar theme. Since this theme has the same name as the theme in the styles.xml file in the res\values directory, this theme overrides the theme in the res\values directory. As a result, API 21 and later use the Material theme instead of the Holo theme.

In this figure, the three theme customizations apply to all APIs. However, if you want to add customizations that apply only to certain API ranges, you can add the customization to the appropriate style element.

The styles.xml file in the...

res\values directory

```xml
<resources>
    <style name="AppTheme"
            parent="AppBaseTheme">

        <!-- Theme customizations NOT specific to a particular API -->
        <item name="android:textViewStyle">@style/TextView</item>
        <item name="android:editTextStyle">@style/EditText</item>
        <item name="android:buttonStyle">@style/Button</item>
    </style>
    <style name="AppBaseTheme"
            parent="android:Theme.Holo.Light.DarkActionBar">
        <!-- API 15-20 theme customizations go here -->
    </style>

    <!-- These styles are applied automatically -->
    <style name="TextView"
            parent="@android:style/Widget.TextView">
        <item name="android:textSize">20sp</item>
    </style>

    <style name="EditText" parent="@android:style/Widget.EditText">
        <item name="android:textSize">20sp</item>
    </style>

    <style name="Button" parent="@android:style/Widget.Button">
        <item name="android:textSize">20sp</item>
        <item name="android:textStyle">bold</item>
    </style>

    <!-- This style needs to be applied manually -->
    <style name="TextView.Label">
        <item name="android:textStyle">bold</item>
        <item name="android:padding">10dp</item>
    </style>
</resources>
```

res\values-v21 directory

```xml
<resources>
    <style name="AppBaseTheme"
            parent="android:Theme.Material.Light.DarkActionBar">
        <!-- API 21+ theme customizations go here. -->
    </style>

    <!-- Custom styles go here -->
</resources>
```

Description

- The theme named AppTheme inherits the theme named AppBaseTheme.
- The theme named AppBaseTheme inherits a theme that's appropriate for the API level.
- Within each theme, you can specify custom styles for different APIs.

Figure 7-9   How to modify a theme depending on the API

How to modify the text appearance for a theme

Android uses a series of built-in styles to control the appearance of the text for a theme. For example, by default, a TextView widget uses the TextAppearance.Small style that's displayed in figure 7-10. If you want to change the text appearance for a single widget, you can set its textAppearance property to a new TextAppearance style. However, if you want to modify the text appearance for multiple widgets, you may want to customize the TextAppearance styles for a theme as shown in this figure.

To start, this figure lists some of the built-in text appearance styles. Android primarily uses these styles to control the size and color of the text. Here, the Small, Medium, and Large styles can be appended to the TextAppearance style to change the size of the text. Or, the Inverse style can be applied to the TextAppearance style to use the inverse color scheme.

The styles.xml file in this figure works much like the styles.xml file in figure 7-7. However, it uses the TextAppearance styles instead of the styles for the different types of widgets. Here, the code overrides the TextAppearance. Small and TextAppearance.Small.Inverse styles and sets the textSize property to 20sp. Since the textAppearance property of the TextView widget is set to the TextAppearance.Small style by default, this automatically sets all TextView widgets to a text size of 20sp.

However, the textAppearance property of other widgets might not be set to the TextAppearance.Small style by default. As a result, if you want one of these widgets to use the TextAppearanceSmall style that you defined, you can manually set the textAppearance property of the widget. In this figure, for instance, the last example sets the textAppearance property to the TextAppearanceSmall style that's available from the styles.xml file shown in this figure.

Some built-in styles for controlling text appearance

Name	Description
`TextAppearance`	Displays standard text appearance and color scheme.
`TextAppearance.Inverse`	Displays standard text appearance with inverse color scheme.
`TextAppearance.Small`	Displays small font size.
`TextAppearance.Small.Inverse`	Displays small font size and inverse color scheme.
`TextAppearance.Medium`	Displays medium font size.
`TextAppearance.Large`	Displays large font size.

A styles.xml file that modifies the TextAppearance styles

```
<resources>
    <style name="AppTheme" parent="Theme.AppCompat.Light.DarkActionBar">
        <!-- Set two new text appearance styles in the theme -->
        <item name="android:textAppearanceSmall">
            @style/TextAppearanceSmall</item>
        <item name="android:textAppearanceSmallInverse">
            @style/TextAppearanceSmallInverse</item>
    </style>

    <!-- These text appearance styles are applied automatically -->
    <style name="TextAppearanceSmall"
            parent="@android:style/TextAppearance.Small">
        <item name="android:textSize">20sp</item>
    </style>
    <style name="TextAppearanceSmallInverse"
            parent="@android:style/TextAppearance.Small.Inverse">
        <item name="android:textSize">20sp</item>
    </style>
</resources>
```

An EditText widget that uses a TextAppearance style

```
<EditText
    android:id="@+id/billAmountEditText"
    android:layout_width="wrap_content"
    android:layout_height="wrap_content"
    android:inputType="numberDecimal"
    android:text="@string/bill_amount"
    android:textAppearance="@style/TextAppearanceSmall">
</EditText>
```

Description

- Android themes use a series of styles to control the appearance of the text on all widgets in the theme.

Figure 7-10   How to modify the text appearance for a theme

A summary of built-in themes

Figure 7-11 lists some of the built-in themes that are available from Android. If you want to view a complete list of all built-in themes for an activity, display the activity in the graphical layout editor. Then, click on the Themes button available from the toolbar above the graphical editor to display a dialog that allows you to browse through all available themes. This list includes the themes shown in this figure as well as many other themes, including any user-defined themes that are available.

If you select a theme from the graphical layout editor, it allows you to preview a theme for your project. However, this doesn't change the theme that's used when you run the app on a device or emulator. To do that, you typically need to change the theme that's specified in the project's Android manifest or styles.xml files as shown in the next figure.

Android 4.0 (API 14) introduced a family of DeviceDefault themes. These themes provide a way for manufacturers to inherit and override a built-in Android theme without changing the built-in theme itself. For example, Google's Nexus devices use the DeviceDefault themes to override the Holo themes. As a result, the unmodified Holo themes are available on the device, and the customized DeviceDefault themes are also available on the device.

This provides the app developer with a choice. On one hand, the app developer can choose the DeviceDefault themes. This makes the appearance of the app more consistent with other apps on that device. However, the appearance of the app may vary when viewed on different devices.

On the other hand, the app developer can choose one of the Holo or AppCompat themes. This makes the appearance of the app more consistent across multiple devices. However, the appearance of the app might not be consistent with other apps on the same device since those apps might use one of the DeviceDefault themes.

Some old themes

Name	Displays the activity...
Theme.Light	With a white background.
Theme.Black	With a black background.
Theme.Light.NoTitleBar	With a white background and no title bar.
Theme.Black.NoTitleBar	With a black background and no title bar.
Theme.Dialog	As a dialog box with a black background.

Some Holo themes

Name	Displays the activity....
Theme.Holo	With a black background.
Theme.Holo.Dialog	With a black background as a dialog.
Theme.Holo.NoActionBar	With a black background and no action bar.
Theme.Holo.Light	With a white background.
Theme.Holo.Light.Dialog	With a white background as a dialog.
Theme.Holo.Light.NoActionBar	With a white background and no action bar.
Theme.Holo.Light.DarkActionBar	With a white background and a dark action bar.

Some Material themes

Name	Displays the activity...
Theme.Material	With a black background.
Theme.Material.Dialog	With a black background as a dialog.
Theme.Material.NoActionBar	With a black background and no action bar.
Theme.Material.Light	With a white background.
Theme.Material.Light.Dialog	With a white background as a dialog.
Theme.Material.Light.NoActionBar	With a white background and no action bar.
Theme.Material.Light.DarkActionBar	With a white background and a dark action bar.

Description

- To view a list of all built-in themes available to an activity, display the activity in the graphical layout editor and use the drop-down list that's available from the Themes button to view the themes.

- Android 4.0 (API 14) introduced the family of DeviceDefault themes. The DeviceDefault themes provide a way for manufacturers to provide a default theme for a device without having to modify the other built-in Android themes such as the Holo theme.

Figure 7-11   A summary of built-in themes

How to apply themes

You can use the AndroidManifest.xml file to apply themes to an entire applicaton or to a specific activity. To apply a theme to the entire application, you use the theme attribute of the application element to specify the theme. For instance, the first example in figure 7-12 shows how to specify a custom theme named AppTheme for the entire application.

To apply a theme to a specific activity, you use the theme attribute of the activity element to specify the theme. For instance, the second example specifies a built-in theme named Holo.Light.Dialog for an activity. This displays the activity as a dialog box like the one shown in this figure.

The third example in this figure works similarly to the first. However, it specifies a custom theme named DialogTheme. This custom theme uses the built-in theme named Holo.Light.Dialog for APIs prior to Android 5.0 (API 21), and it uses a newer built-in theme named Material.Light.Dialog for Android 5.0 (API 21) and later. For API 21 and later, this displays the activity as a dialog box like the one shown in this figure, but using the Material theme instead of the Holo theme.

Once again, for these themes to work, your activity needs to extend the Activity class rather than AppCompatActivity class. Otherwise, your application will crash when it attempts to use these built-in themes.

When you specify a theme at both the application and activity levels, the theme at the activity level overrides the theme at the application level. In fact, it's common to specify a default theme at the application level that applies to most of the activities in the app. Then, you can specify a theme at the activity level for any activities that need to override the default theme for the app.

When you apply a style to a widget, that style's properties override the properties that are specified at the application or activity level. Similarly, if you set a property directly on a widget, it overrides any properties that are specified by a style.

Theme.Holo.Light.Dialog

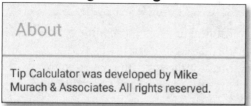

An AndroidManifest.xml file that's used to...

Apply a user-defined theme to the entire application

```
<application
    android:icon="@drawable/ic_launcher"
    android:label="@string/app_name"
    android:theme="@style/AppTheme"
    android:allowBackup="true">
```

Apply a built-in theme to a single activity

```
<activity
    android:name="com.murach.dialogtest.DialogActivity"
    android:theme="@android:style/Theme.Holo.Light.Dialog"
    android:label="@string/about_title" >
```

Apply a user-defined theme to a single activity

```
<activity
    android:name="com.murach.dialogtest.DialogActivity"
    android:theme="@style/DialogTheme"
    android:label="@string/about_title" >
```

Two styles.xml files for a user-defined theme

In the res\values directory (prior to API 21)

```
<style name="DialogTheme" parent="android:Theme.Holo.Light.Dialog"/>
```

In the res\values-v21 directory (API 21 and later)

```
<style name="DialogTheme" parent="android:Theme.Material.Light.Dialog"/>
```

Description

- You can use the AndroidManifest.xml file to apply a theme to the entire application or to a specific activity.

- To apply a theme to the entire application, use the theme attribute of the application element to specify the theme as shown in figure 7-2.

- To apply a theme to a specific activity, use the theme attribute of the activity element to specify the theme.

- You can use the theme attribute to specify a built-in theme or a custom theme.

- When you specify a theme for the application and activity levels, the theme at the activity level overrides the theme at the application level.

Figure 7-12   How to apply themes

How to work with colors

As you work with themes and styles, you'll find that some properties allow you to specify colors. For example, you can specify a color for the textColor and background properties of most widgets. Most of the time, the colors in the default theme are appropriate for your app. However, if you need to modify the colors in your app, figure 7-13 shows how.

How to define colors

If you want to work with colors, it's generally considered a best practice to begin by adding a colors.xml file to the res/values directory. Then, you can add a color element for each color that you want to define as shown in the first example. Within the color element, you can use the name attribute to specify a name for the color. In this figure, for example, the colors.xml file creates colors named primary, secondary, tertiary, background, dark, and light. Together these colors provide a color scheme that you can use for your app.

Within the body of a color element, you can use a *hexadecimal*, or *hex*, value to specify an RGB value for the color. This works very similarly to using hex values to specify RGB colors with HTML and CSS. If you aren't familiar with how this works, you can search the Internet to learn more about it.

How to apply colors

The second example shows three ways to apply a color to the textColor property of a widget. First, you can use a hex value to specify a color. Second, you can use a name for a custom color. Third, you can use a name for a built-in Android color. Of these techniques, using a custom color can make it easier to apply colors consistently, and it can make your app easier to maintain if you decide to change colors later.

The third example shows how to use a style to apply a color. To do that, you use an item element within a style to set the color for a property. When you use this technique, you can use hex values or named styles, though it's generally preferable to use named styles whenever possible.

The fourth example shows how to apply a color by overriding the default colors of a theme. To do that, you use an item element within a theme to override one of the color attributes for the theme. When you use this technique, you must use a named style, not a hex value.

If you want to learn more about the colors that Android uses in its built-in themes, you can view the attrs.xml and colors.xml files that define the default attributes and colors for Android. To find these files, you can look in the data\res\ values directories for the Android APIs on your computer.

A colors.xml file in the res/values directory

```xml
<?xml version="1.0" encoding="utf-8"?>
<resources>
    <color name="primary">#141315</color>
    <color name="secondary">#736C6B</color>
    <color name="tertiary">#DDE0CE</color>
    <color name="background">#A6D39D</color>
    <color name="dark">#000000</color>
    <color name="light">#FFFFFF</color>
</resources>
```

How to apply colors to a widget

Using a hexadecimal value

```
android:textColor="#141315"
```

Using a name from the colors.xml file

```
android:textColor="@color/primary"
```

Using the name of a built-in Android color

```
android:textColor="@android:color/darker_gray"
```

How to apply colors to a style

```xml
<style name="TextView" parent="@android:style/Widget.TextView">
    <item name="android:textColor">@color/primary</item>
</style>
```

How to apply colors to a theme

```xml
<style name="AppTheme" parent="AppBaseTheme">
    <!-- Set new colors for the theme -->
    <item name="android:windowBackground">@color/background</item>
    <item name="android:textColorPrimary">@color/primary</item>
    <item name="android:textColorSecondary">@color/secondary</item>
</style>
```

The colors.xml file for Android 6.0 (API 23)

```
\android-sdks\platforms\android-23\data\res\values\colors.xml
```

Description

- To define names for colors, you can add a colors.xml file to the res/values directory.

- To specify a color, you can use *hexadecimal*, or *hex*, values to specify an RGB value.

- To apply a color, you can use a hex value, a name for a user-defined color, or a name for a built-in Android color.

- To learn more about built-in Android colors, view the colors.xml file for one of the Android SDKs that are installed on your system.

Figure 7-13   How to work with colors

Perspective

Now that you've finished this chapter, you should understand how to work with themes and styles. For small apps, you may decide that the default themes and styles are adequate. In that case, you can manually format each widget on the app.

For larger apps, you may want to create a style sheet to apply formatting. Or, you may want to use a custom theme to override some properties of a built-in theme. Both of these techniques separate the design from the content, which is generally considered a best practice. In addition, they help to apply formatting consistently, reduce code duplication, and make it easier to develop and maintain an app.

Although this chapter shows you the basics for working with colors, there's more to learn about them. For example, you can use XML to define color gradients. In addition, you can specify colors that include a level of transparency.

Terms

style	style sheet
theme	hexadecimal
action bar	hex

Summary

- A *style* is a collection of properties that specify formatting for a widget.
- A *theme* is a collection of styles that apply to an entire activity or app. Android includes several built-in themes.
- With Android 4.0 (API 14) and later, an activity can include an *action bar*, which is a title bar that can also display buttons.
- A style element can specify a theme or a style.
- The AndroidManifest.xml file uses its application element to specify the theme for the application.
- A *style sheet* is a collection of styles that can be applied throughout an application.
- To define names for colors, you can add a colors.xml file to the res/values directory.
- To specify a color, you can use *hexadecimal*, or *hex*, values to specify an RGB value.
- To apply a color, you can use a hex value, a name for a user-defined color, or a name for a built-in Android color.

Exercise 7-1   Use built-in themes

In this exercise, you'll modify the Tip Calculator app so it uses built-in themes to display a dark background.

1. Open the project named ch07_ex1_TipCalculator.

2. Open the styles.xml files in the values and values-v21 directories. These files each specify a built-in theme for the API level.

3. Open the AndroidManifest.xml file and review its code. It sets the custom theme named AppTheme as the theme for the application.

4. Open the layout for the app and view it in the graphical layout editor. Use the Android Version button to view the layout for API levels 16 and 23. This theme should look different for both of these APIs.

5. Edit the two styles.xml files so the app uses the theme named Theme.Holo prior to API 21 and uses Theme.Material for API 21 and later.

6. Open the layout for the app and view it in the graphical layout editor. These themes should display a dark background for all APIs.

Exercise 7-2   Use styles

In this exercise, you'll modify the styles for the Tip Calculator app presented in this chapter.

1. Open the project named ch07_ex2_TipCalculator.

2. Open the Gradle build script for this file. Note that it includes a dependency for the v7 appcompat library.

3. Open the class for the activity. Note that it extends the AppCompatActivity class.

4. Open the styles.xml file in the values directory. Note that this style uses the AppCompat theme. Note also that it defines several custom styles.

5. Open the layout for the activity and view its XML. Each widget applies one of the custom styles specified in the styles.xml file.

6. For the first TextView widget, delete the style attribute that applies the style. This should remove the style's formatting from that widget. Then, restore this style attribute.

7. For the last TextView widget, change the style from TextView.Label to TextView.Label.Indent. This should indent the Total label. Then, change the style back to the TextView.Label style.

8. In the styles.xml file, change the textSize attribute of the TextView style to 16sp. This should change the text size for all widgets on the layout.

9. In the styles.xml file, modify the TextView style so it sets the layout_width, layout_height, and padding attributes for TextView and EditText widgets to the values that are currently stored in the XML file for the layout. Then, delete these attributes from the XML file for the layout. This should reduce code duplication.

10. In the styles.xml file, change the padding attribute of the TextView style to 10dp. This should set the space between the widgets.

11. In the styles.xml file, modify the Button style so it sets the layout_width and layout_height attributes to 40dp. Then, switch to the XML for the layout and delete these attributes from that file. This should make both buttons a little smaller, and it should reduce code duplication.

12. Add two new TextView widgets to the form for a Per Person label and amount. Set the id, text, and alignment attributes for these widgets appropriately. Then, apply the TextView.Label style to the label and the TextView style to the amount. This is an easy way to apply consistent formatting to these new widgets.

Exercise 7-3   Modify a theme

In this exercise, you'll modify the theme that's used by the Tip Calculator app presented in this chapter.

1. Open the project named ch07_ex3_TipCalculator.

2. Open the styles.xml file in the values directory. This file modifies the theme to automatically apply custom styles for the TextView, EditText, and Button widgets. This file also includes a custom style named TextView.Label that can be applied manually.

3. Open the layout for the app and view its XML. This XML only applies the TextView.Label style.

4. Add a RadioGroup widget to the layout. Note that the radio buttons within the group use a different text size than the other widgets in the layout.

5. Modify the styles.xml file so it modifies the TextAppearance styles for the theme instead of modifying the styles for individual widgets. This should set the textSize property to 20sp for the TextAppearance style and its Small, Medium, and Inverse variations. When you're done, all widgets should use a textSize of 20sp.

8

How to work with menus and preferences

An Android app often includes menus that allow users to perform common tasks and preferences that allow users to customize its settings. Often, a menu includes an item that allows users to display the preferences screen for changing settings. In this chapter, you'll learn how to add menus and preferences to an app.

How to work with menus

A *menu* can contain one or more *items*. These items often provide a way for the user to perform actions and navigate through an app. Since a menu is hidden until the user activates it, menus can improve an app by hiding functionality that isn't typically needed. This frees screen space and reduces clutter on the screen.

An introduction to menus

In Android, the most common type of menu is known as the *options menu*. This type of menu can be displayed by an activity as shown in figure 8-1.

If a device has a physical Menu button, the user can use that button to display the options menu across the bottom of the activity. In this figure, for example, the first activity displays a menu with two items across the bottom of the activity.

If a device doesn't have a physical Menu button, an *action overflow icon* is displayed on the right side of the action bar to display the options menu. In this figure, for example, the second activity displays an action overflow icon that provides a way to display the options menu on the top right side of the activity.

An activity with an options menu that has two items

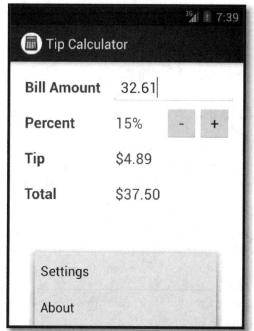

The same options menu displayed from an action overflow icon

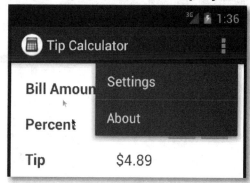

Description

- An activity can include an *options menu* that includes one or more *menu items*.
- If the device has a physical Menu button, the user can use that button to display the options menu across the bottom of the screen.
- If a device doesn't have a physical Menu button, an *action overflow icon* is displayed on the right side of the action bar to display the options menu.

Figure 8-1   An introduction to menus (part 1 of 2)

The action bar can display some or all of the items from the options menu as *action items*. It's generally considered a good design guideline to use action items for a small number of commonly used actions. For example, it's common to add action items such as search, refresh, add, edit, delete, and so on.

Action items can be displayed as text or as icons. In part 2 of figure 8-1, for example, the first activity displays both of the items in the options menu as text. Then, the second activity displays both of the items in the options menu as icons. If you specify an icon for an item, Android uses that icon in the action bar, but it uses text for that item in the options menu.

Any menu items that aren't displayed on the action bar are displayed in the options menu. In this figure, for example, the last activity shows the Settings icon and the overflow icon in the action bar, and it shows the About icon in the options menu.

It's generally considered bad design to put items that the user doesn't commonly use in the action bar. For example, it's not good design to put functions such as settings, about, and help in the action bar. In this chapter, I use these items to show how action items work. However, to improve the design, I eventually move these items back into the options menu.

Although they aren't shown in this figure, Android also provides two other types of menus. First, a *floating context menu* can be displayed when a user performs a long click on a widget. This menu appears as a floating list of menu items. However, it's generally considered a better practice to use *contextual action mode*. This mode displays action items that apply to the selected item or items. Second, a *popup menu* can be displayed when a user clicks on a widget or action item.

A task bar with two action items displayed as text

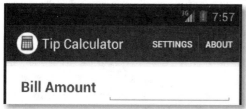

A task bar with two action items displayed as icons

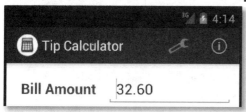

A task bar with one action icon and an overflow action icon

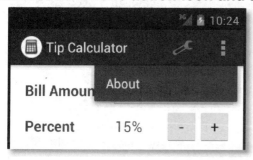

Description

- The action bar can include menu items. Typically, these items are for a small number of actions that are important to the app.

- Any items that aren't displayed on the action bar are displayed in the options menu.

- By default, if you specify an icon for an item, Android uses that icon in the action bar, but it uses text for that item in the options menu.

- A *floating context menu* appears as a floating list of menu items when a user performs a long click on the widget. Alternately, *contextual action mode* can display action items that apply to the selected item or items.

- A *popup menu* usually appears as a list below a widget or action item.

Figure 8-1   An introduction to menus (part 2 of 2)

How to define a menu

Figure 8-2 shows how to define a menu. To do that, you add an XML file for the menu to the project's res\menu directory. In this figure, for example, the menu is stored in an XML file with a name that clearly indicates that the menu is for the Tip Calculator activity.

The XML file for the menu begins with a root element named menu. Within the menu element, the XML can specify one or more item elements that define the items in the menu. In this figure, for example, the XML defines a menu with the two items shown in the previous figure.

Both item elements define four attributes. To start, the id attribute specifies an ID that can be used to access the item. Then, the title attribute specifies the text for the item that's displayed in the menu. This text is stored in the strings. xml file. Next, the icon attribute specifies an icon for the item that's usually displayed in the action bar. The file for this icon is stored in the res\drawable-xhdpi directory.

The showAsAction attribute specifies whether the item is shown in the action bar. For the first item, this attribute has been set to a value of "ifRoom". As a result, this item is shown in the action bar if there is room. If there isn't enough room, this item is shown in the options menu. For the second item, this attribute has been set to "never". As a result, this item is never shown in the action bar, even if there is room.

When an item is displayed in the action bar, Android's default behavior is to display the icon for the item. Conversely, when an item is displayed on the options menu, Android's default behavior is to display the text for the item, not the icon. Most of the time, that's what you want. As a result, you don't usually need to override Android's default behavior.

If you want to use your own icons, you should start by copying the file for the icon into the appropriate res\drawable directory in your project. In this figure, for example, both icons are for extra-high density screens, so they are stored in the res\drawable-xhdpi directory. Here, both icons are PNG files, and their names begin with a prefix of "ic_" to indicate that they are icons. This prefix isn't required, but it's commonly used.

To find an icon for your item, you can search through the standard icons that are available from the different versions of Android. To do that, you can look through the Android SDKs that are installed on your system, find the Android platform you want to use, and browse through its data\res\drawable folders. Or, you can search the Internet for the appropriate icons.

For this project, I have only supplied icons for the drawable-xhdpi folder. This works because Android automatically scales the icons for other densities. However, for a production app, you typically provide icons that are sized appropriately for all screen densities (ldpi, mdpi, hdpi, and xhdpi).

The file that contains the XML for the menu

```
res\menu\activity_tip_calculator.xml
```

The XML for the menu

```xml
<menu xmlns:android="http://schemas.android.com/apk/res/android">
    <item android:id="@+id/menu_settings"
            android:title="@string/menu_settings"
            android:icon="@drawable/ic_settings"
            android:showAsAction="ifRoom" />
    <item android:id="@+id/menu_about"
            android:title="@string/menu_about"
            android:icon="@drawable/ic_about"
            android:showAsAction="never" />
</menu>
```

Some attributes of a menu item

Name	Description
`title`	Specifies the text for the item.
`icon`	Specifies the icon for the item.
`showAsAction`	Specifies whether the item is shown on the action bar. Typically this attribute is set to a value of "always", "never", or "ifRoom".
`orderInCategory`	Specifies an int value for the sequence of the item where the lowest number is displayed first.

The location of the icon files for the items

```
res\drawable-xhdpi\ic_settings.png
res\drawable-xhdpi\ic_about.png
```

A directory that has standard icons for Android 6.0 (API 23)

```
\sdk\platforms\android-23\data\res\drawable-xhdpi
```

Description

- To provide an icon for a menu item, copy the icon file into the appropriate res\drawable directories in your project. Then, use the icon attribute to identify the name of the icon file.

Figure 8-2   How to define a menu

How to display an options menu

The first code example in figure 8-3 shows how to display an options menu. To do that, this code overrides the onCreateOptionsMenu method. This method is a lifecycle method of an activity that Android calls before it needs to display the menu for the first time.

Within the onCreateOptionsMenu method, the first statement gets a MenuInflater object and uses that object to convert, or *inflate*, the XML for the menu items into Java objects. This stores these Java objects in the Menu parameter. Then, the second statement returns a true value to display the menu.

Android only calls the onCreateOptionsMenu method the first time it displays the options menu. As a result, you can't use this method to update the menu each time it's displayed. For most apps, you don't need to do that anyway. However, if you need to update the menu every time it's displayed, you can override the onPrepareOptionsMenu method to do that. To learn more about that method, you can look it up in the API documentation or you can search the Internet.

How to handle option menu events

The second code example in figure 8-3 also shows how to handle the events that occur when a user selects an item from the options menu. To do that, this code overrides the onOptionsItemSelected method. Android calls this method when the user selects an item from the options menu.

Within the onOptionsItemSelected method, this code uses a switch statement to determine which menu item was selected. In this figure, for example, the switch statement gets the ID of the menu item and uses it to determine which menu item was selected.

If the ID of the menu item matches one of the menu items shown earlier in this chapter, this code displays a toast that indicates the name of the item. In addition, it returns a true value. This indicates that the method has consumed the event and stops further processing of the event.

If the ID of the menu item doesn't match one of the menu items shown earlier in this chapter, this code passes the menu item to the onOptionsItem-Selected method of the superclass. This allows the superclass to processes the menu item.

The code that displays the menu

```
@Override
public boolean onCreateOptionsMenu(Menu menu) {
    getMenuInflater().inflate(R.menu.activity_tip_calculator, menu);
    return true;
}
```

The code that handles the menu item events

```
@Override
public boolean onOptionsItemSelected(MenuItem item) {
    switch (item.getItemId()) {
        case R.id.menu_settings:
            Toast.makeText(this, "Settings", Toast.LENGTH_SHORT).show();
            return true;
        case R.id.menu_about:
            Toast.makeText(this, "About", Toast.LENGTH_SHORT).show();
            return true;
        default:
            return super.onOptionsItemSelected(item);
    }
}
```

Description

- To display an options menu, you can override the onCreateOptionsMenu method.
- Within the onCreateOptionsMenu method, you typically use a MenuInflater object to convert, or *inflate*, the XML for the menu items into Java objects and store them in the Menu parameter.
- The onCreateOptionsMenu method must return a true value to display the menu.
- To handle the event that's generated when a user selects an item from the options menu, you can override the onOptionsItemSelected method.
- Within the onOptionsItemSelected method, you can use a switch statement to determine which menu item was selected.
- The onOptionsItemSelected method can return a true value to indicate it has consumed the event and to stop further processing. Or, it can return a false value to allow processing to continue.
- Android only calls the onCreateOptionsMenu method the first time it displays the options menu. To update the menu every time it is displayed, you can override the method named onPrepareOptionsMenu.

Figure 8-3 How to display an options menu and handle its events

How to start a new activity

When an app contains more than one activity, menus are often used to navigate between those activities. To do that, you need to create an *intent*, which is an object that provides a description of an operation to be performed. In this case, you need to create an intent for an activity. Then, you can pass this intent to the startActivity method to start the activity as shown in figure 8-4.

The first code example shows two techniques for starting a new activity. The first technique uses two statements. Here, the first statement creates an Intent object by passing two arguments to the constructor of the Intent class. The first argument specifies the application context, and the second argument specifies the name of the class for the activity. Then, the second statement passes the Intent object to the startActivity method that's available from the Activity class.

The second technique for starting an activity works the same as the first, but it uses a single statement. Since this yields shorter code that's still easy enough to read, the second technique is commonly used.

The second code example shows how to use menu items to start activities. Here, the first item starts the Settings activity. This activity allows the user to change the settings for an app, and you'll learn about it later in this chapter. Then, the second item starts an About activity. This activity uses a dialog box to display some information about the app. For this to work, both of these activities must be declared in the project's AndroidManifest.xml file, just as the Tip Calculator activity is declared in this file.

Code that starts a new activity

Two statements

```
Intent settings =
    new Intent(getApplicationContext(), SettingsActivity.class);
startActivity(settings);
```

One statement

```
startActivity(new Intent(getApplicationContext(),
    SettingsActivity.class));
```

Code that uses menu items to start new activities

```
@Override
public boolean onOptionsItemSelected(MenuItem item) {
    switch (item.getItemId()) {
        case R.id.menu_settings:
            startActivity(new Intent(getApplicationContext(),
                SettingsActivity.class));
            return true;
        case R.id.menu_about:
            startActivity(new Intent(getApplicationContext(),
                AboutActivity.class));
            return true;
        default:
            return super.onOptionsItemSelected(item);
    }
}
```

Description

- An *intent* provides a description of an operation to be performed. Intents are commonly used with the startActivity method to start activities.

- To create an intent for an activity within the app, pass two arguments to the constructor of the Intent class. The first argument specifies the application context, and the second argument specifies the name of the class for the activity.

- To start a new activity, create an Intent object for the activity and pass that object to the startActivity method.

Figure 8-4   How to start a new activity

How to work with preferences

Preferences, or *settings*, provide a way for users to customize how an app works. Android provides Preference APIs that allow you to build an interface that's consistent with the user experience in other Android apps including built-in apps such as the system settings.

An introduction to preferences

Figure 8-5 shows a Settings activity that provides two preferences. The first preference has a name, a description, and a box that allows you to check or uncheck the preference. The second preference also has a name and a description. If you click on this preference, it displays a dialog box like the one shown in this figure that allows you to select one of three options.

The Settings activity

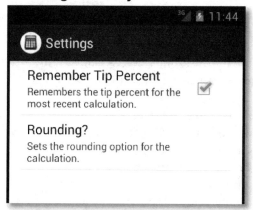

The dialog for the "Rounding?" item

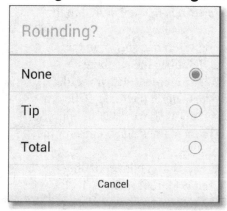

Description

- An app often allows the user to change *preferences*, or *settings*, for the app.
- Android provides Preference APIs that allow you to build an interface that's consistent with the user experience in other Android apps including built-in apps such as the system settings.

Figure 8-5   An introduction to preferences

How to define preferences

For most activities, you use subclasses of the View class to build the user interface. In this book, for example, the Tip Calculator activity uses the TextView, EditText, and Button classes to build its user interface. When working with preferences, you use subclasses of the Preference class to build the user interface as shown in figure 8-6.

To start, you add an XML file to the res\xml directory. This XML file is typically named preferences.xml, but you can use any name you want. In this figure, the root element for the XML file is a PreferenceScreen element. This element defines a screen that displays one or more preferences.

Within the PreferenceScreen element, the CheckBoxPreference element defines the first preference shown in the previous figure. Here, the key attribute specifies the key that's used to access the preference. Then, the next three attributes specify the name, description, and default value for the preference. The values for the title and summary attributes are strings that are stored in the strings.xml file.

The ListPreference element defines the second preference shown in the previous figure. Here, the key, title, summary, and defaultValue attributes work the same as they do for the CheckBoxPreference element. However, the ListPreference element includes three more attributes. Here, the dialogTitle attribute specifies the title for the dialog box that's used to display the list of options. Then, the entries and entryValues attributes specify the names and values for the list. These arrays are also stored in the strings.xml file.

When you use the Preference API, each preference has a corresponding key-value pair that Android saves in a default shared preferences file, which is available to other activities in the app. When a user changes a setting, Android automatically updates this file. Then, you can read these settings from other activities and use them to modify the behavior of the app.

Although this chapter only shows how to use the CheckBoxPreference and ListPreference classes, other subclasses of the Preference class exist. For example, it's common to use an EditTextPreference element to allow the user to enter text such as a username or password. For more information, about these subclasses, look up the Preference class in the API documentation and follow the links to view its subclasses.

The file that contains the XML for the preferences

```
res\xml\preferences.xml
```

The XML for the preferences

```xml
<?xml version="1.0" encoding="utf-8"?>
<PreferenceScreen
        xmlns:android="http://schemas.android.com/apk/res/android">
    <CheckBoxPreference
        android:key="pref_remember_percent"
        android:title="@string/remember_percent_title"
        android:summary="@string/remember_percent_summary"
        android:defaultValue="true" />
    <ListPreference
        android:key="pref_rounding"
        android:title="@string/rounding_title"
        android:summary="@string/rounding_summary"
        android:dialogTitle="@string/rounding_title"
        android:entries="@array/rounding_keys"
        android:entryValues="@array/rounding_values"
        android:defaultValue="@string/rounding_default" />
</PreferenceScreen>
```

Some attributes that apply to all Preference elements

Name	Description
key	Specifies the ID that's used to access the preference.
title	Specifies the title for the preference.
summary	Specifies the summary for the preference.
defaultValue	Specifies the default value for the preference.

Some attributes that apply to a ListPreference element

Name	Description
dialogTitle	Specifies the title of the dialog box that sets the value of the preference.
entries	Specifies the names of the entries in the list.
entryValues	Specifies the values of the entries in the list.

Description

- Instead of using View objects to build the user interface, settings are built using various subclasses of the Preference class that you declare in an XML file. This file is typically named preferences.xml, but you can use any name you want.

- For a list of commonly used Preference objects, see the API documentation for the Preference class.

- Each preference has a corresponding key-value pair that the system saves in the default shared preferences file. Whenever the user changes a setting, the system automatically updates that file.

Figure 8-6   How to define preferences

How to display preferences

In the early days of Android, it was common to display preferences directly within an activity. As of Android 3.0 (API 11), this technique has been deprecated. As a result, for most modern apps, you should use the technique described in figure 8-7 to display preferences in a *fragment*, which is a class that you can use to define part of a user interface. Then, you can use an activity to display one or more fragments. Also, if necessary, you can reuse a fragment in multiple activities.

Fragments were introduced in Android 3.0 (API 11). They make it possible to provide flexible user interfaces that can work on both smaller screens available from phones and larger screens available from tablets and other devices.

The first code example shows a fragment named SettingsFragment that inherits the PreferenceFragment class. This class only includes a single method, the onCreate method. Within this method, the second statement calls the addPreferencesFromResource method of the Fragment class. This method adds the preferences defined in the XML file shown in the previous figure to the fragment defined by this class.

The second code example shows how to add a fragment to an activity. Here, the fragment named SettingsFragment is added to the activity named SettingsActivity. To do that, this code uses the getFragmentManager method to get a FragmentManager object. From that object, this code calls the beginTransaction, replace, and commit methods. These methods start a transaction, replace the content of the activity with the fragment, and commit the transaction. Here, the replace method uses this Android resource:

```
android.R.id.content
```

to identify the content of the activity, and it creates a new SettingsFragment object to replace that content.

The SettingsFragment class

```
package com.murach.tipcalculator;

import android.os.Bundle;
import android.preference.PreferenceFragment;

public class SettingsFragment extends PreferenceFragment {

    @Override
    public void onCreate(Bundle savedInstanceState) {
        super.onCreate(savedInstanceState);

        // Load the preferences from an XML resource
        addPreferencesFromResource(R.xml.preferences);
    }
}
```

The SettingsActivity class

```
package com.murach.tipcalculator;

import android.app.Activity;
import android.os.Bundle;

public class SettingsActivity extends Activity {

    @Override
    public void onCreate(Bundle savedInstanceState) {
        super.onCreate(savedInstanceState);

        // Display the fragment as the main content
        getFragmentManager().beginTransaction()
                .replace(android.R.id.content, new SettingsFragment())
                .commit();
    }
}
```

Description

- You can use a *fragment* to define part of the user interface for an activity.
- An activity can display one or more fragments.
- A fragment can be reused in multiple activities.
- Fragments are available from Android 3.0 (API 11) and higher.
- For a user-interface that's built using Preference objects instead of View objects, a fragment needs to extend the PreferenceFragment class.
- From a fragment that extends the PreferenceFragment class, you can use the addPreferencesFromResource method to add the preferences defined in the XML file to the fragment.
- From an activity, you can use a FragmentManager object to replace the content of the activity with the specified fragment.
- You must declare all activities in the AndroidManifest.xml file.

Figure 8-7   How to display preferences

How to get preferences

Figure 8-8 shows how to get preferences that have automatically been written to a file by the Preference API. In the Tip Calculator activity, for example, you need to be able to get preferences so you can use them to modify that app's behavior.

The first example shows how to define the instance variables for the preferences. Here, the first instance variable is for the SharedPreferences object, and the next two are for the Remember Tip Percent and Rounding preferences.

The second example shows how to set the default values in the preferences file the first time the app starts on a device. To do that, this code calls the setDefaultValues method of the PreferenceManager class. This writes the default values that are stored in the XML file that defines the preferences to the default file for shared preferences. Here, the third parameter controls whether to reset the default values that are specified in the XML file. If false, the default values are set only if this method has never been called on this device, which is usually what you want. Since you only need to run this code once, it's often stored in the onCreate method of the activity.

The third example shows how to get the SharedPreferences object. To do that, this code calls the getDefaultSharedPreferences method of the PreferenceManager class. Here, the parameter uses the *this* keyword to specify the current activity as the context. This code is often stored in the onCreate method of an activity.

The fourth example shows how to get the values for the preferences from the SharedPreferences object. To do that, this code calls the appropriate getXxx method from the object and passes a key for the preference that matches the key specified in the XML file that defines the preferences. In addition, this code specifies a default value that's only used if the preference can't be retrieved. To make sure that your preferences are current, you typically store this code in the activity's onResume method.

In the fourth example, the first statement calls the getBoolean method and passes the key for the Remember Tip Percent preference. In most cases, this returns a Boolean value that indicates whether this preference is checked or unchecked. However, if this method isn't able to get a value from the object, the default value of true is returned.

Then, the second statement calls the getString method and passes the key for the Rounding preference. In most cases, this returns a string that indicates the selected option. Since the Tip Calculator activity uses an int value to keep track of rounding, this code converts the string for the rounding option to an int value.

Step 1: Define the instance variables for the preferences

```
private SharedPreferences prefs;
private boolean rememberTipPercent = true;
private int rounding = ROUND_NONE;
```

Step 2: Set the default values in the preferences file (onCreate)

```
PreferenceManager.setDefaultValues(this, R.xml.preferences, false);
```

Step 3: Get the SharedPreferences object (onCreate)

```
prefs = PreferenceManager.getDefaultSharedPreferences(this);
```

Step 4: Get the preferences (onResume)

```
rememberTipPercent = prefs.getBoolean("pref_remember_percent", true);
rounding = Integer.parseInt(prefs.getString("pref_rounding", "0"));
```

Some get methods of the SharedPreferences object

Name	Description
`getBoolean`(key, default)	Gets a Boolean value for the specified key. If the value doesn't exist, it gets the specified default value.
`getString`(key, default)	Gets a String value for the specified key.
`getInt`(key, default)	Gets an int value for the specified key.
`getLong`(key, default)	Gets a long value for the specified key.
`getFloat`(key, default)	Gets a float value for the specified key.

Description

- You can use the default SharedPreferences object to get preferences that have been automatically saved by the Preferences API.

- To set the default values the first time the app starts on a device, you can use the setDefaultValues method of the PreferenceManager class. Here, the third parameter controls whether to reset the default values that are specified in the XML file. If false, the default values are set only if this method has never been called in the past.

- To get the default SharedPreferences object, you can call the getDefaultShared-Preferences method of the PreferenceManager class.

- To get a preference from a SharedPreferences object, you can use the appropriate getXxx method for that preference. When you do, you specify a key that corresponds with the key in the XML file for the preferences. In addition, you specify a default value that's used if no value is retrieved for that preference.

Figure 8-8   How to get preferences

How to use preferences

Once you get preferences, you can use them to change the way your app works as shown in figure 8-9. Typically, you begin by using an if/else statement to check the value for the preference. Then, you execute the code that's appropriate for the value of the preference.

The first code example checks the value of the Remember Tip Percent preference, which has been stored in a variable named rememberTipPercent. If this variable contains a true value, this code remembers the tip percent by reading the last saved tip percent from a SharedPreferences object named prefs. Otherwise, it does not remember the tip percent by setting the tip percent to a default value of 15%.

The second code example checks the value of the Rounding preference, which has been stored in a variable named rounding. If this variable is equal to the constant named ROUND_NONE, the code calculates the tip and total without any rounding. If this variable is equal to the constant named ROUND_TIP, the code rounds the tip. If this variable is equal to the constant named ROUND_TOTAL, the code rounds the total.

The code in the second example may change the tip percent when the tip or total is rounded. However, this code shouldn't change the tip percent that's specified by the user. To provide for that, this code uses a variable named tipPercent-ToDisplay to store the tip percent that's displayed by the calculation. That way, the tip percent that's specified by the user is stored separately in the tipPercent variable. When there is no rounding, these variables contain the same value, but they may contain different values after the rounding is applied.

Use the "Remember Tip Percent" preference in the onResume method

```
if (rememberTipPercent) {
    tipPercent = prefs.getFloat("tipPercent", 0.15f);
}
else {
    tipPercent = 0.15f;
}
```

Use the "Rounding" preference in the calculateAndDisplay method

```
float tipPercentToDisplay = 0;
if (rounding == ROUND_NONE) {
    tipAmount = billAmount * tipPercent;
    totalAmount = billAmount + tipAmount;
    tipPercentToDisplay = tipPercent;
}
else if (rounding == ROUND_TIP) {
    tipAmount = StrictMath.round(billAmount * tipPercent);
    totalAmount = billAmount + tipAmount;
    tipPercentToDisplay = tipAmount / billAmount;
}
else if (rounding == ROUND_TOTAL) {
    float tipNotRounded = billAmount * tipPercent;
    totalAmount = StrictMath.round(billAmount + tipNotRounded);
    tipAmount = totalAmount - billAmount;
    tipPercentToDisplay = tipAmount / billAmount;
}
```

Description

- Once you get the preferences from the default preference file, you can use them to change the way your app works.

Figure 8-9   How to use preferences

More skills for working with preferences

So far in this chapter, you have learned some basic skills for working with preferences. These skills apply to most apps. Now, you're ready to learn some more skills for working with preferences. These skills may apply to apps that have more preferences, or apps that have more complex requirements for how their preferences work.

How to group preferences

If your app has a large number of preferences, you may want to organize them in groups as shown in figure 8-10. Here, the Settings activity has three preferences that have been grouped in two categories. The first category has a title of TIP PERCENT, and it contains the Remember Tip Percent and the Default Tip Percent preference. The second category has a title of ROUNDING, and it contains the Rounding preference.

The code example shows how to create the TIP PERCENT category. To do that, this code nests a CheckBoxPreference element and a ListPreference element within a PreferenceCategory element. These preferences are for the Remember Tip Percent and Default Tip Percent preferences. Here, the PreferenceCategory has a title attribute that specifies the title for the category. As usual, this string is stored in the project's strings.xml file.

A Settings activity that uses categories

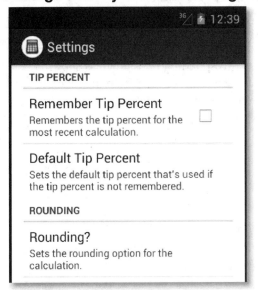

The XML for the preferences

```xml
<?xml version="1.0" encoding="utf-8"?>
<PreferenceScreen
    xmlns:android="http://schemas.android.com/apk/res/android">
    <PreferenceCategory
        android:title="@string/percent_category_title"
        android:key="pref_percent_category">
        <CheckBoxPreference
            android:key="pref_remember_percent"
            android:title="@string/forget_remember_title"
            android:summary="@string/forget_remember_summary"
            android:defaultValue="false" />
        <ListPreference
            android:key="pref_default_percent"
            android:title="@string/default_percent_title"
            android:summary="@string/default_percent_summary"
            android:dialogTitle="@string/default_percent_title"
            android:entries="@array/default_percent_keys"
            android:entryValues="@array/default_percent_values"
            android:defaultValue="@string/default_percent_default" />
    </PreferenceCategory>
    ...
</PreferenceScreen>
```

Description

- You can group preferences by nesting one or more Preference elements within a PreferenceCategory element.

Figure 8-10   How to group preferences

How to enable and disable preferences

For some apps, you may want to disable a preference when a corresponding CheckBoxPreference element is unchecked. To do that, you can use a preference's dependency attribute to link it to the corresponding preference as shown in figure 8-11. Here, the Settings activity has two preferences where the second preference is disabled when the first preference is unchecked. Conversely, the second preference is enabled when the first preference is checked.

A Settings activity that uses dependencies

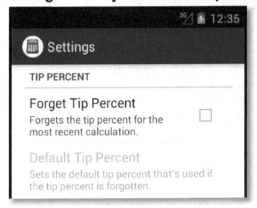

The XML for the preferences

```xml
<?xml version="1.0" encoding="utf-8"?>
<PreferenceScreen
    xmlns:android="http://schemas.android.com/apk/res/android">
    <PreferenceCategory
        android:title="@string/percent_category_title"
        android:key="pref_percent_category">
        <CheckBoxPreference
            android:key="pref_forget_percent"
            android:title="@string/forget_percent_title"
            android:summary="@string/forget_percent_summary"
            android:defaultValue="false" />
        <ListPreference
            android:key="pref_default_percent"
            android:title="@string/default_percent_title"
            android:summary="@string/default_percent_summary"
            android:dependency="pref_forget_percent"
            android:dialogTitle="@string/default_percent_title"
            android:entries="@array/default_percent_keys"
            android:entryValues="@array/default_percent_values"
            android:defaultValue="@string/default_percent_default" />
    </PreferenceCategory>
    ...
</PreferenceScreen>
```

The dependency attribute

Name	Description
dependency	Specifies the key of the preference that the current preference depends upon.

Description

- You can enable or disable a preference by using its dependency attribute to link the preference to a corresponding CheckBoxPreference element.

Figure 8-11   How to enable or disable a preference

How to use Java to work with preferences

So far in this chapter, you've used the default behavior provided by the Preference API. Now, figure 8-12 presents a Java class that shows how to work with preferences. This is often useful if you need to override the default behavior of the Preference API. For example, the last figure showed how to use the default behavior to disable the Default Tip Percent preference when the Forget Tip Percent preference is unchecked. Now, this figure shows how to disable the Default Tip Percent preference when the Remember Tip Percent preference is checked.

The SettingsFragment class in this figure handles the event that occurs when the user changes any of the preferences defined by the XML file. To do that, this class implements the OnSharedPreferenceChangeListener. Then, it registers this listener in the onResume method, and it unregisters this listener in the onPause method. As a result, Android executes the onSharedPreferenceChanged method whenever the user changes a preference.

The onCreate method gets the default SharedPreferences object. Then, the onResume method uses this object to get the Boolean value for the Remember Tip Percent preference. Finally, the onResume method passes this Boolean value to the setDefaultPercentPreference method.

The setDefaultPercentPreference method enables or disables the preference. To start, the first statement uses the findPreference method that's available from the current class to get a Preference object for the Default Tip Percent preference. Then, an if/else statement checks whether the rememberPercent parameter is true. If so, this statement uses the setEnabled method to disable the Default Tip Percent preference. Otherwise, it uses the setEnabled method to enable this preference.

The onSharedPreferenceChanged method begins by checking if the Remember Tip Percent preference was changed. If so, this code gets the value of this preference and passes it to the setDefaultPercentPreference method, which enables or disables the Default Tip Percent preference.

A class that works with preferences

```
package com.murach.tipcalculator;

import android.content.SharedPreferences;
import android.content.SharedPreferences.OnSharedPreferenceChangeListener;
import android.os.Bundle;
import android.preference.Preference;
import android.preference.PreferenceFragment;
import android.preference.PreferenceManager;

public class SettingsFragment extends PreferenceFragment
implements OnSharedPreferenceChangeListener {

    private SharedPreferences prefs;
    private boolean rememberPercent;

    @Override
    public void onCreate(Bundle savedInstanceState) {
        super.onCreate(savedInstanceState);
        addPreferencesFromResource(R.xml.preferences);
        prefs = PreferenceManager.getDefaultSharedPreferences(getActivity());
    }

    @Override
    public void onResume() {
        super.onResume();
        rememberPercent = prefs.getBoolean("pref_remember_percent", true);
        this.setDefaultPercentPreference(rememberPercent);
        prefs.registerOnSharedPreferenceChangeListener(this);
    }

    private void setDefaultPercentPreference(boolean rememberPercent) {
        Preference defaultPercent = findPreference("pref_default_percent");
        if (rememberPercent) {
            defaultPercent.setEnabled(false);
        } else {
            defaultPercent.setEnabled(true);
        }
    }

    @Override
    public void onPause() {
        prefs.unregisterOnSharedPreferenceChangeListener(this);
        super.onPause();
    }

    @Override
    public void onSharedPreferenceChanged(SharedPreferences prefs,
            String key) {
        if (key.equals("pref_remember_percent")) {
            rememberPercent = prefs.getBoolean(key, true);
        }
        this.setDefaultPercentPreference(rememberPercent);
    }
}
```

Figure 8-12   How to use Java to work with preferences

Perspective

The skills presented in this chapter should be enough to get you started with menus and preferences. However, Android provides many more features for working with menus and preferences. If necessary, you can use the Menu and Preference APIs to create more complex menus and preferences. With menus, for example, you can group menu items, and you can use Java code to work with menus dynamically. With preferences, you can create a preference that leads to another screen of preferences, and you can build custom preferences by extending one of the Preference classes.

Terms

menu	contextual action mode
items	inflate
options menu	intent
action overflow icon	preferences
action items	settings
floating context menu	fragment
popup menu	

Summary

- An activity can include an *options menu* that includes one or more *menu items*.
- If the device doesn't have a physical Menu button, an *action overflow icon* is displayed on the right side of the action bar to display the options menu.
- The action bar can include menu items.
- Any items that aren't displayed on the action bar are displayed in the options menu.
- By default, if you specify an icon for an item, Android uses that icon in the action bar, but it uses text for that item in the options menu.
- Within the onCreateOptionsMenu method, you typically use a MenuInflater object to convert, or *inflate*, the XML for the menu items into Java objects and store them in the Menu parameter.
- To handle the event that's generated when a user selects an item from the options menu, you can override the onOptionsItemSelected method and use a switch statement to determine which menu item was selected.
- An *intent* provides a description of an operation to be performed. Intents are commonly used with the startActivity method to start activities.
- Android provides Preference APIs that allow you to build an interface that's consistent with the user experience in other Android apps.

- Each preference has a corresponding key-value pair that the system saves in the default shared preferences file. Whenever the user changes a setting, the system automatically updates that file.

- You can use a *fragment* to define part of the user interface for an activity.

- You can use the default SharedPreferences object to get preferences that have been automatically saved by the Preferences API.

- You can group preferences by nesting one or more Preference elements within a PreferenceCategory element.

- You can enable or disable a preference by using its dependency attribute to link the preference to a corresponding CheckBoxPreference element.

Exercise 8-1   Experiment with menus and settings

In this exercise, you'll experiment with the menus and settings that are available from the Tip Calculator app.

1. Open the project named ch08_ex1_TipCalculator.

2. Run the app. If the device or emulator has a physical Menu button, use it to display the menu. Otherwise, use the action overflow icon to display the menu. This menu should have two items: Settings and About.

3. Select the Settings item to display the Settings activity. Then, click the Back button to return to the main activity.

4. Select the About item to display the About activity. This activity should appear as a dialog box. Then, click outside of the dialog box to return to the main activity.

5. Open the XML file in the menu directory and modify it so the Settings item is displayed in the action bar if there is enough room. This should display an icon for the Settings item in the action bar, but the About item should remain in the options menu.

6. Run the app. Then, use the main activity to set the tip percent to 17%.

7. Select the Settings action item to display the Settings activity. Then, remove the check from the Remember Tip Percent box. Next, click the Back button to return to the main activity. The tip percent should be reset to its default value of 15%.

8. Use the main activity to set the tip percent to 17%.

9. Change the orientation of the activity. Since the app isn't remembering the tip percent, this should reset the tip percent to 15%.

10. Select the Settings action item to display the Settings activity.

11. Use the Settings activity to select the "Remember Tip Percent" check box.

12. Use the Settings activity to change the default rounding for the app so that the total is always rounded.

13. Use the main activity to increase or decrease the tip. To do that, you may need to click on the Increase (+) and Decrease (-) buttons several times. This should automatically round the total and adjust the tip percent accordingly.

14. Open the XML file for the menu and delete the icon attribute from the Settings item. This should display text for the Settings item in the action bar.

15. View the files in the res\drawable-xhdpi directory of the project. This directory should include a second Settings icon.

16. Open the XML file for the menu and modify it so the Settings item uses the new Settings icon. This should display the new Settings icon in the action bar.

Exercise 8-2   Work with menus

In this exercise, you'll modify the menus used by the Tip Calculator app presented in this chapter.

1. Open the project named ch08_ex2_TipCalculator.

2. Open the XML file in the menu directory. Then, add a Help item to the end of this menu.

3. Open the class for the Tip Calculator activity. Then, modify the code so it displays a toast that indicates that the Help feature has *not* been implemented.

4. Add a Refresh item to the beginning of the menu. This item should display the icon named ic_refresh that's in the project's res\drawable-xhdpi directory, but only if there's enough room on the action bar.

5. Modify the class for the Tip Calculator activity so the Refresh item calls the calculateAndDisplay method.

6. Modify the onClick method so it does *not* call the calculateAndDisplay method. Instead, add some code that updates and displays the tip percent, but not the tip amount or total. That way, you need to use the Refresh item to update the tip amount and total after you click on the buttons.

7. Create an XML file for a menu for the Preferences activity. This menu should have two items: Tip Calculator and About.

8. Modify the Preferences activity so it displays the menu.

9. Modify the Preferences activity so it handles the two items. The Tip Calculator item should display the Tip Calculator activity, and the About item should display the About activity.

Exercise 8-3   Work with preferences

In this exercise, you'll add another setting to the Tip Calculator app presented in this chapter.

1. Open the project named ch08_ex3_TipCalculator.

2. Open the preferences.xml file in the xml directory. Add a ListPreference for a setting named Default Tip Percent. This setting should allow the user to specify a default tip percent of 10%, 15%, or 20%.

3. Open the class for the Tip Calculator activity. Modify this code so it gets the default tip percent and uses it if the user has decided not to remember the tip percent.

4. Switch to the preferences.xml file. Modify this code so it puts the Remember Tip Percent and Default Tip Percent preferences in a category named Tip Percent. Also, create a category named Rounding for the Rounding preference.

5. Run the app. It should work correctly. However, the default tip percent is only used if the app does *not* remember the tip percent. As a result, there's no need to set the default tip percent unless you uncheck the Remember Tip Percent box.

6. Open the class for the Settings fragment. Modify this code so it disables the Default Tip Percent preference when the Remember Tip Percent preference is unchecked.

9

How to work with fragments

Android introduced fragments in version 3.0 (API 11). As a result, they're available on most modern Android devices. You can use fragments to create user interfaces that work well on both small and large screens. For example, fragments allow you to create a user interface that works well on a small screen such as a phone while also taking advantage of the extra space that's available from a large screen such as a tablet.

An introduction to fragments

A *fragment* is a class that you can use to define part of a user interface. Then, you can use an activity to display one or more fragments. In this context, fragments are sometimes referred to as *panes*.

Single-pane and multi-pane layouts

On a small screen, an activity typically only displays a single fragment. This is known as a *single-pane layout*. In the first example of figure 9-1, for instance, the Tip Calculator activity displays the Tip Calculator fragment, and the Settings activity displays a Settings fragment. At this point, the app looks and acts as if it was not using fragments.

However, on a large screen, an activity can display multiple fragments. This is known as a *multi-pane layout*. In the second example, for instance, a large screen is in portrait and landscape orientation. As a result, the Tip Calculator activity displays both the Tip Calculator and Settings fragments.

In landscape orientation, the Tip Calculator activity displays the Tip Calculator fragment on the left side of the screen. Then, it displays the Settings fragment on the right side of the screen.

In portrait orientation, the Tip Calculator activity displays the Tip Calculator fragment at the top of the screen. Then, it displays the Settings fragment below the Tip Calculator fragment. This makes both fragments immediately available to the user.

On both of the large screens in this figure, there's plenty of room to display the soft keyboard. As a result, Android often displays this keyboard automatically. If there is enough room, Android displays an enhanced version of the soft keyboard that includes a separate keypad for some special characters. This makes it easier for the user to enter the bill amount.

Two activities displaying two fragments

One activity displaying two fragments in both landscape and portrait orientation

Description

- You can use a *fragment* to define part of the user interface for an activity. A fragment is sometimes referred to as a *pane*.

- On a small screen, an activity typically only displays a single fragment. This is known as a *single-pane layout*.

- On a large screen, an activity can display multiple fragments. This is known as a *multi-pane layout*.

- A fragment can be reused in multiple activities.

Figure 9-1   Single-pane and multi-pane layouts

The lifecycle methods of a fragment

The Fragment class defines methods that are called by the Android operating system at different points in the lifecycle of an activity. This is shown by the diagram in figure 9-2. Most of these methods are similar to the lifecycle methods of an activity. For example, a fragment includes the onCreate, onPause, and onResume methods that are available from an activity.

The lifecycle of the activity that contains the fragment affects the lifecycle of the fragment. In other words, when Android calls a lifecycle method for an activity, it also calls a similar lifecycle method for each fragment within that activity. For example, when Android calls the onPause method of an activity, it also calls the onPause method for each fragment within that activity. However, a fragment has some extra lifecycle methods that aren't available from an activity. Of these methods, the most commonly used method is the onCreateView method.

Android calls the onCreateView method when it's time to display the layout for the fragment. This occurs after the onCreate method but before the onStart method. Conversely, Android calls the onDestroyView method when it's time to remove the fragment's layout from the activity. This occurs after the onStop method but before the onDestroy method.

The onAttach and onDetach methods are another pair of lifecycle methods that are occasionally used to work with a fragment. Android calls the onAttach method when the fragment has been associated with an activity, and it calls the onDetach method when the fragment has been disassociated from the activity.

The onActivityCreated method does not have a corresponding lifecycle method. Android calls this method after it has returned from its call to the onCreate method of the activity.

When an activity is in the resumed state, it can add and remove fragments. However, when the activity leaves the resumed state, any fragments within the activity must respond to the lifecycle methods of the activity such as the onPause, onStop, and onDestroy methods.

The lifecycle methods of a fragment

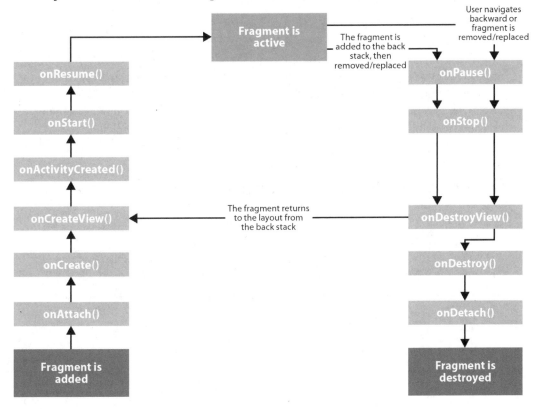

Description

- A fragment has lifecycle methods that are similar, but not identical to the lifecycle methods of an activity.

Figure 9-2   The lifecycle methods of a fragment

How to use single-pane layouts for small screens

Now that you understand how fragments work, you're ready to learn how to use them in a single-pane layout. This type of layout is useful for small screen devices such as phones. Since many devices have small screens, you often begin by creating a single-pane layout. Then, once you have that layout working correctly, you can add one or more multi-pane layouts for devices that have larger screens as shown later in this chapter.

How to create the layout for a fragment

Figure 9-3 shows how to create the layout for a fragment. To start, the XML code for a fragment works the same as the XML code for an activity. As a result, if you want to convert an existing layout for an activity to a fragment, you can rename the XML file. Then, you can use the graphical editor to work with the layout of a fragment just as you can use the graphical editor to work with the layout for an activity. In this figure, for instance, the graphical editor shows the layout for the Tip Calculator fragment.

The layout for a fragment

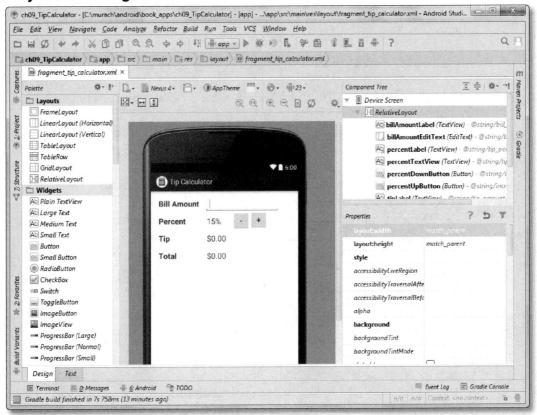

Description

- The XML for a fragment works the same as the XML for an activity.
- You can use the graphical layout editor to work with the layout for a fragment.

Figure 9-3   How to create the layout for a fragment

How to create the class for a fragment

Figure 9-4 shows how to create the class for a fragment. In general, this class works much like the class for an activity. However, there are a few differences. To start, the class for a fragment extends the Fragment class, not the Activity class.

Within the class for the fragment, the onCreate method typically initializes components of the fragment that aren't related to the layout for the user interface. In this figure, for example, the first two statements work with the preferences for the app. Both of these statements were described in the previous chapter, so you should understand how they work. Then, the third statement uses the setHasOptionsMenu method to indicate that this fragment has an options menu with items that should be added to the options menu for the activity. This is necessary because an activity that contains multiple fragments may need to combine the menu items from multiple fragments in its options menu.

The onCreateView method for a fragment typically contains the code that's used to create and initialize the layout for the fragment. Here, the first parameter of this method is a LayoutInflater object that you can use to inflate a layout. And the second paramter is a ViewGroup object for the parent layout. The layout for the fragment should be attached to this layout.

Within the method, the first statement calls the inflate method from this object to return a View object for the layout. To do that, this statement passes three arguments to the inflate method. The first argument specifies the ID for the layout of the fragment, the second argument specifies the container parameter of the method, and the third argument specifies a false value to indicate that the layout is not the root layout.

After inflating the layout, the code in the onCreateView method can get references to the widgets on the layout. To do that, you can call the findViewById method from the View object that was created by the first statement. In this figure, for example, the second statement gets a reference to the EditText widget for the bill amount.

The onCreateView method can also set listeners for widgets. This code works like the code in the onCreate method of an activity. As a result, you should already understand how it works.

If you need to convert an existing activity to a fragment, you can usually start by cutting code from the activity class and pasting it into the fragment class. For example, to convert the Tip Calculator activity described in the previous chapter to the Tip Calculator fragment described in this chapter, you can begin by creating the class for the Tip Calculator fragment. Then, you can cut and paste the instance variables and methods from the activity into the fragment. Next, you can modify the onCreate and onCreateView methods so they work as shown in this figure.

Method of the Fragment class

Method	Description
`setHasOptionsMenu(bool)`	Set to true to indicate that this fragment has an options menu and would like to participate in populating the activity's option menu by having Android call this fragment's onCreateOptionsMenu method.

The declaration for the TipCalculatorFragment class

```
public class TipCalculatorFragment extends Fragment
implements OnEditorActionListener, OnClickListener
```

The onCreate method

```
@Override
public void onCreate(Bundle savedInstanceState) {
    super.onCreate(savedInstanceState);

    // set the default values for the preferences
    PreferenceManager.setDefaultValues(getActivity(),
        R.xml.preferences, false);

    // get default SharedPreferences object
    prefs = PreferenceManager.getDefaultSharedPreferences(getActivity());

    // turn on the options menu
    setHasOptionsMenu(true);
}
```

The onCreateView method

```
@Override
public View onCreateView(LayoutInflater inflater, ViewGroup container,
        Bundle savedInstanceState) {

    // inflate the layout for this fragment
    View view = inflater.inflate(R.layout.fragment_tip_calculator,
        container, false);

    // get references to the widgets
    billAmountEditText = (EditText)
            view.findViewById(R.id.billAmountEditText);

    // set the listeners
    billAmountEditText.setOnEditorActionListener(this);

    // return the View for the layout
    return view;
}
```

Description

- The Java code for a fragment works much like the Java code for an activity. However, there are several differences, especially in the onCreate and onCreateView methods.

Figure 9-4   How to create the class for a fragment

How to display a fragment in an activity

Figure 9-5 shows how to display a fragment in an activity. To start, you can create an XML file that contains the layout for the activity. In this figure, for example, that file is named activity_main.xml. This XML begins by defining a linear layout that matches the width and height of the parent. Within the LinearLayout element, this code contains a single fragment element. The name attribute of this element specifies the name of the package and the class for the Tip Calculator fragment shown in the previous figure.

After creating the XML file for the layout, you can use the activity class to display the layout as shown in the second example. In this figure, for instance, the activity class displays the layout named activity_main that was created in the first example. Since the event handling has been moved into the fragment class, this activity class just needs to display the layout.

The activity_main.xml file

```
<LinearLayout xmlns:android="http://schemas.android.com/apk/res/android"
    android:orientation="horizontal"
    android:layout_width="match_parent"
    android:layout_height="match_parent">

    <fragment android:name="com.murach.tipcalculator.TipCalculatorFragment"
            android:id="@+id/main_fragment"
            android:layout_weight="1"
            android:layout_width="0dp"
            android:layout_height="match_parent" />

</LinearLayout>
```

The TipCalculatorActivity class

```
package com.murach.tipcalculator;

import android.app.Activity;
import android.os.Bundle;

public class TipCalculatorActivity extends Activity {

    @Override
    public void onCreate(Bundle savedInstanceState) {
        super.onCreate(savedInstanceState);
        setContentView(R.layout.activity_main);
    }
}
```

Description

- To add a fragment to a layout, add a fragment element and use its name attribute to specify the fully qualified name for the class that defines the fragment.

Figure 9-5   How to display a fragment in an activity

How to create a preference fragment

Figure 9-6 shows how to create a fragment that displays preferences. Since this user interface is built using Preference objects instead of View objects, your fragment needs to extend the PreferenceFragment class instead of the Fragment class.

The first code example shows a fragment named SettingsFragment that inherits the PreferenceFragment class. This code is the same as the code for the SettingsFragment class shown in the previous chapter.

How to display a preference fragment in an activity

Once you've created a preference fragment like the one shown in the first example, it's easy to display it in an activity. To do that, you can define a layout for the activity that uses a fragment element to display the fragment as shown in the second example. Then, you can define an activity class that displays that layout as shown in the third example.

The SettingsFragment class

```
package com.murach.tipcalculator;

import android.os.Bundle;
import android.preference.PreferenceFragment;

public class SettingsFragment extends PreferenceFragment {

    @Override
    public void onCreate(Bundle savedInstanceState) {
        super.onCreate(savedInstanceState);

        // Load the preferences from an XML resource
        addPreferencesFromResource(R.xml.preferences);
    }
}
```

The activity_settings.xml file

```
<LinearLayout xmlns:android="http://schemas.android.com/apk/res/android"
    android:orientation="horizontal"
    android:layout_width="match_parent"
    android:layout_height="match_parent">

    <fragment android:name="com.murach.tipcalculator.SettingsFragment"
              android:id="@+id/settings_fragment"
              android:layout_weight="1"
              android:layout_width="0dp"
              android:layout_height="match_parent" />

</LinearLayout>
```

The SettingsActivity class

```
package com.murach.tipcalculator;

import android.app.Activity;
import android.os.Bundle;

public class SettingsActivity extends Activity {

    @Override
    public void onCreate(Bundle savedInstanceState) {
        super.onCreate(savedInstanceState);

        // set the view for the activity using XML
        setContentView(R.layout.activity_settings);
    }
}
```

Description

- Since the user interface is built using Preference objects instead of View objects, your fragment needs to extend the PreferenceFragment class instead of the Fragment class. Then, it can use the addPreferencesFromResource method to add the preferences defined in the XML file to the fragment.

Figure 9-6 How to work with preference fragments

How to use multi-pane layouts for large screens

Now that you understand how to use fragments in a single-pane layout, you're ready to learn how to use them in a multi-pane layout. This type of layout is commonly used for devices that have large screens such as tablets.

How to add multiple fragments to a layout

Figure 9-7 shows how to add multiple fragments to a layout. To do that, you can add two or more fragment elements to a layout file.

The first code example is designed to display two fragments when the screen is in landscape mode. To do that, this code uses a linear layout with its orientation set to horizontal. Then, it uses two fragment elements to display the Tip Calculator and Settings fragments. Both of these fragments have their layout_weight attributes set to a value of 1. As a result, they split the width of the screen equally.

The second code example is designed to display two fragments when the screen is in portrait mode. To do that, this code uses a linear layout with its orientation set to vertical. Then, it uses two fragment elements to display the Tip Calculator and Settings fragments. Both of these fragments have their layout_weight attributes set to a value of 1. As a result, they split the height of the screen equally.

Two layout files for the main activity

res\layout\activity_main_twopane_land.xml

```xml
<LinearLayout xmlns:android="http://schemas.android.com/apk/res/android"
    android:orientation="horizontal"
    android:layout_width="match_parent"
    android:layout_height="match_parent">

    <fragment android:name="com.murach.tipcalculator.TipCalculatorFragment"
            android:id="@+id/main_fragment"
            android:layout_weight="1"
            android:layout_width="0dp"
            android:layout_height="match_parent" />

    <fragment android:name="com.murach.tipcalculator.SettingsFragment"
            android:id="@+id/settings_fragment"
            android:layout_weight="1"
            android:layout_width="0dp"
            android:layout_height="match_parent" />

</LinearLayout>
```

res\layout\activity_main_twopane_port.xml

```xml
<LinearLayout xmlns:android="http://schemas.android.com/apk/res/android"
    android:orientation="vertical"
    android:layout_width="match_parent"
    android:layout_height="match_parent">

    <fragment android:name="com.murach.tipcalculator.TipCalculatorFragment"
            android:id="@+id/main_fragment"
            android:layout_weight="1"
            android:layout_height="0dp"
            android:layout_width="match_parent" />

    <fragment android:name="com.murach.tipcalculator.SettingsFragment"
            android:id="@+id/settings_fragment"
            android:layout_weight="1"
            android:layout_width="match_parent"
            android:layout_height="0dp" />

</LinearLayout>
```

Description

- To add more than one fragment to a layout, add two or more fragment elements to a layout file.

Figure 9-7   How to add multiple fragments to a layout

How to detect screen width

When the Tip Calculator app starts, it launches the Tip Calculator activity, which displays the activity_main layout that's in the layout directory. However, for large screens, you probably want to display one of the two-pane layouts described in the previous figure. To do that, you can detect large screens by creating a values directory that includes a smallest-width qualifier as shown in figure 9-8.

In this figure, the first example is stored in this directory:

```
res\values-sw600dp-land
```

This directory uses two qualifiers. First, it uses a smallest-width qualifier of "sw600dp" to indicate that this layout should be used for screens with a minimum width of 600dp. This is typically the smallest width of a 7" tablet. Second, it uses the land qualifier to indicate that this directory contains a layout for devices that are in landscape orientation.

Within this directory, the layout.xml file contains code that specifies which layout to use. To do that, it uses an item element to create an *alias*, which is a file that points to another resource file. You can often use an alias to reduce code duplication.

To specify an alias for a layout, code an item element. Then, set the name attribute to the name of the alias and set the type attribute of the item to a value of "layout". Next, use the body of the element to point to the correct layout that's in the project's layout directory. In the first example, the alias named activity_main points to the layout named activity_main_twopane_land. As a result, if the screen is large and in landscape orientation, the Tip Calculator app displays the layout named activity_main_twopane_land, instead of the activity_main layout, which is what you want.

The second example works the same as the first example. However, the values directory uses the port qualifier. Then, the alias points to a layout that's appropriate for portrait orientation. As a result, this code is only used when the screen is in portrait orientation.

If necessary, you can define several smallest-width qualifiers like this:

```
res\values-sw480dp
```

and this:

```
res\values-sw600dp
```

and this:

```
res\values-sw720dp
```

Then, Android selects the layout based on the screen's smallest width. For example, a screen that's 700dp x 500dp would use the the layout specified by the values-sw480dp directory. Similarly, a screen that's 1024dp x 600dp would use the layout specified by the values-sw600dp directory.

Smallest-width qualifiers examples

Qualifier	Typical screen size	Typical device
sw480dp	640dp x 480dp	5" tablet
sw600dp	1024dp x 600dp	7" tablet
sw720dp	960dp x 720dp	10" tablet

The layout files for devices with a minimum screen size

res\values-sw600dp-land\layout.xml

```xml
<?xml version="1.0" encoding="utf-8"?>
<resources>
    <item name="activity_main" type="layout">
        @layout/activity_main_twopane_land
    </item>
</resources>
```

res\values-sw600dp-port\layout.xml

```xml
<?xml version="1.0" encoding="utf-8"?>
<resources>
    <item name="activity_main" type="layout">
        @layout/activity_main_twopane_port
    </item>
</resources>
```

Description

- For devices with small screens, Android uses a layout in the layout directory.

- To detect large screens, you can create a values directory that uses the smallest-width qualifier. For example, a qualifier of sw600dp specifies a device with a smallest width of at least 600dp.

- To detect landscape and portrait orientations, you can create a values directory that uses the land or port qualifiers.

- Within a values directory, you can use an *alias* to point to a layout file that's stored in a layout directory. This provides a way to avoid duplicating code in multiple layout files.

- To specify an alias for a layout, code an item element. Then, set its name attribute to the name of the layout that you want to replace and set its type attribute to a value of "layout". Next, use the body of the element to point to the correct XML file in the project's layout directory.

- The smallest-width qualifier was introduced in Android 3.2 (API 13). As a result, it's available on most modern Android devices.

- With Android Studio, you can create a layout.xml file and its directory by right-clicking on the values directory, selecting the New→Values Resource item, and using the resulting dialog to enter the name of the file and directory.

Figure 9-8   How to detect screen width

How to control the soft keyboard

In figure 9-9, the EditText element contains a requestFocus element. As a result, when the app starts on a device with a large screen, Android moves the focus into this EditText widget. Then, if there is enough room, Android displays the soft keyboard for that EditText widget. If that isn't what you want, you can delete the requestFocus element.

Another common problem is that Android doesn't always display the correct action button on the soft keyboard by default. For example, it may display a Next button instead of a Done button. To fix this problem, you can use the imeOptions attribute of an EditText element to specify the correct action button for the soft keyboard. In this figure, for example, the imeOptions attribute specifies a Done button for the EditText widget.

Some values for the imeOptions attribute

Setting	The EditText widget displays a...
actionDone	Done button
actionNext	Next button
actionPrevious	Previous button
actionGo	Go button
actionSeach	Search button

Attributes of an EditText widget that can control the soft keyboard

```
<EditText
    android:id="@+id/billAmountEditText"
    android:layout_width="wrap_content"
    android:layout_height="wrap_content"
    android:layout_alignBaseline="@+id/billAmountLabel"
    android:layout_marginLeft="5dp"
    android:layout_toRightOf="@+id/billAmountLabel"
    android:ems="8"
    android:inputType="numberDecimal"
    android:imeOptions="actionDone"
    android:text="@string/bill_amount"
    android:textSize="20sp" >

    <requestFocus />
</EditText>
```

Description

- If you don't want Android to display the soft keyboard when the app starts, you can delete the requestFocus element from the body of the EditText element.

- If Android does not display the correct action button on the soft keyboard, you can use the imeOptions attribute of the EditText element to specify the correct action button on the soft keyboard.

Figure 9-9   How to control the soft keyboard

Other skills for working with fragments

So far, this chapter has presented the skills that you typically need for working with fragments. However, in some situations, you may need the skills presented in the next two figures to work with fragments.

How to get a reference to a fragment

Figure 9-10 shows how to get a reference to a fragment object. Once you have a reference to a fragment object, you can call any of its public methods. Or, if you aren't able to get a reference, you can take the appropriate action such as displaying the appropriate menu.

The onCreateOptionsMenu of the Tip Calculator fragment displays the appropriate options menu depending on whether the Settings fragment has been created. To do that, the first statement calls the getFragmentManager method to return a FragmentManager object. This method can be called from within an activity or a fragment. In this case, it's called from within a fragment.

After getting the FragmentManager object, this code calls the findFragmentById method from that object to find the Settings fragment. Then, it casts the Fragment object that's returned to the SettingsFragment type.

After the first statement, this code checks whether the manager was able to get the fragment. If so, this code displays a menu that includes a Settings item. Otherwise, it displays a menu that does not include a Settings item, which is usually what you want if the Settings fragment is already displayed.

The onSharedPreferenceChanged method of the Settings fragment calls a method from the Tip Calculator fragment. To do that, the first statement attempts to get the Tip Calculator fragment. Since this code works like the code in the first example, you shouldn't have much trouble understanding it.

After the first statement, this code checks whether the manager was able to get the fragment. If so, it calls the onResume method of this fragment to refresh the user interface so it reflects the latest changes to the settings. Without this statement, the user interface of the Tip Calculator fragment won't refresh when you use the Settings fragment to change settings.

Method of the Activity and Fragment classes

Method	Description
getFragmentManager()	Returns a FragmentManager object.

Method of the FragmentManager class

Method	Description
findFragmentById(id)	Returns a Fragment object for the fragment with the specified ID. If a Fragment object doesn't exist for that ID, this method returns a null value.

The onCreateOptionsMenu method of the Tip Calculator fragment

```java
@Override
public void onCreateOptionsMenu(Menu menu, MenuInflater inflater) {

    // attempt to get the fragment
    SettingsFragment settingsFragment = (SettingsFragment)
            getFragmentManager()
            .findFragmentById(R.id.settings_fragment);

    // if the fragment is null, display the appropriate menu
    if (settingsFragment == null) {
        inflater.inflate(R.menu.fragment_tip_calculator, menu);
    } else {
        inflater.inflate(R.menu.fragment_tip_calculator_twopane, menu);
    }
}
```

The onSharedPreferenceChanged method of the Settings fragment

```java
@Override
public void onSharedPreferenceChanged(SharedPreferences prefs,
        String key) {

    // attempt to get the fragment
    TipCalculatorFragment tipFragment =
        (TipCalculatorFragment) getFragmentManager()
            .findFragmentById(R.id.main_fragment);

    // if the fragment is not null, call a method from it
    if (tipFragment != null) {
        tipFragment.onResume();
    }
}
```

Description

- You can use the FragmentManager object to get a reference to a Fragment object.

Figure 9-10   How to get a reference to a fragment

How to replace one fragment with another

Figure 9-11 shows how to use Java code to replace the content of a container with a fragment. Here, the fragment named SettingsFragment replaces any other fragments that might be in the content container for the activity.

To do that, this code uses the getFragmentManager method to get a FragmentManager object. From that object, this code calls the beginTransaction, replace, and commit methods. These methods start a transaction, replace the content of the activity with the fragment, and commit the transaction. Here, the replace method uses this Android resource:

```
android.R.id.content
```

to identify the container for the content of the activity, and it creates a new SettingsFragment object to replace that content.

The code in this figure accomplishes the same task as the code shown in the second and third examples of figure 9-6. So, which approach is better? This is largely a matter of personal choice. In this book, I use the approach taken in figure 9-6 because I think XML is easier to understand when you're first getting started. However, this figure shows you often have a choice between using XML or Java to accomplish the same task. As your applications grow more complex, you may want or need to use Java to manage your fragments.

Method of the FragmentManager class

Method	Description
`beginTransaction()`	Begins a transaction and returns a FragmentTransaction object.

Methods of the FragmentTransaction class

Method	Description
`replace(id, fragment)`	Replaces the fragment in the container with the specified ID with an instance of the specified Fragment object.
`commit()`	Finishes the transaction.

The SettingsActivity class

```
package com.murach.tipcalculator;

import android.app.Activity;
import android.os.Bundle;

public class SettingsActivity extends Activity {

    @Override
    public void onCreate(Bundle savedInstanceState) {
        super.onCreate(savedInstanceState);

        // display the fragment as the main content
        getFragmentManager().beginTransaction()
                .replace(android.R.id.content, new SettingsFragment())
                .commit();
    }
}
```

Description

- You can use a FragmentManager object to replace one fragment with another fragment.

Figure 9-11   How to replace one fragment with another

Perspective

The skills presented in this chapter should be enough to get you started with fragments. However, Android apps can use fragments in many ways. As you progress through this book, you'll encounter other examples of apps that use fragments. For example, chapter 14 shows how to use fragments in a way that's common for many apps. As a result, studying that chapter should help to broaden your understanding of fragments.

Terms

fragment	multi-pane layout
panes	support library
single-pane layout	alias

Summary

- You can use a *fragment* to define part of the user interface for an activity. A fragment is sometimes referred to as a *pane*.

- On a small screen, an activity typically only displays a single fragment. This is known as a *single-pane layout*.

- On a large screen, an activity can display multiple fragments. This is known as a *multi-pane layout*.

- Fragments are available from Android 3.0 (API 11) and higher.

- The XML and Java code for a fragment works much like the XML and Java code for an activity.

- To add a fragment to a layout, add a fragment element and use its name attribute to specify the fully qualified name for the class that defines the fragment.

- If the user interface for a fragment uses Preference objects instead of View objects, the fragment needs to extend the PreferenceFragment class instead of the Fragment class.

- To add more than one fragment to a layout, add two or more fragment elements to a layout file.

- To detect large screens, you can create a values directory that uses the smallest-width qualifier to select screens whose smallest width is at least as wide as specified width.

- To detect landscape and portrait orientations, you can create a values directory that uses the land or port qualifiers.

- Within a values directory, a layout can use an *alias* to point to a layout file that's stored in a layout directory. This provides a way to avoid duplicating code in multiple layout files.

- From a fragment or activity, you can call the getFragmentManager method to get a FragmentManager object that you can use to find and work with the fragments in your app.

Exercise 9-1   Test the app

In this exercise, you'll experiment with the fragments that are available from the Tip Calculator app.

1. Open the project named ch09_ex1_TipCalculator.

2. In the Project view, expand the src directory. This directory should contain the source code for three activities and two fragments.

3. Expand the res\layout directory. This directory should contain five activity layouts. Of these layouts, three are for the "main" activity, the Tip Calculator activity. These layouts include both two-pane layouts described in this chapter.

4. Expand the res\values directory. Then, expand the layout.xml file. This should show two versions of the layout.xml file. Both use the smallest-width modifier.

5. Open the layout.xml file that uses the sw600dp and port modifiers. This file should use an alias to display the two-pane portrait layout for the Tip Calculator activity.

6. Run the app on a phone or an emulator for a phone. This should use the Tip Calculator activity to display the Tip Calculator fragment, and it should use the Settings activity to display the Settings fragment.

7. Run the app on a tablet or an emulator for a tablet. If necessary, you can create an emulator for a tablet as described in chapter 4. This should use the Tip Calculator activity to display both the Tip Calculator and Settings fragments.

8. Switch the orientation. This should still display both fragments, but it should use a different layout.

9. Test the app to make sure it works correctly when displayed in a tablet. It should. If it doesn't, fix any problems you encounter.

Exercise 9-2   Create a new fragment

In this exercise, you'll create a third fragment (the About fragment) and display it.

1. Open the project named ch09_ex2_TipCalculator.

2. In the res\layout directory add a layout named fragment_about. Then, copy the XML from the activity_about layout into the layout for the fragment.

3. Add a class to the src directory named AboutFragment. Then, write the code for this activity so it displays the fragment_about layout. This class should be a regular fragment, not a preference fragment.

4. Open the activity_about layout in the res\layout directory and modify it so it uses a fragment element to display the AboutFragment class. This fragment element's height and width should wrap the content.

5. Run the app. It should work just as it did before.

6. Open the activity_main_twopane_port layout in the res\layout directory and modify it so it displays the About fragment after the other two fragments.

7. Run the app on a tablet that has a large screen. In portrait orientation, the app should display all three fragments.

8. Open the activity_main_twopane_land layout and modify it so it displays the About fragment below the Settings fragment. To do that, you can nest both of these fragments in a linear layout that has vertical orientation.

9. Run the app on a tablet. The app should display all three fragments in both orientations. In addition, both orientations should allow you to access an Options menu that contains an About item.

10. Open the Java code for the TipCalculatorFragment. In the onCreateOptions-Menu method, modify the code so it doesn't inflate a menu if you're using one of the "two-pane" layouts.

Exercise 9-3   Use the fragment manager

In this exercise, you'll use a different technique to display a fragment in an activity.

1. Open the project named ch09_ex3_TipCalculator.

2. Open the class for the Tip Calculator activity. Modify this code so it uses the FragmentManager object to display the Tip Calculator fragment.

3. Run the app to make sure it still works correctly.

4. Open the class for the Settings activity. Modify this code so it uses the FragmentManager object to display the Settings fragment.

5. Run the app to make sure it still works correctly.

Section 3

The News Reader app

In sections 1 and 2, you learned the essential skills for developing a simple Android app, the Tip Calculator app. Now, this section shows how to develop another app, the News Reader app.

Along the way, this section presents many essential Android skills that can be applied to other apps. To start, chapter 10 shows how to build a simple version of a News Reader app that uses threads, files, a simple adapter, and intents. Then, chapter 11 shows how to modify this app so it uses services and notifications. Finally, chapter 12 shows how to modify this app so it uses broadcast receivers.

10

How to work with threads, files, adapters, and intents

In this chapter, you'll learn how to create a News Reader app that downloads news items from the web and displays them so they can be read by the user. To be able to create such an app, you need to learn several new skills including how to work with threads, files, adapters, and intents.

An introduction
to the News Reader app

The skills presented in this chapter are necessary to create the News Reader app. To put these skills in context, this chapter begins by showing the user interface for this app. Then, it shows the XML file that contains the data for this app.

The user interface

Figure 10-1 shows the user interface for the News Reader app. When the app starts, the top of the Items activity displays the title of the news feed. Below that, the Items activity displays a list of items in the news feed. Here, each item has a publication date and a title.

If the user clicks on an item in the Items activity, the app displays the Item activity, which displays more information about the clicked item. Here, each item has a title, a publication date, a description, and a link to the original article that says "Read more on the web".

If the user clicks on this link, the app starts the default web browser for the device. Then, it displays the link in that browser. Although this exits the News Reader app, the user can easily use the Back button to navigate back to the News Reader app.

The Items activity

The Item activity

Description

- When the app starts, the Items activity displays the title of the news feed followed by a list of items in the news feed where each item has a publication date and a title.

- If the user clicks on a news item, the Item activity displays more information about that item, including a link that can display the original article in a web browser.

Figure 10-1   The user interface for the News Reader app

The XML for an RSS feed

Figure 10-2 shows the URL for the *RSS (Rich Site Summary) feed* that's used by the News Reader app. A website can use an RSS feed to publish frequently updated works, such as blog entries and news headlines. Then, an app can read that RSS feed. Since RSS uses a standardized XML file format, an app that can read one RSS feed can usually read other types of RSS feeds with little or no modification.

The RSS feed that's returned by the URL in this figure is actually much more complex than the XML shown in this figure. However, the XML shown in this figure shows the parts of the RSS feed that are used by the News Reader app.

To start, there are title and pubDate elements that come before any item elements. These elements store the title and publication data for the feed.

After the elements for the feed, there are a series of item elements. Each item element contains title, link, description, and pubDate elements. These elements store the title, link, description, and publication date for each item in the feed.

The URL for the RSS feed

```
http://rss.cnn.com/rss/cnn_tech.rss
```

Simplified XML for the RSS feed

```
<rss xmlns:media="http://search.yahoo.com/mrss/"
     xmlns:feedburner="http://rssnamespace.org/feedburner/ext/1.0"
     version="2.0">
<channel>
  <title>CNN.com - Technology</title>
  <pubDate>Tue, 15 Sep 2015 17:18:54 GMT</pubDate>
  <item>
    <title>Prosthetic hand 'tells' the brain what it is touching</title>
    <link>http://rss.cnn.com/c/35492/f/676960/s/story01.htm</link>
    <description>Research on prosthetic hands has come a long way, but
    most of it has focused on improving the way the body controls the
    device.</description>
    <pubDate>Tue, 15 Sep 2015 17:18:08 GMT</pubDate>
  </item>
  <item>
    <title>They're here! New Horizons' best shots of Pluto</title>
    <link>http://rss.cnn.com/c/35492/f/676960/story01.htm</link>
    <description>&lt;br clear='all'/&gt;</description>
    <pubDate>Mon, 14 Sep 2015 18:59:52 GMT</pubDate>
  </item>
  <item>
    <title>5 big Mario moments</title>
    <link>http://rss.cnn.com/c/35492/story01.htm</link>
    <description>&lt;br clear='all'/&gt;</description>
    <pubDate>Mon, 14 Sep 2015 18:53:03 GMT</pubDate>
  </item>
  ...
  ...
</channel>
</rss>
```

Description

- An *RSS* (*Rich Site Summary*) *feed* can be used to publish frequently updated works, such as blog entries and news headlines.

- Since RSS uses a standardized XML file format, the feed can be published once and viewed by many different apps.

Figure 10-2   The XML for an RSS feed

How to work with threads

The News Reader app reads data from the Internet, it writes data to a file, and it reads data from that file. In an Android app, you should perform tasks like these in their own threads.

How threads work

As figure 10-3 explains, a *thread* is a single flow of execution through an app. By default, an Android app uses a single thread, called the *UI thread*. This thread displays the user interface for the app and listens for events that occur when the user interacts with the user interface.

The user interface for an app should always be responsive to the user. As a result, you don't want to do any processing on the UI thread that takes more than a second or so. For example, downloading data from the Internet may take several seconds or even minutes depending on the amount of data and speed of the connection. As a result, you should never perform this type of processing on the UI thread. Instead, you should create a second thread.

Similarly, if you want to perform a file I/O operation such as reading or writing a file, you should do that processing in a separate thread. That's because I/O operations are thousands of times slower than CPU operations. So any program that reads data from a disk spends much of its time waiting for that information to be retrieved.

The first diagram shows how an app might work when executed as a single thread. First, the UI thread displays the user interface and listens for events. Then, an event occurs that causes the program to perform an I/O operation. Since this I/O operation runs in the UI thread, the UI thread can't listen for events and respond to the user while the I/O thread is running. In other words, if the user clicks on a button while the I/O operation is running, the UI thread can't respond to that event. When the I/O operation completes, the user interface becomes responsive again.

The second diagram shows how this app benefits from using a separate thread to perform the I/O operation. This allows the two tasks to overlap. As a result, the UI thread can continue listening for events while the I/O operation runs. The result is a user interface that's always responsive.

How using threads improves user interface responsiveness

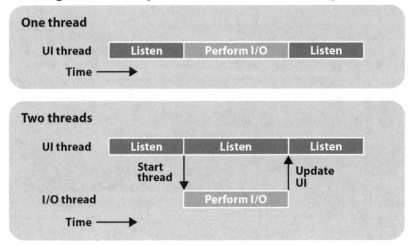

Description

- A *thread* is a single sequential flow of control within a program. A thread often completes a specific task.

- By default, an Android app uses a single thread, called the *UI thread*, to display the user interface. Any task that can slow or stop the responsiveness of the UI thread should be run in a separate thread.

Figure 10-3   How threads work

How to execute asynchronous tasks

An *asynchronous task* is a task that runs in a separate thread in the background and does not need to be synchronized with other threads. Since Android apps often need to perform asynchronous tasks, the Android framework includes a class named AsyncTask that makes it easy to perform these types of tasks. This class handles the code that creates and manages the thread for you. As a result, you just need to create a class that inherits the AsyncTask class and override the appropriate methods.

Since an asynchronous class is usually closely related to an activity class, it's usually coded as a nested inner class of the activity class. In part 1 of figure 10-4, for example, the DownloadFeed class defines an asynchronous task that's closely related to the ItemsActivity class. As a result, the DownloadFeed class is nested within the ItemsActivity class.

The ItemsActivity class begins by defining a constant for a string that contains the URL for the RSS feed described earlier in this chapter. Then, this class defines an onCreate method. This method starts by displaying the layout for the news items. Then, it creates an instance of the DownloadFeed object and calls its execute method to start its thread. In addition, this code uses the execute method to pass the URL string to the DownloadFeed object.

The DownloadFeed class extends the AsyncTask class. This class uses *generics*, which is a feature of Java that allows a class to operate on various types of objects. In particular, the declaration of the AsyncTask class allows it to work with three generic types: (1) parameters, (2) progress, and (3) result. As the programmer, you can decide what types of objects you want to use with this class. In this figure, the DownloadFeed class uses an AsyncTask class like this:

```
AsyncTask<String, Void, String>
```

to indicate that it accepts one or more string parameters, doesn't handle progress values, and returns a string.

Within the DownloadFeed class, the doInBackground method contains the code that runs in the background thread. This method accepts the array of String objects as specified by the first generic type. Since this activity only passes one string value, the first statement in this method gets that string from the params variable. At this point, you are ready to download the data from the feed specified by the URL. You'll learn how to do that later in this chapter. Finally, this method returns a message in a String object as specified by the third generic type.

After the doInBackground method finishes, the onPostExecute method runs. This method accepts a String object as specified by the third generic type. This method contains code that updates the user interface. For now, this method uses a toast to display the result string that's returned by the doInBackground method. Later in this chapter, you'll see how this method can be used to update the user interface.

An activity with a nested AsyncTask class

```
package com.murach.newsreader;

import android.os.AsyncTask;
import android.os.Bundle;
import android.app.Activity;
import android.content.Context;
import android.widget.Toast;

public class ItemsActivity extends Activity {

    private static String URL_STRING =
            "http://rss.cnn.com/rss/cnn_tech.rss";

    @Override
    protected void onCreate(Bundle savedInstanceState) {
        super.onCreate(savedInstanceState);
        setContentView(R.layout.activity_items);

        new DownloadFeed().execute(URL_STRING);
    }

    class DownloadFeed extends AsyncTask<String, Void, String> {
        @Override
        protected String doInBackground(String... params) {
            // get the parameter
            String urlString = params[0];

            // download the feed and write it to a file

            // return a message
            return "Feed downloaded";
        }

        @Override
        protected void onPostExecute(String result) {
            Context context = ItemsActivity.this;
            Toast.makeText(context, result, Toast.LENGTH_LONG).show();
        }
    }
}
```

Description

- An *asynchronous task* is a task that runs in a separate thread in the background and does not need to be synchronized with other threads.

- It's common to use a nested inner class to create an asynchronous thread. Within the inner class, you can access the context for the activity by coding the name of the outer class, a period, and the *this* keyword.

Figure 10-4   How to execute asynchronous tasks (part 1 of 2)

When you nest one class within another, the inner class can access the object for the outer class by coding the name of the outer class, a period, and the *this* keyword. In this figure, for example, the onPostExecute method of the inner class uses this code:

```
ItemsActivity.this
```

to get the ItemsActivity object. Then, it stores the ItemsActivity object in a Context object, which is later used by the makeText method of the Toast object.

Part 2 of this figure shows more details about how the AsyncTask class works. To start, it shows several examples for how you could declare a class that extends the AsyncTask class.

The first example declares a DownloadFeed class that extends the AsyncTask class. This class accepts one or more string parameters, uses int values to update the progress of the thread, and returns a string. In this case, the doInBackground method would accept an array of String objects and return a String object. Similarly, the onPostExecute method would accept the String object returned by the doInBackground method. For primitive types such as the int type, you must use the corresponding wrapper class such as the Integer class.

The second example accepts a URL object as a parameter, doesn't update the progress of the thread, and returns a string. This shows that you can use any object type as a generic type. In addition it shows that you can use the Void keyword if your class doesn't use one of the generic types.

The third example shows that you can use the Void keyword for all three generic types. This can make sense when the class for the asynchronous task works with instance variables. In that case, it often doesn't need to accept parameters or return data. In this case, the doInBackground method would accept an array of Void objects and return a Void object. To do that, this method can return a null value.

The fourth example declares a ReadFeed class that extends the AsyncTask class. This class accepts one or more string parameters, uses int values to update the progress of the thread, and returns an RSSFeed object. The code for the RSSFeed class is shown later in this chapter.

This figure also shows the four methods of the AsyncTask class that you can override. Of these four methods, the code within the doInBackground method executes in a background thread. As a result, you should put any network or I/O code in this thread.

The other three methods execute in the UI thread. As a result, you can use them to update the user interface. Typically, you use the onPreExecute method to display a ProgressBar widget that displays the progress of the background thread. Then, you can use the onProgressUpdate method to update the ProgressBar widget every time the background thread uses the publishProgress method to update its progress. Finally, you can use the onPostExecute method to update the user interface when the background thread finishes.

The generic types for the AsyncTask class

```
AsyncTask<Params, Progress, Result>
```

Possible class declarations

```
class DownloadFeed extends AsyncTask<String, Integer, String> { ... }
class DownloadFeed extends AsyncTask<URL, Void, String> { ... }
class DownloadFeed extends AsyncTask<Void, Void, Void> { ... }
class ReadFeed extends AsyncTask<String, Integer, RSSFeed> { ... }
```

The AsyncTask class

Method	Is executed...
`onPreExecute()`	On the UI thread before the task is executed.
`doInBackground(Params...)`	On the background thread immediately after the onPreExecute method finishes. An array of parameters is passed to this method. This method returns a result that's passed to the onPostExecute method. If this method uses the publishProgress method, it passes progress values to the onProgressUpdate method.
`onProgressUpdate(Progress...)`	On the UI thread after a call to the publishProgress method is made in the doInBackground method.
`onPostExecute(Result)`	On the UI thread after the doInBackground method finishes.

Description

- The AsyncTask class uses *generics* to allow a class to operate on various types of objects. The AsyncTask class provides for three generic types: (1) parameters, (2) progress, and (3) result.

- The AsyncTask class provides an easy way to perform a background task and publish results on the UI thread without having to manually manipulate threads.

Figure 10-4   How to execute asynchronous tasks (part 2 of 2)

How to execute timed tasks

Android apps sometimes need to start a separate thread to execute a timed task, which is a task that executes after a specified delay or at a specified interval. For the News Reader app, for example, you may want to check for updates every hour. For a Stopwatch app, you may want to update the user interface every second. To do that, you can use the Timer and TimerTask classes as shown in figure 10-5. These Java classes provide a way to start a thread that executes tasks at a specified interval.

The code example in this figure shows a startTimer method that you can use to start a timed task. This method begins by creating a variable named startMillis that stores the number of milliseconds for the current time. To get these milliseconds, this code calls the currentTimeMillis method from the System class.

After getting the starting time, this code creates a new TimerTask object named task. To be able to create this object, you must override the abstract run method that's declared in the TimerTask class. Then, you can store the code that executes the task within this method. In this figure, for example, the run method contains two statements. The first statement gets the number of milliseconds that have elapsed since the timer was started. Then, the second statement passes the number of elapsed milliseconds to a method named updateView. This method updates the user interface and is shown in the next figure.

After creating the TimerTask object, this code creates a Timer object and uses it to execute the TimerTask object. To start, the first statement creates a Timer object that starts a *daemon thread*, which is a thread that ends when the app ends. This is the most common type of thread for Android apps. However, you can also start a regular thread that would continue running even if the app ended. Then, the second statement calls the schedule method of the Timer object to execute the task immediately (after 0 milliseconds) and to continue executing it every second (1000 milliseconds). As a result, the task continues to execute every second until the timer is cancelled or until the app ends.

Sometimes, you may want to execute a task only once. To do that, you don't specify a repeat interval for the schedule method. For example, to execute a task once after a delay of 2 hours, you could use this code:

```
task.schedule(task, 1000 * 60 * 60 * 2);
```

This code calculates the number of milliseconds for 2 hours by multiplying the number of milliseconds in 1 second (1000) by the number of seconds in a minute (60) by the number of minutes in an hour (60) by the number of hours (2).

The classes used to work with timed tasks

```
java.util.Timer
java.util.TimerTask
```

A method that starts a timed task

```java
private void startTimer() {
    final long startMillis = System.currentTimeMillis();
    TimerTask task = new TimerTask() {

        @Override
        public void run() {
            long elapsedMillis =
                System.currentTimeMillis() - startMillis;
            updateView(elapsedMillis);
        }
    };
    Timer timer = new Timer(true);
    timer.schedule(task, 0, 1000);   // execute every second
}
```

The TimerTask class

Constructor/Method	Description
`TimerTask()`	Creates a new TimerTask object.
`run()`	An abstract method of the TimerTask class. You can override this method and execute the code for the task within it.

The Timer class

Constructor/Method	Description
`Timer(isDaemon)`	Creates a Timer object and specifies whether its thread is a *daemon thread* that ends when the app ends.
`schedule(task, delay)`	Executes the specified TimerTask object once, after the specified delay in milliseconds.
`schedule(task, delay, interval)`	Executes the specified TimerTask object repeatedly, after the specified delay in milliseconds at the specified interval in milliseconds.
`cancel()`	Cancels this timer, discarding any currently scheduled tasks.

Description

- You can use the TimerTask and Timer classes to create a thread that executes tasks after a specified delay or at a specified interval.

Figure 10-5   How to execute timed tasks

How to update the UI thread

If you create a timed task, that task runs in its own thread in the background. As the background thread runs, you need a way to update the UI thread. To do that, you can use the post method that's available from a View object as shown in figure 10-6. This method accepts any object that implements the Runnable interface.

The updateView method in this figure begins by declaring that it accepts a parameter for a long value. This parameter contains the number of elapsed milliseconds. In addition, this parameter must be declared as final so it can be used within the UI thread.

Within this method, the first statement calls the post method on a TextView widget. Within the parentheses for this method, this code uses an anonymous class to create an instance of a Runnable object.

This anonymous class begins by calculating the number of elapsed seconds by dividing the number of elapsed milliseconds by 1000. Then, the anonymous class implements the run method of the Runnable interface. This method contains a single statement that uses the setText method of a TextView widget to update the text that's displayed on the widget. This text includes the number of seconds that have elapsed since the timed task was started.

A method that updates the UI thread

```
private void updateView(final long elapsedMillis) {
    // UI changes need to be run on the UI thread
    messageTextView.post(new Runnable() {

        int elapsedSeconds = (int) elapsedMillis/1000;

        @Override
        public void run() {
            messageTextView.setText("Seconds: " + elapsedSeconds);
        }
    });
}
```

The View class

Method	Description
post(runnable)	Executes the specified Runnable object.

The Runnable interface

Method	Description
run()	Contains the statements that are run within the thread.

Description

- You can use the post method of a View object to execute any Runnable object on the UI thread.

Figure 10-6   How to update the UI thread

How to work with files

Now that you know how to create and execute a thread, you're ready to learn how to write the code that's executed by the thread. This code often reads and writes files. For example, the News Reader app needs to read an RSS feed from the Internet and write that data to an XML file. Then, it needs to read that XML file and parse its data into a series of Java objects.

How to download a file from the Internet

Figure 10-7 shows how to download data from the Internet and store it in a file on an Android device. To do that, you get an input stream from the Internet, you get an output stream to a file on your device, and you use standard Java I/O methods to read data from the input stream and write that data to the output stream.

The code example begins by declaring a constant that stores the name, but not the path, of the XML file that stores the data for the news feed. Then, this code uses a try/catch statement to handle any I/O exceptions that may be thrown by the I/O code. If an IOException is thrown, the catch clause handles the exception by displaying it in the LogCat view.

Within the try clause, the first statement creates a URL object from a string that points to the URL for the RSS feed described earlier in this chapter. Then, the second statement calls the openStream method on the URL object to return an InputStream object that you can use to read the input stream.

To be able to read data from the Internet, you must include the INTERNET permission in the manifest file for your app. To do that, you can add a permission like the one shown in this figure at the same indentation level as the application element.

After getting the InputStream object, this code calls the openFileOutput method to get a FileOutputStream object. Calling the openFileOutput method is an application-level operation. However, you can call this method from an activity and Android automatically calls the method from the application level. This is true even if you call this method from an inner class of an activity.

The openFileOutput method accepts two parameters. The first parameter specifies the filename, but not the path, of the file. You can't specify a path because Android automatically uses the default path for the app. This path uses the package name of the app to give each app its own directory for storing files. If the filename doesn't exist in this directory, Android automatically creates the file for you and opens it. Otherwise, it opens the existing file.

The second parameter specifies the mode that Android uses when it opens the file. For this parameter, you can use any of the MODE_XXX constants that are available from the Context class. However, the MODE_PRIVATE constant is used in most cases. This constant only allows the current app to work with the file, which is what you want for a file like news_feed.xml. In addition, if a file with the specified name already exists, this mode overwrites the existing file, which is also what you want for a file like news_feed.xml.

The classes used to download a file from the Internet

```
java.net.URL
android.content.Context
```

How to download a file

```java
final String FILENAME = "news_feed.xml";
try{
    // get the input stream
    URL url = new URL("http://rss.cnn.com/rss/cnn_tech.rss");
    InputStream in = url.openStream();

    // get the output stream
    FileOutputStream out =
            openFileOutput(FILENAME, Context.MODE_PRIVATE);

    // read input and write output
    byte[] buffer = new byte[1024];
    int bytesRead = in.read(buffer);
    while (bytesRead != -1)
    {
        out.write(buffer, 0, bytesRead);
        bytesRead = in.read(buffer);
    }
    out.close();
    in.close();
}
catch (IOException e) {
    Log.e("News reader", e.toString());
}
```

The URL class

Method	Description
openStream()	Returns an InputStream object for the specified URL.

The INTERNET permission in the AndroidManifest.xml file

```xml
<uses-permission android:name="android.permission.INTERNET" />
```

The Context class

Method	Description
openFileOutput(filename, mode**)**	Returns a FileOutputStream object for the specified file. If the file doesn't already exist, this method creates it. For the second parameter, use MODE_PRIVATE to overwrite any existing files.

Description

- You can use Java and Android APIs to download a file by reading input from the Internet and writing output to the file system.

Figure 10-7   How to download a file from the Internet

The MODE_PRIVATE constant stores a value of 0. As a result, the code in figure 10-7 can also be written like this:

```
FileOutputStream out = openFileOutput(FILENAME, 0);
```

Although this code is shorter, using the MODE_PRIVATE constant is a better practice because it makes the code easier to read and because it will work even if the value of the constant changes in a future version of Android.

How to parse an XML file

Figure 10-8 shows how to read the XML file that was written in the previous figure and how to use *SAX* (*Simple API for XML*) to parse this XML into a series of Java objects. To do that, the code in this figure uses Java classes from the javax.xml.parsers and org.xml.sax packages.

The code example in this figure begins by declaring a constant for the name of the XML file that stores the data for the news feed. Then, the second statement declares an RSSFeed object named feed. The RSSFeed class is a custom class that's used to store the data for the news feed. The code for this class is shown later in this chapter. For now, all you need to know is that an RSSFeed object can store the data for a news feed.

After declaring the RSSFeed object, this code uses a try/catch statement to handle any exceptions that may be thrown by the code. If an exception is thrown, the catch block displays it in the LogCat view.

Within the try block, the first three statements get an XMLReader object that can be used to read an XML file. This is boilerplate code that you can cut and paste to get an XMLReader object. Then, this code creates an object from the RSSFeedHandler class. This class is another custom Java class that's shown later in this chapter. For now, all you need to know is that it contains the code that's used to parse the XML file. After creating the RSSFeedHandler object, the next statement sets this object as the content handler in the XMLReader object.

After setting up the content handler, this code uses the openFileInput method to get an input stream for the file. This method works much like the openFile-Output method described in the previous figure. However, it only requires one parameter, the filename.

Now that the XMLReader object and the input stream have been created, this code can parse the data. To do that, it creates an InputSource object from the FileInputStream object. Then, it calls the parse method from the XMLReader object and passes the InputSource object to it. This parses the data in the XML file and stores it in an RSSFeed object. Finally, this code calls the getFeed method of the RSSHandler class to return the RSSFeed object.

The classes used to work with SAX

```
javax.xml.parsers.SAXParser
javax.xml.parsers.SAXParserFactory

org.xml.sax.InputSource
org.xml.sax.XMLReader
```

How to parse an XML file

```
final String FILENAME = "news_feed.xml";
RSSFeed feed;
try {
    // get the XML reader
    SAXParserFactory factory = SAXParserFactory.newInstance();
    SAXParser parser = factory.newSAXParser();
    XMLReader xmlreader = parser.getXMLReader();

    // set content handler
    RSSFeedHandler theRssHandler = new RSSFeedHandler();
    xmlreader.setContentHandler(theRssHandler);

    // get the input stream
    FileInputStream in = openFileInput(FILENAME);

    // parse the data
    InputSource is = new InputSource(in);
    xmlreader.parse(is);

    // get the content handler and return it
    feed = theRssHandler.getFeed();
}
catch (Exception e) {
    Log.e("News reader", e.toString());
}
```

The Context class

Method	Description
openFileInput(filename)	Returns a FileInputStream object for the specified file.

Description

- You can use *SAX* (*Simple API for XML*) to parse XML files.

Figure 10-8   How to parse an XML file

The RSSFeedHandler class

Figure 10-9 shows the RSSFeedHandler class. This class begins by extending the DefaultHandler class, which contains methods that you can override to handle the events that occur when the XML file is parsed. In particular, the DefaultHandler class defines the startDocument, endDocument, startElement, endElement, and characters methods that are overridden in the RSSFeedHandler class.

Within the RSSFeedHandler class, the first two statements define instance variables for the RSSFeed and RSSItem objects that are used to store the data for the feed and for the items within the feed. The code for the RSSFeed and RSSItem classes is presented in the next two figures. For now, all you need to know is that both classes have a default constructor as well as get and set methods that allow you to store data.

The next six statements define the Boolean instance variables that this class uses to determine when the various elements of the XML file are being parsed.

The getFeed method returns the RSSFeed object. This method can be called after the parser finishes parsing the XML file.

The startDocument method is executed when the parser starts reading the XML document. The code in this method creates the RSSFeed and RSSItem objects. As a result, you can use these objects in the other methods of this class.

The endDocument method is executed when the parser finishes reading this document. In this class, this method doesn't include any statements. However, if you needed to perform some processing when the parser finishes reading the document, you could add it here.

The startElement method is executed after the parser reads a start element such as <item>. The third parameter of this method contains the qualified name of the element. Within this method, an if/else statement checks whether the element is needed by the News Reader app. If so, this code takes the appropriate action. For the element named item, this code creates a new RSSItem object that can be used to store the data for the item. Then, it exits this method. For the element named title, this code sets isTitle variable to true. Then, it exits the method. And so on.

The RSSFeedHandler class   Page 1

```java
package com.murach.newsreader;

import org.xml.sax.helpers.DefaultHandler;
import org.xml.sax.*;

public class RSSFeedHandler extends DefaultHandler {
    private RSSFeed feed;
    private RSSItem item;

    private boolean feedTitleHasBeenRead = false;
    private boolean feedPubDateHasBeenRead = false;

    private boolean isTitle = false;
    private boolean isDescription = false;
    private boolean isLink = false;
    private boolean isPubDate = false;

    public RSSFeed getFeed() {
        return feed;
    }

    public void startDocument() throws SAXException {
        feed = new RSSFeed();
        item = new RSSItem();
    }

    public void endDocument() throws SAXException { }

    public void startElement(String namespaceURI, String localName,
            String qName, Attributes atts) throws SAXException {

        if (qName.equals("item")) {
            item = new RSSItem();
            return;
        }
        else if (qName.equals("title")) {
            isTitle = true;
            return;
        }
        else if (qName.equals("description")) {
            isDescription = true;
            return;
        }
        else if (qName.equals("link")) {
            isLink = true;
            return;
        }
        else if (qName.equals("pubDate")) {
            isPubDate = true;
            return;
        }
    }
```

Figure 10-9   The RSSFeedHandler class (part 1 of 2)

The endElement method is executed after the parser reads an end element such as </item>. Like the startElement method, the third parameter contains the qualified name of the element. Within this method, an if statement checks whether the element is named item. If so, it adds the current RSSItem object to the RSSFeed object.

The characters method is executed when the parser reads the characters within an element. This method contains three parameters. Within this method, the first statement converts these three parameters into a string. Then, a nested if/else statement stores this string in the appropriate RSSFeed or RSSItem object.

This code is a little tricky because both the feed and each item have elements named title and pubDate. As a result, the code for the title element begins by checking whether the feed title has already been read. If not, it sets the title in the RSSFeed object. Then, it sets the appropriate Boolean variable to indicate that the feed title has been read. Otherwise, this code sets the title in the RSSItem object. Finally, this code sets the isTitle variable to a false value to indicate that the title element is no longer being parsed.

The code for the pubDate element works similarly to the code for the title element.

The code for the link and description elements are simpler since they only exist in the item element. However, many description elements don't store simple text. Instead, they store links to other URLs. For these elements, this code sets the description to indicate that it isn't available directly from the feed.

The RSSFeedHandler class   **Page 2**

```java
    public void endElement(String namespaceURI, String localName,
            String qName) throws SAXException {
        if (qName.equals("item")) {
            feed.addItem(item);
            return;
        }
    }

    public void characters(char ch[], int start, int length) {
        String s = new String(ch, start, length);
        if (isTitle) {
            if (feedTitleHasBeenRead == false) {
                feed.setTitle(s);
                feedTitleHasBeenRead = true;
            }
            else {
                item.setTitle(s);
            }
            isTitle = false;
        }
        else if (isLink) {
            item.setLink(s);
            isLink = false;
        }
        else if (isDescription) {
            if (s.startsWith("<")) {
                item.setDescription("No description available.");
            } else{
                item.setDescription(s);
            }
            isDescription = false;
        }
        else if (isPubDate) {
            if (feedPubDateHasBeenRead == false) {
                feed.setPubDate(s);
                feedPubDateHasBeenRead = true;
            }
            else {
                item.setPubDate(s);
            }
            isPubDate = false;
        }
    }
}
```

Figure 10-9   The RSSFeedHandler class (part 2 of 2)

The RSSFeed class

Figure 10-10 shows the code for the RSSFeed class. This class stores the data for the RSS feed. This data includes the feed's title and publication date as well as an ArrayList object that stores RSSItem objects. Most of the code for this class is standard Java code that lets you set and get data. As a result, you shouldn't have much trouble understanding it.

However, the code within the getPubDateMillis method uses the SimpleDateFormat object to parse the string for the date into a Date object. Here, the parse method accepts a string in this format:

```
Fri, 18 Dec 2015 14:11:56 EST
```

and converts that string into a Date object. Then, this code calls the getTime method of the Date object to return the milliseconds for the time.

The RSSItem class

Figure 10-11 shows the code for the RSSItem class. This class stores the data for each item in the RSS feed. This data includes the item's title, description, link, and publication date. Most of the code for this class is standard Java code that lets you set and get this data. As a result, you shouldn't have much trouble understanding it.

However, the code within the getPubDateFormatted method uses two SimpleDateFormat objects to change the format for the date. To start, the parse method of the first SimpleDateFormat object converts the string for the date into a Date object. Then, the format method of the second SimpleDateFormat object formats the Date object so it uses this format:

```
Friday, 5:56 AM (Dec 18)
```

As a result, the News Reader app can use this user-friendly date format instead of the unwieldy date format that's stored in the XML file.

The RSSFeed class

```
package com.murach.newsreader;

import java.text.ParseException;
import java.text.SimpleDateFormat;
import java.util.ArrayList;
import java.util.Date;

public class RSSFeed {
    private String title = null;
    private String pubDate = null;
    private ArrayList<RSSItem> items;

    private SimpleDateFormat dateInFormat =
            new SimpleDateFormat("EEE, dd MMM yyyy HH:mm:ss Z");

    public RSSFeed() {
        items = new ArrayList<RSSItem>();
    }

    public void setTitle(String title) {
        this.title = title;
    }

    public String getTitle() {
        return title;
    }

    public void setPubDate(String pubDate) {
        this.pubDate = pubDate;
    }

    public long getPubDateMillis() {
        try {
            Date date = dateInFormat.parse(pubDate.trim());
            return date.getTime();
        }
        catch (ParseException e) {
            throw new RuntimeException(e);
        }
    }

    public int addItem(RSSItem item) {
        items.add(item);
        return items.size();
    }

    public RSSItem getItem(int index) {
        return items.get(index);
    }

    public ArrayList<RSSItem> getAllItems() {
        return items;
    }
}
```

Figure 10-10   The RSSFeed class

The RSSItem class

```java
package com.murach.newsreader;

import java.text.ParseException;
import java.text.SimpleDateFormat;
import java.util.Date;

public class RSSItem {

    private String title = null;
    private String description = null;
    private String link = null;
    private String pubDate = null;

    private SimpleDateFormat dateOutFormat =
        new SimpleDateFormat("EEEE h:mm a (MMM d)");

    private SimpleDateFormat dateInFormat =
        new SimpleDateFormat("EEE, dd MMM yyyy HH:mm:ss Z");

    public void setTitle(String title) {
        this.title = title;
    }

    public String getTitle() {
        return title;
    }

    public void setDescription(String description) {
        this.description = description;
    }

    public String getDescription() {
        return description;
    }

    public void setLink(String link) {
        this.link = link;
    }

    public String getLink() {
        return link;
    }

    public void setPubDate(String pubDate) {
        this.pubDate = pubDate;
    }

    public String getPubDate() {
        return pubDate;
    }
```

Figure 10-11   The RSSItem class (part 1 of 2)

The RSSItem class

```java
    public String getPubDateFormatted() {
        try {
            Date date = dateInFormat.parse(pubDate.trim());
            String pubDateFormatted = dateOutFormat.format(date);
            return pubDateFormatted;
        }
        catch (ParseException e) {
            throw new RuntimeException(e);
        }
    }
}
```

Figure 10-11   The RSSItem class (part 2 of 2)

How to work with adapters

After the News Reader app reads the RSS feed and parses it into a series of Java objects, it needs to display the data that's stored in those objects. To do that, you can store the data in an adapter. Then, you can use a ListView widget to display the data that's stored in the adapter.

How to create the layout for a list view

Figure 10-12 shows the activity_items layout as it's displayed in the Graphical Layout editor. This layout begins by using a TextView widget to display the title of the news feed. Then, it uses a ListView widget to display the items of the news feed.

This figure also shows the listview_item layout as it's displayed in the Graphical Layout editor. This layout uses two TextView widgets to display the publication date and title for each item. The first TextView widget displays the item's publication date in the default font size. The second TextView widget displays the item's title with a larger font size of 24sp. In addition this layout sets the margins for these widgets to control the alignment and spacing of these widgets.

The activity_items layout

The listview_item layout

The ListView widget in the activity_items layout

```
<ListView
    android:id="@+id/itemsListView"
    android:layout_width="match_parent"
    android:layout_height="match_parent" />
```

The listview_item layout

```
<?xml version="1.0" encoding="utf-8"?>
<LinearLayout xmlns:android="http://schemas.android.com/apk/res/android"
    android:layout_width="match_parent"
    android:layout_height="match_parent"
    android:orientation="vertical" >

    <TextView
        android:id="@+id/pubDateTextView"
        android:layout_width="wrap_content"
        android:layout_height="wrap_content"
        android:layout_marginLeft="10dp"
        android:layout_marginTop="5dp"
        android:text="@string/item_pub_date" />

    <TextView
        android:id="@+id/titleTextView"
        android:layout_width="wrap_content"
        android:layout_height="wrap_content"
        android:layout_marginLeft="10dp"
        android:layout_marginRight="10dp"
        android:text="@string/item_title"
        android:textSize="24sp" />

</LinearLayout>
```

Figure 10-12   How to create the layout for a list view

How to use an adapter to display data in a list view

In chapter 5, you learned how to use an adapter to display an array of strings in a Spinner widget. Now, figure 10-13 shows how to use another type of adapter to display data in a ListView widget. More specifically, it shows how to use the SimpleAdapter class to display data in the ListView widget described in the previous figure.

The code example begins by calling the getAllItems method of the RSSFeed object to get an ArrayList object that contains RSSItem objects. Then, this code creates an ArrayList object that contains HashMap objects. Here, each HashMap object stores the data for each item in the list. In this figure, for example, each HashMap object stores the publication date and title for each item.

After storing the data for the adapter in the ArrayList variable named data, this code gets the ID for the listview_item layout and stores it in the variable named resource. This resource defines the layout for each item in the ListView widget.

Then, this code creates an array of strings that contain the names of the keys that get the data from the HashMap objects. This is where the data comes from. These keys correspond with the keys that are used to store data in the HashMap objects.

Next, it creates an array of int values that contain the IDs of the widgets in the listview_item layout. This is where the data goes to. As a result, these IDs should exist in the listview_item layout.

After setting up all of the necessary variables, this code creates the SimpleAdapter object. Here, the first parameter uses the *this* keyword to specify the context for the adapter. Then, the next four parameters use the data, resource, from, and to variables that were set up earlier.

Finally, the last statement calls the setAdapter method to set the SimpleAdapter object in the ListView widget. This causes the ListView widget to display the data stored in the SimpleAdapter object.

As you review this figure, you may notice that the adapter accepts a List object that contains Map objects. However, the code in this figure uses an ArrayList object that contains HashMap objects. This works because an ArrayList object is a type of List object, and a HashMap object is a type of Map object.

Code that creates and sets the adapter

```
// get the items for the feed
ArrayList<RSSItem> items = feed.getAllItems();

// create a List of Map<String, ?> objects
ArrayList<HashMap<String, String>> data =
        new ArrayList<HashMap<String, String>>();
for (RSSItem item : items) {
    HashMap<String, String> map = new HashMap<String, String>();
    map.put("date", item.getPubDateFormatted());
    map.put("title", item.getTitle());
    data.add(map);
}

// create the resource, from, and to variables
int resource = R.layout.listview_item;
String[] from = {"date", "title"};
int[] to = {R.id.pubDateTextView, R.id.titleTextView};

// create and set the adapter
SimpleAdapter adapter =
    new SimpleAdapter(this, data, resource, from, to);
itemsListView.setAdapter(adapter);
```

The constructor for the SimpleAdapter class

Parameter	Description
context	The context of the View associated with this SimpleAdapter object.
data	A List object that contains Map objects. The Map objects contain the data for the items in the list. The Map objects should include all keys specified in the parameter named from.
resource	The ID of a layout for each item in the list. The layout file should include the views defined in the parameter named to.
from	An array of column names that are in the Map objects.
to	An array of IDs for the widgets that should display the columns in the parameter named from.

Description

- You can use the SimpleAdapter class to display data in a ListView widget.

Figure 10-13   How to use an adapter to display data in a list view

How to handle events for an adapter

Figure 10-14 shows how to handle the events that occur when the user clicks on one of the items in a list. To do that, you implement the OnItemClickListener interface that's nested in the AdapterView class. Then, you wire this event handler to the appropriate widget. In this figure, for example, the onItemClick method is executed when the user clicks on one of the items in the ListView widget.

Within the onItemClick method, the third parameter provides the position of the selected item. In most cases, this is the only parameter that you need. However, if necessary, the first parameter provides the AdapterView object for the list of items, the second parameter provides a View object for the selected item, and the fourth parameter provides the ID of the selected item.

Within the onItemClick method, the first statement uses the position parameter to get the RSSItem object at the specified position. At this point, you can write additional code that processes the data stored in the RSSItem object. In this next figure, for example, you'll learn how to pass this data to another activity so it can be displayed.

Step 1: Import the interface for the listener

```
import android.widget.AdapterView.OnItemClickListener;
```

Step 2a: Implement the interface for the listener

```
public class ItemsActivity extends Activity
implements OnItemClickListener {
```

Step 2b: Implement the interface for the listener

```
@Override
public void onItemClick(AdapterView<?> parent, View v,
        int position, long id) {

    // get item at position
    RSSItem item = feed.getItem(position);
}
```

Step 3: Set the listeners

```
itemsListView.setOnItemClickListener(this);
```

Description

- The onItemClick method is executed when the user clicks on one of the items in the ListView widget.

- Within the onItemClick method, the third parameter provides the position of the selected item.

Figure 10-14   How to handle events for an adapter

How to work with intents

In the News Reader app, the Items activity displays a list of items. If you click on one of these items, you need to display the data for that item in the Item activity. To do that, you can use an intent to pass data from the Items activity to the Item activity.

Within the Item activity, you can click on a link to display it in the default browser on your device. To do that, you can use an intent to tell your device that you want to view the link in a web browser.

How to pass data between activities

As you learned in chapter 8, an *intent* is an object that provides a description of an operation to be performed. When an intent specifies a specific component within an app such as an activity, it can be referred to as an *explicit intent*. If you want, you can use an explicit intent to pass data from one component to another.

In figure 10-15, for example, the code in the ItemsActivity class creates an explicit intent for the ItemActivity class. Then, it uses the putExtra method of the Intent object to store some data in the intent. Here, it stores a string named title and an int value named position. Next, this code uses the startActivity method to start the activity specified by the intent.

The code in the ItemActivity class begins by using the getIntent method to get the Intent object that was passed to it. Then, it uses the getStringExtra method to get the string named title, and it uses the getIntExtra method to get the int value named position.

Although the code example only shows how to work with a string and an int value, the putExtra method of the Intent class works for most primitive types and arrays of primitive types. In addition, the Intent class includes an appropriate getXxxExtra method for most primitive types. As a result, you can use an Intent object to pass most types of data between activities.

Code in the ItemsActivity class

```
// create the intent
Intent intent = new Intent(this, ItemActivity.class);

// put data in the intent
intent.putExtra("title", item.getTitle());
intent.putExtra("position", position);

// start the intent
this.startActivity(intent);
```

Code in the ItemActivity class

```
// get the intent
Intent intent = getIntent();

// get data from the intent
String pubDate = intent.getStringExtra("pubDate");
int position = intent.getIntExtra("position", 0);
```

The Intent class

Constructor/Method	Description
Intent(context, class)	Creates an intent for the specific class in the project.
putExtra(name, value)	Stores the specified value or array of values with the specified name.
getStringExtra(name)	Gets the string value with the specified name.
getIntExtra(name, default)	Gets the int value with the specified name. If no int value exists with the specified name, this method returns the specified default value.

Description

- An *explicit intent* specifies a component such as an activity.
- You can use an explicit intent to pass data from one activity to another.
- The Intent class provides getXxxExtra methods for most primitive types and arrays of primitive types.

Figure 10-15   How to pass data from one activity to another

How to view a URL in a web browser

Figure 10-16 shows how to view a URL in a web browser. To do that, you can use an *implicit intent*, which is an intent that specifies the action you want to perform. Then, Android determines the best app to perform that action.

The first example in this figure begins by creating a string for a URL that links to a news article on the Internet. Then, it uses the parse method of the Uri class to create a Uri object for this link.

After creating the Uri object, this example creates an Intent object. This code passes the ACTION_VIEW constant as the first argument of the constructor, and it passes the Uri object as the second argument. As a result, the Intent object specifies that you want to view the specified URI in a web browser. Then, after you use the startActivity method to start the Intent object, Android determines which web browser to start and uses that browser to view the specified URI.

How to dial or call a phone number

This figure also shows how to use an implicit intent to dial or call a phone number. This is similar to viewing a page in a web browser.

However, there are three primary differences. First, the string that specifies the URI must begin with "tel:", not "http:". Second, the first parameter that's passed to the constructor for the Intent must be the ACTION_DIAL or ACTION_CALL constant. Third, if you use the ACTION_CALL constant, you must add the CALL_PHONE permission to the AndroidManifest.xml file for your project. As with all permissions, this permission must be coded at the same indentation level as the application element.

How to view a URL in a web browser

```
// create a Uri object for the link
String link = "http://rss.cnn.com/~r/rss/cnn_tech/~3/N9m_DSAe5rY/";
Uri uri = Uri.parse(link);

// create the intent and start it
Intent viewIntent = new Intent(Intent.ACTION_VIEW, uri);
startActivity(viewIntent);
```

How to call a phone number

```
// get the Uri for the phone number
String number = "tel:800-111-1111";
Uri callUri = Uri.parse(number);

// create the intent and start it
Intent callIntent = new Intent(Intent.ACTION_DIAL, callUri);
startActivity(callIntent);
```

The Intent class

Constructor		Description
Intent(action, uri)		Create an intent with the specified action constant and the specified data Uri.

Constant	Target	Action
ACTION_VIEW	Activity	View specified Uri in a web browser.
ACTION_DIAL	Activity	Places the specified phone number in the dialer, but lets the user decide whether to make the call.
ACTION_CALL	Activity	Call the specified phone number. This method requires the CALL_PHONE permission.

The CALL_PHONE permission in the AndroidManifest.xml file

```
<uses-permission android:name="android.permission.CALL_PHONE" />
```

Description

- An *implicit intent* specifies the action you want to perform. Then, Android determines the best app to perform that action.
- You can use an implicit intent to view a URL in a web browser or to call a phone number.

Figure 10-16   How to view a URL in a web browser

The News Reader app

Now that you've learned the skills necessary to create the News Reader app, you're ready to see how all the pieces fit together.

The activity_items layout

Figure 10-17 shows the XML for the activity_items layout. This layout uses a linear layout with vertical orientation. It begins by displaying a TextView widget for the title of the feed. Then, it uses a ListView widget to display the items of the feed. In turn, the ListView widget uses the listview_layout to display each item. Since this was described earlier in this chapter, you shouldn't have much trouble understanding how it works.

The activity_items layout

```xml
<?xml version="1.0" encoding="utf-8"?>
<LinearLayout xmlns:android="http://schemas.android.com/apk/res/android"
    android:layout_width="match_parent"
    android:layout_height="match_parent"
    android:orientation="vertical" >

    <TextView
        android:id="@+id/titleTextView"
        android:layout_width="match_parent"
        android:layout_height="wrap_content"
        android:background="#FFAC83"
        android:padding="7dp"
        android:text="@string/items_title"
        android:textSize="22sp" />

    <ListView
        android:id="@+id/itemsListView"
        android:layout_width="match_parent"
        android:layout_height="match_parent" />

</LinearLayout>
```

The listview_item layout

```xml
<?xml version="1.0" encoding="utf-8"?>
<LinearLayout xmlns:android="http://schemas.android.com/apk/res/android"
    android:layout_width="match_parent"
    android:layout_height="match_parent"
    android:orientation="vertical" >

    <TextView
        android:id="@+id/pubDateTextView"
        android:layout_width="wrap_content"
        android:layout_height="wrap_content"
        android:layout_marginLeft="10dp"
        android:layout_marginTop="5dp"
        android:text="@string/item_pub_date" />

    <TextView
        android:id="@+id/titleTextView"
        android:layout_width="wrap_content"
        android:layout_height="wrap_content"
        android:layout_marginLeft="10dp"
        android:layout_marginRight="10dp"
        android:text="@string/item_title"
        android:textSize="24sp" />

</LinearLayout>
```

Figure 10-17   The activity_items layout

The ItemsActivity class

Figure 10-18 shows the Java code for the ItemsActivity class. This class uses most of the skills presented in this chapter.

The onCreate method begins by creating a FileIO object from the FileIO class that's shown in the next figure. Then, it gets references to the TextView and ListView widgets for the Items activity. Next, it sets the listener for the ListView widget. Finally, it creates a new object from the inner class named DownloadFeed, and it calls the execute method from that object to start it.

The inner class named DownloadFeed uses a background thread to download the XML file for the RSS feed. After the background thread finishes executing, this class displays a message in the LogCat view. Then, it creates a new object from the inner class named ReadFeed, and it calls the execute method from that object to start it.

The inner class named ReadFeed uses a background thread to read the XML file for the RSS feed and parse it into an RSSFeed object that contains multiple RSSItem objects. After the background thread finishes executing, this class displays a message in the LogCat view. Then, it calls the updateDisplay method of its outer class to update the user interface.

Both of these inner classes specify the Void type for all three generic types of the AsyncTask class. As a result, the doInBackground and onPostExecute methods both use the Void type.

The updateDisplay method uses an adapter to display the items in the feed in the ListView widget. When it finishes, it displays a message in the LogCat view.

The onItemClick method handles the event that occurs when the user clicks on one of the items in the ListView widget. To handle this event, this code starts the Item activity and uses an explicit intent to pass all of the data for the item to that activity.

The ItemsActivity class

```java
package com.murach.newsreader;

import java.util.ArrayList;
import java.util.HashMap;

import android.os.AsyncTask;
import android.os.Bundle;
import android.app.Activity;
import android.content.Intent;
import android.util.Log;
import android.view.View;
import android.widget.AdapterView;
import android.widget.ListView;
import android.widget.SimpleAdapter;
import android.widget.TextView;
import android.widget.AdapterView.OnItemClickListener;

public class ItemsActivity extends Activity
implements OnItemClickListener {

    private RSSFeed feed;
    private FileIO io;

    private TextView titleTextView;
    private ListView itemsListView;

    @Override
    protected void onCreate(Bundle savedInstanceState) {
        super.onCreate(savedInstanceState);
        setContentView(R.layout.activity_items);

        io = new FileIO(getApplicationContext());

        titleTextView = (TextView) findViewById(R.id.titleTextView);
        itemsListView = (ListView) findViewById(R.id.itemsListView);

        itemsListView.setOnItemClickListener(this);

        new DownloadFeed().execute();
    }

    class DownloadFeed extends AsyncTask<Void, Void, Void> {
        @Override
        protected Void doInBackground(Void... params) {
            io.downloadFile();
            return null;
        }

        @Override
        protected void onPostExecute(Void result) {
            Log.d("News reader", "Feed downloaded");
            new ReadFeed().execute();
        }
    }
```

Figure 10-18   The ItemsActivity class (part 1 of 3)

The ItemsActivity class

```java
class ReadFeed extends AsyncTask<Void, Void, Void> {
    @Override
    protected Void doInBackground(Void... params) {
        feed = io.readFile();
        return null;
    }

    @Override
    protected void onPostExecute(Void result) {
        Log.d("News reader", "Feed read");

        // update the display for the activity
        ItemsActivity.this.updateDisplay();
    }
}

public void updateDisplay()
{
    if (feed == null) {
        titleTextView.setText("Unable to get RSS feed");
        return;
    }

    // set the title for the feed
    titleTextView.setText(feed.getTitle());

    // get the items for the feed
    ArrayList<RSSItem> items = feed.getAllItems();

    // create a List of Map<String, ?> objects
    ArrayList<HashMap<String, String>> data =
            new ArrayList<HashMap<String, String>>();
    for (RSSItem item : items) {
        HashMap<String, String> map = new HashMap<String, String>();
        map.put("date", item.getPubDateFormatted());
        map.put("title", item.getTitle());
        data.add(map);
    }

    // create the resource, from, and to variables
    int resource = R.layout.listview_item;
    String[] from = {"date", "title"};
    int[] to = {R.id.pubDateTextView, R.id.titleTextView};

    // create and set the adapter
    SimpleAdapter adapter =
        new SimpleAdapter(this, data, resource, from, to);
    itemsListView.setAdapter(adapter);

    Log.d("News reader", "Feed displayed");
}
```

Figure 10-18   The ItemsActivity class (part 2 of 3)

The ItemsActivity class   **Page 3**

```
@Override
public void onItemClick(AdapterView<?> parent, View v,
        int position, long id) {

    // get the item at the specified position
    RSSItem item = feed.getItem(position);

    // create an intent
    Intent intent = new Intent(this, ItemActivity.class);

    intent.putExtra("pubdate", item.getPubDate());
    intent.putExtra("title", item.getTitle());
    intent.putExtra("description", item.getDescription());
    intent.putExtra("link", item.getLink());

    this.startActivity(intent);
}
}
```

Figure 10-18   The ItemsActivity class (part 3 of 3)

The FileIO class

Figure 10-19 shows the Java code for the FileIO class. This class begins by defining a constant named URL_STRING for a string that specifies the URL for the RSS feed. Then, it defines a constant named FILENAME that specifies the name of the file that stores the current RSS feed. Next, it defines a Context object that can store the context for the app.

The constructor for the FileIO class accepts a Context object. Within the constructor, a single statement sets the variable named context for this class to the Context object parameter. That way, the context variable can be used by the downloadFile and readFile methods of this class.

The downloadFile method downloads the XML for the RSS feed from the Internet and stores it in the specified file. Conversely, the readFile method reads the specified file and returns an RSSFeed object. To do that, these methods use the techniques described earlier in this chapter for downloading, writing, and reading a file. However, both of these methods use the FILENAME constant defined earlier in this class. As a result, you can be sure that both of these methods are working with the same file, which is what you want.

The activity_item layout

Figure 10-20 shows the XML for the activity_item layout. This layout uses a scroll view to make sure that the user can scroll up and down if the item is too tall to fit on the screen. Then, it uses a linear layout with vertical orientation to display four TextView widgets for the title, publication date, description, and link of the item. This code sets the color of the fourth TextView widget to blue to indicate to the user that it is a link that can be clicked.

The FileIO class   **Page 1**

```java
package com.murach.newsreader;

import java.io.FileInputStream;
import java.io.FileOutputStream;
import java.io.IOException;
import java.io.InputStream;
import java.net.URL;

import javax.xml.parsers.SAXParser;
import javax.xml.parsers.SAXParserFactory;
import org.xml.sax.InputSource;
import org.xml.sax.XMLReader;

import android.content.Context;
import android.util.Log;

public class FileIO {

    private final String URL_STRING = "http://rss.cnn.com/rss/cnn_tech.rss";
    private final String FILENAME = "news_feed.xml";
    private Context context = null;

    public FileIO (Context context) {
        this.context = context;
    }

    public void downloadFile() {
        try{
            // get the URL
            URL url = new URL(URL_STRING);

            // get the input stream
            InputStream in = url.openStream();

            // get the output stream
            FileOutputStream out =
                context.openFileOutput(FILENAME, Context.MODE_PRIVATE);

            // read input and write output
            byte[] buffer = new byte[1024];
            int bytesRead = in.read(buffer);
            while (bytesRead != -1)
            {
                out.write(buffer, 0, bytesRead);
                bytesRead = in.read(buffer);
            }
            out.close();
            in.close();
        }
        catch (IOException e) {
            Log.e("News reader", e.toString());
        }
    }
```

Figure 10-19   The FileIO class (part 1 of 2)

The FileIO class

```
public RSSFeed readFile() {
    try {
        // get the XML reader
        SAXParserFactory factory = SAXParserFactory.newInstance();
        SAXParser parser = factory.newSAXParser();
        XMLReader xmlreader = parser.getXMLReader();

        // set content handler
        RSSFeedHandler theRssHandler = new RSSFeedHandler();
        xmlreader.setContentHandler(theRssHandler);

        // read the file from internal storage
        FileInputStream in = context.openFileInput(FILENAME);

        // parse the data
        InputSource is = new InputSource(in);
        xmlreader.parse(is);

        // set the feed in the activity
        RSSFeed feed = theRssHandler.getFeed();
        return feed;
    }
    catch (Exception e) {
        Log.e("News reader", e.toString());
        return null;
    }
}
}
```

Figure 10-19   The FileIO class (part 2 of 2)

The activity_item layout

```xml
<?xml version="1.0" encoding="utf-8"?>
<ScrollView xmlns:android="http://schemas.android.com/apk/res/android"
    android:layout_width="fill_parent"
    android:layout_height="fill_parent"
    android:orientation="vertical" >

    <LinearLayout
        android:layout_width="match_parent"
        android:layout_height="wrap_content"
        android:orientation="vertical" >

        <TextView
            android:id="@+id/titleTextView"
            android:layout_width="match_parent"
            android:layout_height="wrap_content"
            android:paddingLeft="7dp"
            android:paddingRight="7dp"
            android:paddingTop="5dp"
            android:text="@string/item_title"
            android:textSize="24sp"
            android:textStyle="bold" />

        <TextView
            android:id="@+id/pubDateTextView"
            android:layout_width="match_parent"
            android:layout_height="wrap_content"
            android:paddingLeft="7dp"
            android:paddingTop="5dp"
            android:text="@string/item_pub_date" />

        <TextView
            android:id="@+id/descriptionTextView"
            android:layout_width="match_parent"
            android:layout_height="wrap_content"
            android:paddingLeft="7dp"
            android:paddingRight="7dp"
            android:paddingTop="5dp"
            android:text="@string/item_description"
            android:textSize="18sp" />

        <TextView
            android:id="@+id/linkTextView"
            android:layout_width="match_parent"
            android:layout_height="wrap_content"
            android:layout_marginTop="10dp"
            android:paddingLeft="7dp"
            android:paddingTop="5dp"
            android:text="@string/item_link"
            android:textColor="@color/blue"
            android:textSize="18sp" />

    </LinearLayout>
</ScrollView>
```

Figure 10-20   The activity_item layout

The ItemActivity class

Figure 10-21 shows the Java code for the ItemActivity class. When this activity is started, its onCreate method gets all necessary data from the Intent object and displays it on the four TextView widgets defined by the layout for this activity.

If the user clicks on the TextView widget for the link, the onClick method is executed. This method starts by creating a Uri object for the link that's stored in the Intent object. Then, it uses an implicit intent to view the URI in a web browser.

The ItemActivity class

```java
package com.murach.newsreader;

import android.net.Uri;
import android.os.Bundle;
import android.view.View;
import android.view.View.OnClickListener;
import android.widget.TextView;
import android.app.Activity;
import android.content.Intent;

public class ItemActivity extends Activity implements OnClickListener {

    @Override
    protected void onCreate(Bundle savedInstanceState) {
        super.onCreate(savedInstanceState);
        setContentView(R.layout.activity_item);

        // get references to widgets
        TextView titleTextView = (TextView)
            findViewById(R.id.titleTextView);
        TextView pubDateTextView = (TextView)
            findViewById(R.id.pubDateTextView);
        TextView descriptionTextView = (TextView)
            findViewById(R.id.descriptionTextView);
        TextView linkTextView = (TextView)
            findViewById(R.id.linkTextView);

        // get the intent and its data
        Intent intent = getIntent();
        String pubDate = intent.getStringExtra("pubdate");
        String title = intent.getStringExtra("title");
        String description =
            intent.getStringExtra("description").replace('\n', ' ');

        // display data on the widgets
        pubDateTextView.setText(pubDate);
        titleTextView.setText(title);
        descriptionTextView.setText(description);

        // set the listener
        linkTextView.setOnClickListener(this);
    }

    @Override
    public void onClick(View v) {
        // get the intent and create the Uri for the link
        Intent intent = getIntent();
        String link = intent.getStringExtra("link");
        Uri viewUri = Uri.parse(link);

        // create the intent and start it
        Intent viewIntent = new Intent(Intent.ACTION_VIEW, viewUri);
        startActivity(viewIntent);
    }
}
```

Figure 10-21   The ItemActivity class

Perspective

Now that you've finished this chapter, you should be able to create an app like the News Reader app that uses threads, files, adapters, and intents. However, there are many possible ways to improve this app. In the next chapter, you'll learn how to use services and notifications to improve this app.

Terms

RSS (Rich Site Summary) feed	daemon thread
thread	SAX (Simple API for XML)
UI thread	intent
asynchronous task	explicit intent
generics	implicit intent

Summary

- An *RSS* (*Rich Site Summary*) *feed* can be used to publish frequently updated works, such as blog entries and news headlines. Since RSS uses a standardized XML file format, the feed can be published once and viewed by many different apps.

- A *thread* is a single sequential flow of control within a program that often completes a specific task.

- By default, an Android app uses a single thread, called the *UI thread*, to display the user interface. Any task that can slow or stop the responsiveness of the UI thread should be run in a separate thread.

- An *asynchronous task* is a task that runs in a separate thread in the background and does not need to be synchronized with other threads.

- The AsyncTask class uses *generics* to allow a class to operate on various types of objects, and provides for three generic types: (1) parameters, (2) progress, and (3) result. It provides an easy way to perform a background task and publish results on the UI thread without having to manually manipulate threads.

- You can use the TimerTask and Timer classes to create a thread that executes tasks after a specified delay or at a specified interval.

- You can use Java and Android APIs to download a file by reading input from the Internet and writing output to the file system.

- You can use *SAX* (*Simple API for XML*) to parse XML files.

- You can use the SimpleAdapter class to display data in a ListView widget.

- An *explicit intent* specifies a component such as an activity, and can be used to pass data from one activity to another.

- An *implicit intent* specifies the action you want to perform, and can be used to view a URL in a web browser or to call a phone number.

Exercise 10-1   Review the News Reader app

In this exercise, you'll take the News Reader app for a test drive to see how it works.

1. Open the project named ch10_ex1_NewsReader that's in the ex_starts directory.

2. Run the app. The Items activity should display a list of news items, and the LogCat view should display messages indicating that the RSS feed has been downloaded, read, and displayed. Scroll down through this list of items to its end.

3. Click on a news item. This should display the news item in the Items activity, but it shouldn't display any new messages in the LogCat view.

4. Click on the "Read more" link. This should start a web browser and display the web page for the news item in that browser.

5. Click on the Back button twice to return to the Items activity. This should not display any new messages in the LogCat view.

6. Change the orientation of the emulator or device. This should display messages in the LogCat view that indicate that the RSS feed has been downloaded, read, and displayed. On the good side, this provides a way for you to refresh the news feed so it has the latest news items. On the other hand, this isn't the best way to check for updates. In the next chapter, you'll learn a better way to handle this issue.

Exercise 10-2   Work with asynchronous tasks

In this exercise, you'll modify the DownloadFeed and ReedFeed classes used by the News Reader app presented in this chapter so they accept parameters and return results.

Modify the DownloadFeed class

1. Open the project named ch10_ex2_NewsReader that's in the ex_starts directory.

2. Open the class for the Items activity. Then, modify the code for the DownloadFeed class so it uses this declaration:

```
class DownloadFeed extends AsyncTask<String, Void, Void>
```

3. Save the file. When you do, Android Studio should display an error that shows that the declaration for the doInBackground method doesn't specify the correct types.

4. Modify the declaration for the doInBackground method so it accepts multiple String objects as a parameter.

5. Modify the code for the doInBackground method so it gets the first String object that's passed to it and uses that string to create the URL object. In other words, this method should not use the URL_STRING constant.

6. Modify the code that creates and executes the DownloadFeed class so it passes the String object for the URL to the doInBackground method.

7.   Run the app to make sure this code works correctly.

Modify the ReadFeed class

8.   Modify the code for the ReadFeed class so it uses this declaration:

```
class ReadFeed extends AsyncTask<Void, Void, RSSFeed>
```

9.   Save the file. When you do, Android Studio should show that there are errors with both the doInBackground and onPostExecute methods.

10. Modify the declaration for the doInBackground method so it returns an RSSFeed object.

11. Modify the declaration for the onPostExecute method so it accepts an RSSFeed object as a parameter.

12. Modify the code for the doInBackground method so the end of the try clause returns an RSSFeed object instead of setting it in the activity. Also, modify the end of the catch clause so it returns a null value.

13. Modify the code for the onPostExecute method so it sets the feed instance variable in the activity to the RSSFeed parameter. Make sure to set the feed before this method calls the updateDisplay method of the activity.

14. Run the app to make sure this code works correctly.

Exercise 10-3   Modify the News Reader app

In this exercise, you'll make some enhancements to the News Reader app so it reads a different RSS feed and formats it differently.

1.   Open the project named ch10_ex3_NewsReader that's in the ex_starts directory.

2.   Change the URL for the RSS feed to:

```
http://rss.cnn.com/rss/cnn_world.rss
```

3.   In the activity_items layout, add a TextView widget for the publication date just below the TextView widget for the title.

4.   In the ItemsActivity class, add code that uses the TextView widget you just added to display the publication date for the feed.

5.   Modify the feed's publication date so it's displayed in this format:

```
Thursday 9:12 AM (Feb 28)
```

To do that, add a getPubDateFormatted method to the RSSFeed class that works like one that's in the RSSItem class.

6.   In the listview_item layout, add a TextView widget to display the description for the feed.

7.   In the ItemsActivity class, modify the code for the adapter so it also stores the description and displays it in the list of items. When you're done, the Items activity should display the publication date, title, and description for each item.

8.   Modify the code that's executed when you click on an item so that clicking on an item displays the item in a web browser. When you're done, the News Reader app should never display the Item activity.

11

How to work with services and notifications

Some types of Android apps need to execute a task in the background, even when the app isn't running. To do that, you can use an Android component known as a service. When a service completes a task, it may want to notify the user. To do that, you can use an Android feature known as a notification.

How to work with the Application object

Before you learn how to work with services, you should learn how to create a custom Application object for your app. This provides two benefits. First, it allows you to store data that applies to the entire app in a central location that's always available to all components of the app including all of the app's activities and services. Second, it allows you to execute code when the application starts, as opposed to executing code each time an activity starts.

How to define the Application object

Figure 11-1 shows how to define a custom Application object. Here, the NewsReaderApp class extends the Application class. Then, it defines an instance variable that can store the milliseconds that correspond with the publication date for the current RSS feed for the app. Next, it defines get and set methods for this variable.

After defining the methods that store the application's data, this class overrides the onCreate method of the Application class. Android calls this method when the app starts, but not when an activity starts. As a result, this method is a good place to put any code that should only be run once, when the app starts. For now, this method just prints a message to the LogCat view. However, later in this chapter, you'll learn how to modify this method so it starts a service.

When storing data in the Application object, keep in mind that the Application object is created when the app starts and remains available as long as any component of the app is running. As a result, you should avoid storing large amounts of data in this object whenever possible. For the News Reader app, for example, you could store an RSSFeed object in the Application object. That way, the current feed would always be available to all components in the app. Since the RSSFeed object only contains text and isn't too large, this might be OK in most cases. However, since the News Reader app only needs to store the milliseconds for the feed, it's probably a better design to only store a long value for the milliseconds as shown in this figure.

A starting point for the NewsReaderApp class

```
package com.murach.newsreader;

import android.app.Application;
import android.util.Log;

public class NewsReaderApp extends Application {

    private long feedMillis = -1;

    public void setFeedMillis(long feedMillis) {
        this.feedMillis = feedMillis;
    }

    public long getFeedMillis() {
        return feedMillis;
    }

    @Override
    public void onCreate() {
        super.onCreate();
        Log.d("News reader", "App started");
    }
}
```

Description

- To store data and methods that apply to the entire application, you can extend the Application class and add instance variables and methods. The Application object is created when the app starts and remains available until the app ends.

- To run code only once when the application starts, you can override the onCreate method of your custom Application class.

Figure 11-1   How to define the Application object

How to register the Application object

When you extend the Application class, you must register that class. Then, when the application starts, Android creates the Application object from your custom Application class. If you don't register the custom Application class, Android creates the Application object from the Application class, not from your custom class.

To register your custom Application class, open the AndroidManifest.xml file. Then, edit the application element so its name attribute specifies the name of your custom Application class. In figure 11-2, for example, the name attribute of the application element specifies the NewsReaderApp class.

How to use the Application object

Once you have registered your custom application class, you can use the Application object. To do that, you can start by using the getApplication method to get a reference to the Application object. In this figure, the first example declares a variable of the NewsReaderApp type. Then, the second example uses the getApplication method to get a reference to the Application object, and it casts this Application object to the NewsReaderApp type.

Once you have a reference to the Application object, you can use it just as you would use any other object. In this figure, the third example calls the getFeedMillis method from the Application object to get the milliseconds for the publication date of the current RSS feed object for the app.

The application element of the AndroidManifest.xml

```
<application
    android:name=".NewsReaderApp"
    android:allowBackup="true"
    android:icon="@drawable/ic_launcher"
    android:label="@string/app_name"
    android:theme="@style/AppTheme" >
```

How to use the Application object

Step 1: Declare a variable for the Application object

```
NewsReaderApp app;
```

Step 2: Get a reference to the Application object

```
app = (NewsReaderApp) getApplication();
```

Step 3: Use the object

```
long feedMillis = app.getFeedMillis();
```

A method of the Context class

Method	Description
`getApplication()`	Returns the Application object for your app. You can cast this object to the custom Application class for your app.

Description

- When you extend the Application class, you must register that class. Then, when the application starts, Android creates the Application object from your custom Application class.

- To register your custom Application class, open the AndroidManifest.xml file and edit the application element so its name attribute specifies the name of your custom Application class.

- To get a reference to the Application object, you can use the getApplication method of the Context class.

- Once you have a reference to the Application object, you can use it just as you would use any other object.

Figure 11-2   How to register and use the Application object

How to work with services

A *service* is an Android component that does not provide a user interface. Instead, it performs a task that runs in the background. Since a service continues to run even if the user switches to another application, services should be used for tasks that should run independently of activities. For example, the News Reader app could use a service to download updates from the network, even when another app is running. Similarly, a Music Player app could use a service to play music even while another app is running.

The lifecycle of a service

Figure 11-3 shows two possible lifecyles of a service. The service on the left is an *unbound service*, which is a service that does not return a result to the caller. This type of service is created and started when a component such as an activity calls the startService method to start it. Then, the service runs until it's stopped.

An unbound service is typically stopped in one of three ways. First, another component such as an activity can call the stopService method to stop it. Second, the service can call the stopSelf method to stop itself. Third, Android can stop the service when the user turns off the device or if the device runs low on memory.

The service on the right is a *bound service*, which is a service that can interact with one or more components. This type of service is started when a component calls the bindService method to bind to it. Then, other components can bind to the service at the same time. The service runs as long as at least one component is bound to it. However, when the last component unbinds, Android stops the service.

Although a service runs in the background, a service does not run in a background thread. On the contrary, it runs in the same thread as the component that started it. As a result, if your service is going to perform any long running tasks such as networking or file I/O, you should create a new thread within the service to perform those tasks. Otherwise, your user interface may become unresponsive.

In this chapter, you'll learn how to use an unbound service, which is the most common type of service. However, if you need a service to interact with an activity, you can search the Internet to learn more about how to do this. Also, a service can be both bound and unbound. To create such a service, you can implement all necessary lifecycle methods. For example, you can implement the onStartCommand and onBind methods. Then, you can call the startService method to run the service indefinitely, and you can call the bindService command to bind components to the service.

The lifecycle of a service

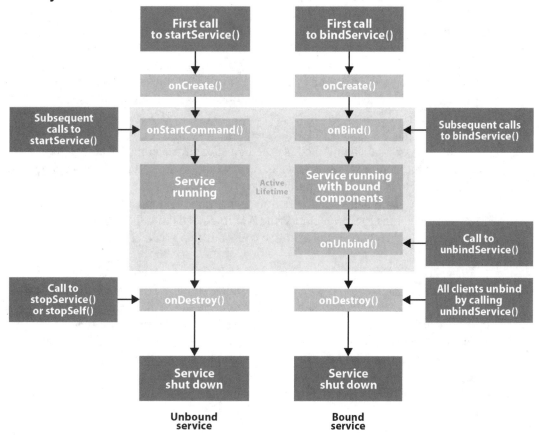

Figure 11-3   The lifecycle of a service

Description

- A *service* performs tasks in the background and does not provide a user interface. Services should be used for tasks that run independently of activities. For example, a service might download updates from the network, play music, and so on.

- A service continues to run even if the user switches to another app.

- An *unbound service* does not interact with other components such as activities. This type of service runs until it's stopped by another component or by itself.

- A *bound service* can interact with components such as activities. This type of service runs only as long as another component is bound to it. Multiple components can bind to the service at once, but when all of them unbind, the service is destroyed.

How to create a service

Figure 11-4 shows the starting code for an unbound service. This service implements all four methods that you typically need for an unbound service.

To start, this code imports the three classes that are needed to implement a service: the Service, Intent, and IBinder classes. Then, the code declares that the NewsReaderService class extends the Service class. Within this class, the onCreate method is called when the service is first created. As a result, it's a good place to put code that's only executed once.

The onStartCommand method is executed each time a component uses the startService method to start this service. As a result, it's a good place to put code that may need to be executed multiple times.

The onStartCommand method returns an int value that tells Android what to do if the device runs low on memory and Android needs to kill the thread for the service. The Service class provides three constants that you can return from this method.

The START_STICKY constant tells Android to leave the service in the started state. Then, when more memory becomes available, restart the service by calling the onStartCommand method with a null intent for the intent parameter. In other words, the "start" for the service is "sticky", so the service should be restarted as soon as possible. This constant is appropriate for services that should always be running in the background. For example, if a service is started once and executes a task at specified intervals, you probably want to restart it.

The START_NOT_STICKY constant tells Android to stop the service and to not restart it. In other words, the "start" is not "sticky", so the service should not be restarted when memory becomes available. This constant is appropriate for a service that doesn't need to be completed and is called by the app at specified intervals. For example, an app might start a service every 15 minutes to perform a task. If the service gets destroyed while doing that task, it's probably best to let it stay stopped since the app will start it again anyway when the app is restarted.

The START_REDELIVER_INTENT constant tells Android to stop the service. However, when more memory becomes available, Android should restart the service by calling the onStartCommand method with the last delivered intent as the intent parameter. In other words, Android calls the onStartCommand method and "redelivers" the last intent. This constant is appropriate for a service that performs a task that needs to be completed. Android continues to redeliver the intent until the service calls its stopSelf method, which indicates that the service has completed.

The onBind method is executed when another component attempts to bind to the service. This method returns an IBinder object that the component can use to interact with the service. For an unbound service, such as the service shown in this chapter, you can return a null value. However, the onBind method is an abstract method that must be implemented. As a result, you can't completely omit this method, even for an unbound service.

The onDestroy method is executed when the service is no longer in use and is being destroyed. As a result, it's a good place to clean up any resources that need to be cleaned up.

The NewsReaderService class with its lifecycle methods implemented

```
package com.murach.newsreader;

import android.app.Service;
import android.content.Intent;
import android.os.IBinder;

public class NewsReaderService extends Service {

    @Override
    public void onCreate() {
        // Code that's executed once when the service is created.
    }

    @Override
    public int onStartCommand(Intent intent, int flags, int startId) {
        // Code that's executed each time another component
        // starts the service by calling the startService method.
        return START_STICKY;
    }

    @Override
    public IBinder onBind(Intent intent) {
        // Code that's executed each time another component
        // binds to the service by calling the bindService method.
        // This method is required.
        // For an unbound service, you can return a null value.
        return null;
    }

    @Override
    public void onDestroy() {
        // Code that's executed once when the service
        // is no longer in use and is being destroyed.
    }
}
```

Some constants of the Service class

Constant	If Android destroys the thread for the service due to lack of memory, Android should...
START_STICKY	Leave the service in the started state. When memory becomes available, restart the service by calling the onStartCommand method with a null value for the intent parameter.
START_NOT_STICKY	Stop the service and do not restart it. As a result, the service will only restart if the app resumes and executes a method that restarts the service.
START_REDELIVER_INTENT	Stop the service. When memory becomes available, restart the service by calling the onStartCommand method with the last delivered intent for the intent parameter.

Figure 11-4   How to create a service

How to register a service

Before you can use a service, you must register it. The first example in figure 11-5 shows how to do that. To start, you open the manifest file and add a service element at the same indentation level as the activity elements. Then, use the name attribute of the service element to specify the name of the class for your service.

By default, a service is public and can be started by other applications. However, if you want to make a service private, you can add an export attribute to the service element and set it to false as shown in the second example.

How to start and stop a service

Before you can start a service, you must create an Intent object for the service. In this figure, for instance, the third example creates an Intent object for the service that's defined by the NewsReaderService class.

You can start a service by calling the startService method and passing the intent for the service to it. In this figure, the fourth example starts the service. Since the startService method belongs to the Context class, you can call it from any component, including an activity or the Application object.

When you use the startService method to start a service, the service runs indefinitely. In some cases, that's what you want. In other cases, you may want to stop the service.

For instance, if you want to allow the user to stop a service, you can provide a menu item that stops the service. Then, when the user selects that menu item, you can use the stopService method to stop the service as shown in the fifth example.

Or, if you have a service that performs a task, you may want to stop the service after it finishes its task. To do that, you can call the stopSelf method from within the class for the service as shown in the sixth example.

A public service in the AndroidManifest.xml file

```
<service
    android:name=".NewsReaderService">
</service>
```

A private service in the AndroidManifest.xml file

```
<service
    android:name=".NewsReaderService"
    android:exported="false">
</service>
```

Code that uses a service

Create the Intent object for the service

```
Intent serviceIntent = new Intent(this, NewsReaderService.class);
```

Start the service

```
startService(serviceIntent);
```

Stop the service from an activity or other component

```
stopService(serviceIntent);
```

Stop the service from within the service

```
stopSelf();
```

Some methods of the Context class

Method	Description
`startService(intent)`	Start the service that's specified by the intent.
`stopService(intent)`	Stop the service that's specified by the intent.

A method of the Service class

Method	Description
`stopSelf()`	Stops the current service.

Description

- Before you can use a service, you must register it.
- To register your service, add a service element to the AndroidManifest.xml file at the same indentation level as the activity elements and use the name attribute of the service element to specify the name of the class for your service.
- To make a service private, so it can only be accessed by the current app, you can add an exported attribute and set it to a value of false.
- If a component starts a service by calling the startService method, the service runs until another component stops it by calling the stopService method, until it stops itself by calling the stopSelf method, or until the device is turned off.

Figure 11-5   How to register, start, and stop a service

How to use threads with services

As mentioned earlier in this chapter, a service does not run in its own thread by default. As a result, if you want a service to perform a long running task such as accessing a network or file I/O, you should create a new thread within the service. If, for example, you want a thread to execute a single task, you can use the AsyncTask class to create a background thread for that class. Or, if you want a thread to execute a task at a specified interval, you can use the TimerTask and Timer classes to create and schedule a thread as shown in figure 11-6.

The NewsReaderService class shown in this figure begins by extending the Service class. Within the class, the first statement declares an instance variable for the Timer class. This instance variable is used to schedule when the timer task is executed.

Android calls the first three methods of this class at various times in the lifecycle of this service. This code sends messages to the LogCat view to show that they have been executed. However, since no components in the app attempt to bind to this service, the onBind method should never be called. Of these metods, the onCreate method calls the startTimer method, and the onDestroy method calls the stopTimer method.

The startTimer method begins by creating a TimerTask object and overriding its run method. This method creates the thread for the task. As a result, all of the long running code for the task should be placed within this method. For now, this example just sends a message to the LogCat view that indicates that the timer task has been executed. After creating the TimerTask object, the startTimer method creates a Timer object and uses it to schedule the TimerTask object to run after a delay of 10 seconds and to continue to run every 10 seconds after that.

The stopTimer method begins by checking to make sure the Timer object is not null. If so, it uses the Timer object to cancel the thread for the timer task.

A Service class that runs a timed task in its own thread

```
package com.murach.newsreader;

import java.util.Timer;
import java.util.TimerTask;

import android.app.Service;
import android.content.Intent;
import android.os.IBinder;
import android.util.Log;

public class NewsReaderService extends Service {

    private Timer timer;

    @Override
    public void onCreate() {
        Log.d("News reader", "Service created");
        startTimer();
    }

    @Override
    public IBinder onBind(Intent intent) {
        Log.d("News reader", "No binding for this service");
        return null;
    }

    @Override
    public void onDestroy() {
        Log.d("News reader", "Service destroyed");
        stopTimer();
    }

    private void startTimer() {
        // create task
        TimerTask task = new TimerTask() {
            @Override
            public void run() {
                Log.d("News reader", "Timer task executed");
            }
        };

        // create and start timer
        timer = new Timer(true);
        int delay = 1000 * 10;      // 10 seconds
        int interval = 1000 * 10;   // 10 seconds
        timer.schedule(task, delay, interval);
    }

    private void stopTimer() {
        if (timer != null) {
            timer.cancel();
        }
    }
}
```

Figure 11-6   How to use threads with services

How to test a service

In the next chapter, you'll learn how to start a service when the device boots. For now, you can test a service by starting it when the app starts. To do that, you can call the startService method from the onCreate method of the Application object as shown in figure 11-7. That way, Android starts the service once when the app starts. For now, this a logical place to start the News Reader service.

Since a service doesn't have a user interface, you can test it by using the Log class to print messages to the LogCat view. Then, after you start the service, you can check the LogCat view to make sure that the service is running and performing its task correctly.

In this figure, the first two messages show that the app has started and the service has been created. Then, the next few messages show that the timer task has been executed. Finally, if the last message is shown, it indicates that the service has been destroyed. This message is typically displayed when the service is stopped because the stopService method has been called by another component, or the stopSelf method has been called by the service after it has completed its task.

The onCreate method of the Application object

```
@Override
public void onCreate() {
    super.onCreate();
    Log.d("News reader", "App started");

    // start service
    Intent service = new Intent(this, NewsReaderService.class);
    startService(service);
}
```

The messages that are displayed in the LogCat view

```
App started
Service created
Timer task executed
Timer task executed
Timer task executed
...
Service destroyed
```

Description

- To start a service when the app starts, you can call the startService method from the onCreate method of the Application object. That way, the service is started once, when the app starts.

- To test a service, you can use the Log class to print messages to the LogCat view.

Figure 11-7   How to test a service

How to view all services

Figure 11-8 shows how to use the Settings app to view all of the apps and services that are running on a device or emulator. To start, you can start the Settings app on the device or emulator. From there, the procedure differs depending on the Android API that's running on the device. This figure shows how to view the apps and services for APIs 23 and 19. However, you should be able to use a similar procedure for most APIs. If necessary, you can search the Internet to learn how to view the apps and services for your device.

In this figure, the screen shows that three apps and two services are running. The Settings app is the app that's currently being used to display the apps and services. However, it doesn't use a service. The News Reader app is also running, and it uses a service. Similarly, the Android Keyboard app is running, and it uses a service as well.

For most devices, it would be unusual for only three apps to be running. To get an idea of how many apps and services typically run on a device, you should take a moment to browse through the apps on a phone or tablet that has been running for a while. Most likely, you'll see a long list of apps and services.

The apps and services that are running on the device

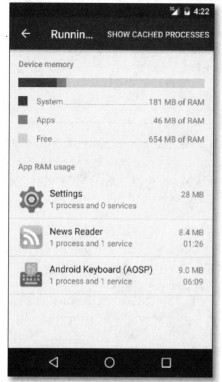

Procedure (API 23)

1. Start the Settings app.
2. Select the Developer Options item.
3. Select the Running Services item.
4. View the apps and services.

Procedure (API 19)

1. Start the Settings app.
2. Select the More tab.
3. Select the Application Manager item.
4. If necessary, display the Running tab. To do that, you may need to scroll to the right.
5. View the apps and services.

Description

- You can use the Settings app to view all of the apps and services that are running on a device or emulator. The procedure for doing this varies depending on the API version.

Figure 11-8   How to view services

How to work with notifications

A *notification* provides a way for a service to display a message even when another app is running. For example, the service for the News Reader app could display a notification that indicates that an updated feed is available.

How notifications work

Figure 11-9 shows how notifications work. To start, when Android first displays a notification, the notification appears as an icon in the *notification area* at the top of the screen. In this figure, the first two screens display a notification in the notification area. Here, the first screen is the News Reader app and the second screen is the Home screen. Both screens display a notification for the News Reader app at the top of the screen.

When one or more notifications have been displayed, the user can pull down on the notification area to open the *notification drawer*. This allows the user to view all notifications. In this figure, the third screen shows the notification drawer when it contains only one notification, a notification that indicates that an updated feed is available from the News Reader app. However, it's possible for the notification area and drawer to display multiple notifications.

Often, a user can perform an action by clicking on the notification in the notification drawer. Typically, this starts the app that displayed the notification. In this figure, if the user selects the notification, it starts the News Reader app and displays the updated feed.

The News Reader app after it has started a notification

A notification displayed in the notification area

The notification drawer

Description

- A *notification* provides a way for a service to display a message even when another app is running.
- When it's first displayed, a notification appears as an icon in the *notification area* at the top of the screen.
- To view a notification, the user can pull down on the notification area to open the *notification drawer*. Then, the user can often execute an action by clicking on the notification. Typically, this starts the app that displayed the notification.

Figure 11-9 How notifications work

How to create a pending intent

To create the action for a notification, you can create a *pending intent*, which is a special type of intent that can be passed to another app and executed by that app at a later time. That way, you can pass this intent to Android, so it can use the intent in its notification system.

Before you learn how to create a pending intent for a notification, you need to understand some terminology. To start, the *back stack* is a stack of all recently used activities. These activities are sorted in the order in which they were used. When the user clicks the Back button, the top activity is removed from the stack, and the next activity is displayed.

All activities in the back stack belong to a task, which is a cohesive unit that can contain multiple activities. For example, the News Reader task contains any Items or Item activities that are in the back stack. When the user clicks the Home button or the Task button to the right of that button, the current task moves to the background. However, all of the activities for that task remain on the back stack for that task. This back stack can be referred to as the task history. That way, if you navigate back to the task, you can continue the task where you left off.

The code examples in figure 11-10 show how to create a pending intent for a notification. The first step is to create a regular intent for an activity as described in the previous chapter. However, when starting an activity from a notification, it's common to use the addFlags method to control how the activity is displayed in the back stack. In most cases, you want to set the "activity clear top" flag as shown in step 1. That way, if the activity is already in the task, the other activities in the task that are on top of the specified activity are cleared from the stack. This brings the specified activity to the top of the stack, which is usually what you want.

If you don't set this flag, Android creates a new instance of the activity and adds it to the top of the stack for the task. If the task already contains this activity, this creates multiple instances of the same activity. Sometimes this is what you want. For example, you may want to place the same activity in the stack twice if it displays different data.

The second step is to create the pending intent. To do that, you typically create a flag that tells Android what to do if an intent with the same name already exists but hasn't yet been executed. In most cases, you want to use an "update current" flag to keep the current intent but update its data with any data in the new intent. (This is the "extra" data that you can set with the putExtra method.) Then, you pass that flag and the intent you created in step 1 to the getActivity method of the PendingIntent class. This returns a PendingIntent object that you can use in a notification. Here, the second parameter specifies a code that isn't currently used by Android. As a result, you can specify any int value you want for this parameter.

The classes used to work with pending intents

```
android.app.PendingIntent
android.content.Intent
```

How to create a pending intent

Step 1: Create the intent

```
Intent notificationIntent = new Intent(this, ItemsActivity.class)
    .addFlags(Intent.FLAG_ACTIVITY_CLEAR_TOP);
```

Step 2: Create the pending intent

```
int flag = PendingIntent.FLAG_UPDATE_CURRENT;
PendingIntent pendingIntent =
        PendingIntent.getActivity(this, 0, notificationIntent, flag);
```

Some constants and methods of the Intent class

Constant	Description
FLAG_ACTIVITY_CLEAR_TOP	If the activity is already running, it closes all activities on top of the specified activity, which brings the specified activity to the top of the task history.
FLAG_ACTIVITY_NEW_TASK	Creates a new task for the activity and places it on top of the task history.
Method	**Description**
addFlags(flags)	Adds flags that control the behavior of the intent.

Some constants and methods of the PendingIntent class

Constant	Description
FLAG_UPDATE_CURRENT	If the pending intent already exists, keep it but replace its extra data with the extra data from the new intent.
Method	**Description**
getActivity(context, code, intent, flags)	Returns a PendingIntent object for the specified activity intent with the specified flags.

Description

- A *pending intent* is an intent that can be passed to other apps so that they can execute the intent at a later time.

- The *back stack* is a stack of all recently used activities. These activities are sorted in the order in which they were used. When the user clicks the Back button, the top activity is removed from the stack, and the next activity is displayed.

- A *task* is a cohesive unit that can contain multiple activities. Every activity belongs to a task. A task can move to the background when the user starts a new task or navigates away from the current task.

- The *task history* is the back stack for a particular task.

Figure 11-10   How to create a pending intent

How to create a notification

Once you've created a pending intent for a notifcation, you can create the notification as shown in figure 11-11. When you do, you set the pending intent in the notification.

The first step in creating a notification is to declare all of the variables that you need. In this figure, the first step declares four variables. These variables specify (1) the icon, (2) the ticker text that's displayed when the notification first appears, (3) the title of the notification when it's displayed in the drawer, and (4) the text for the notification when it's displayed in the drawer. Here, the icon is the same launcher icon that's used for the app. Similarly, the title of the notification is the same as the name of the app.

The second step in creating a notification is to create the Notification object. To do that, you create an object from the Builder class that's nested in the Notification class. Then, you use the various set methods of the Builder object to set the various components of the notification. For example, you can use the setContentIntent method to set the pending intent for the notification. Similarly, you can use the setSmallIcon method to set the icon for the notification. In addition, if you pass a true value to the setAutoCancel method, Android automatically removes the intent if the user selects it in the drawer, which is usually what you want. Finally, you use the build method to build the Notification object. Since all of these methods are designed to be chained, you can chain them together as shown in this figure.

After creating a Notification object like the one shown in this figure, you typically display it. To do that, you use a system service as shown in the next figure.

With Android 4.1 (API 16) and later, you can use the Builder class that's nested within the Notification class to create a Notification object as shown in this figure. In this book, the News Reader app uses this technique. As a result, it specifies a minimum API of 16. Fortunately, that includes most Android devices.

However, if you want to support older versions of Android, you can use the Builder class that's nested within the NotificationCompat class that's available from the v4 support library. To do that, you start by adding a dependency for the v4 support library to your project. Then, you import the NotificationCompat class from the android.support.v4.app package. Finally, you use the Builder class that's nested within the NotificationCompat class to create the Notification object. This Builder class works the same as the Builder class that's nested within the Notification class shown in this figure.

The class used to work with notifications

```
android.app.Notification
```

How to create a notification

Step 1: Create the variables for the notification

```
int icon = R.drawable.ic_launcher;
CharSequence tickerText = "Updated news feed is available";
CharSequence contentTitle = getText(R.string.app_name);
CharSequence contentText = "Select to view updated feed.";
```

Step 2: Create the Notification object

```
Notification notification =
        new Notification.Builder(this)
    .setSmallIcon(icon)
    .setTicker(tickerText)
    .setContentTitle(contentTitle)
    .setContentText(contentText)
    .setContentIntent(pendingIntent)
    .setAutoCancel(true)
    .build();
```

Constructors and methods of the Notification.Builder class

Constructor	Description
`Notification.Builder(context)`	Creates an object from the class with the specified context.

Method	Description
`setSmallIcon(int)`	Set the resource ID of the icon for the notification.
`setTicker(string)`	Set the ticker text that's briefly displayed when the notification is first displayed.
`setContentTitle(string)`	Sets the title of the notification.
`setContentText(string)`	Sets the text of the notification.
`setContentIntent(pendingIntent)`	Sets the intent of the notification.
`setAutoCancel(boolean)`	If true, Android automatically removes the intent after it is selected.
`build()`	Returns a Notification object.

Description

- With Android 4.1 (API 16) and later, you can use the Notification.Builder class to create a Notification object.

- To support older versions of Android, you can use the NotificationCompat.Builder class that's available from the v4 support library.

Figure 11-11   How to create a notification

How to work with system services

A *system service* is a service that's provided by the Android operating system. Android provides system services that allow you to work with notifications, locations, network connections, soft keyboards, and so on.

In chapter 6, for example, you learned how to use the input method service to get an InputMethodManager object, and you learned how to use that object to hide a soft keyboard. In this chapter, you'll learn how to use two services: the notification service and the network connection service. Both of these system services can be used by the News Reader app.

How to display or remove a notification

Figure 11-12 shows how to use a system service to display or remove a notification. To start, you can call the getSystemService method to get a NotificationManager object. Since this method is available from the Context class, you can call it directly from any component such as an activity or service.

Once you have a NotificationManager object, you can call its notify method to display the notification. But first, you should declare an ID for the notification so you can pass it to the notify method. This ID should be unique within your app. That way, your app can uniquely identify that notification later. For example, if you want to remove a notification later, you can pass this ID to the cancel method of the NotificationManager object.

Some constants and methods of the Context class

Method	Description
`getSystemService(name)`	Returns the appropriate object for the specified system-level service.

Constant	Description
`NOTIFICATION_SERVICE`	Returns a NotificationManager object for displaying and removing notifications.
`LOCATION_SERVICE`	Returns a LocationManager object for working with GPS updates.
`CONNECTIVITY_SERVICE`	Returns a ConnectivityManager object for working with network connections.
`WIFI_SERVICE`	Returns a WifiManager object for working with Wi-Fi connectivity.
`INPUT_METHOD_SERVICE`	Returns an InputMethodManager object for displaying and hiding soft keyboards.

The classes used to work with notifications

```
android.app.NotificationManager
```

Some methods of the NotificationManager class

Method	Description
`notify(id, notification)`	Displays the specified notification and sets its ID to the specified ID. This ID should be unique within your application.
`cancel(id)`	Cancels the notification with the specified ID.

How to get a NotificationManager object

```
NotificationManager manager = (NotificationManager)
        getSystemService(NOTIFICATION_SERVICE);
```

How to display a notification

```
final int NOTIFICATION_ID = 1;
manager.notify(NOTIFICATION_ID, notification);
```

How to remove a notification

```
manager.cancel(NOTIFICATION_ID);
```

Description

- A *system service* is a service that's provided by the Android operating system.
- You can use the NotificationManager object to display or remove a notification.

Figure 11-12   How to display and remove notifications

How to check if a network connection is available

Figure 11-13 shows how to use a system service to check if a network connection is available to the device. This is often helpful since attempting to connect to the network when the network isn't available uses the battery unnecessarily. As a result, most apps should check if a connection is available before connecting to the network. In particular, the News Reader app should check if a connection is available before attempting to download the news feed from the Internet.

The first statement in the example calls the getSystemService method of the Context class to get a ConnectivityManager object. Then, the second statement calls the getActiveNetworkInfo method to get a NetworkInfo object that contains data about the active network for the device. If this object is not null, the example calls its isConnected method to determine whether it's possible for the device to connect to a network. If so, it downloads the news feed from the Internet.

The classes used to work with connectivity

```
android.app.ConnectivityManager
android.net.NetworkInfo
```

How to check if you have a connection to the network

```
// get NetworkInfo object
ConnectivityManager connectivityManager = (ConnectivityManager)
    getSystemService(Context.CONNECTIVITY_SERVICE);
NetworkInfo networkInfo =
    connectivityManager.getActiveNetworkInfo();

// if network is connected, download feed
if (networkInfo != null && networkInfo.isConnected()) {
    new DownloadFeed().execute();
}
```

A method of the ConnectivityManager class

Method	Description
getActiveNetworkInfo()	Returns a NetworkInfo object that contains info about the active network for the device.

Some methods of the NetworkInfo class

Method	Description
isConnected()	Returns a boolean value that indicates whether it's possible for the device to connect to the network.
isRoaming()	Returns a boolean value that indicates whether the device is roaming on the network and whether the connection may incur additional costs.
getTypeString()	Returns a string for the type of the network such as "WIFI" or "MOBILE".

The permission in the AndroidManifest.xml file

```
<uses-permission android:name="android.permission.ACCESS_NETWORK_STATE" />
```

Description

- You can use a ConnectivityManager object to check if a network connection is available to the device.

Figure 11-13   How to check if a network connection is available

The News Reader app

At this point, you've learned the skills necessary to make three improvements to the News Reader app. First, you can add a service that periodically downloads a new feed and checks for an updated feed. Second, you can have that service display a notification whenever an updated feed is available. Third, you can add a custom Application object to store the milliseconds for the current news feed and to start the service.

To make these improvements, you can add a custom Application class like the one shown in figure 11-1. In addition, you can add a NewsReaderService class like the one shown in figure 11-14. Finally, you can make some changes to the ItemsActivity and FileIO classes as shown in figures 11-15 and 11-16.

The NewsReaderService class

Figure 11-14 shows the code for the NewsReaderService class. Part 1 shows the lifecycle methods for the service. Here, the onCreate method gets references to the Application and FileIO objects and calls the startTimer method that starts the thread that performs the tasks for the service. The onStartCommand doesn't perform any tasks, but it returns the START_STICKY constant so Android can automatically restart the service if necessary. The onBind method isn't used, so it returns a null value. And the onDestroy method stops the thread for the service.

Part 2 begins by showing the startTimer method. This method creates the thread that executes the task for the service. This thread begins by downloading the file for the feed from the Internet. Then, it reads the file and parses it into an RSSFeed object.

After reading the RSSFeed object, this code checks whether the publication date of the new feed is newer than the current publication date that's stored in the Application object. If so, this code updates the Application object with the milliseconds for the publication date of the new feed. Then, it uses the sendNotification method to display a notification that indicates that an updated feed is available.

The stopTimer method stops the timer thread for the service. Since this method is called from the onDestroy method, this only occurs if the service is being destroyed. And since this app doesn't call the stopService or stopSelf methods, the service is typically only destroyed when the user turns the device off.

Part 3 shows the sendNotification method. This method creates a pending intent for the notification. It creates the variables for the notification. It creates the notification. And it uses a system service to display the notification.

This class contains many statements that display messages to the LogCat view. That provides a way for you to verify that the service is working as expected. Once you're sure the service is working correctly, you can remove some or all of these statements.

The NewsReaderService class   **Page 1**

```java
package com.murach.newsreader;

import java.util.Timer;
import java.util.TimerTask;

import android.app.Notification;
import android.app.NotificationManager;
import android.app.PendingIntent;
import android.app.Service;
import android.content.Intent;
import android.os.IBinder;
import android.util.Log;

public class NewsReaderService extends Service {

    private NewsReaderApp app;
    private Timer timer;
    private FileIO io;

    @Override
    public void onCreate() {
        Log.d("News reader", "Service created");
        app = (NewsReaderApp) getApplication();
        io = new FileIO(getApplicationContext());
        startTimer();
    }

    @Override
    public int onStartCommand(Intent intent, int flags, int startId) {
        Log.d("News reader", "Service started");
        return START_STICKY;
    }

    @Override
    public IBinder onBind(Intent intent) {
        Log.d("News reader", "Service bound - not used!");
        return null;
    }

    @Override
    public void onDestroy() {
        Log.d("News reader", "Service destroyed");
        stopTimer();
    }
```

Figure 11-14   The NewsReaderService class (part 1 of 3)

The NewsReaderService class

```java
    private void startTimer() {
        TimerTask task = new TimerTask() {

            @Override
            public void run() {
                Log.d("News reader", "Timer task started");

                io.downloadFile();
                Log.d("News reader", "File downloaded");

                RSSFeed newFeed = io.readFile();
                Log.d("News reader", "File read");

                // if new feed is newer than old feed
                if (newFeed.getPubDateMillis() > app.getFeedMillis()) {
                    Log.d("News reader", "Updated feed available.");

                    // update app object
                    app.setFeedMillis(newFeed.getPubDateMillis());

                    // display notification
                    sendNotification("Select to view updated feed.");
                }
                else {
                    Log.d("News reader", "Updated feed NOT available.");
                }

            }
        };

        timer = new Timer(true);
        int delay = 1000 * 60 * 60;      // 1 hour
        int interval = 1000 * 60 * 60;   // 1 hour
        timer.schedule(task, delay, interval);
    }

    private void stopTimer() {
        if (timer != null) {
            timer.cancel();
        }
    }
```

Figure 11-14   The NewsReaderService class (part 2 of 3)

The NewsReaderService class   **Page 3**

```java
private void sendNotification(String text) {
    // create the intent for the notification
    Intent notificationIntent = new Intent(this, ItemsActivity.class)
        .addFlags(Intent.FLAG_ACTIVITY_NEW_TASK);

    // create the pending intent
    int flags = PendingIntent.FLAG_UPDATE_CURRENT;
    PendingIntent pendingIntent =
            PendingIntent.getActivity(this, 0, notificationIntent, flags);

    // create the variables for the notification
    int icon = R.drawable.ic_launcher;
    CharSequence tickerText = "Updated news feed is available";
    CharSequence contentTitle = getText(R.string.app_name);
    CharSequence contentText = text;

    // create the notification and set its data
    Notification notification =
            new Notification.Builder(this)
        .setSmallIcon(icon)
        .setTicker(tickerText)
        .setContentTitle(contentTitle)
        .setContentText(contentText)
        .setContentIntent(pendingIntent)
        .setAutoCancel(true)
        .build();

    // display the notification
    NotificationManager manager = (NotificationManager)
            getSystemService(NOTIFICATION_SERVICE);
    final int NOTIFICATION_ID = 1;
    manager.notify(NOTIFICATION_ID, notification);
}
}
```

Figure 11-14   The NewsReaderService class (part 3 of 3)

The ItemsActivity class

Figure 11-15 shows the ItemsActivity class. For the most part, this class is the same as the ItemsActivity class described in the previous chapter. As a result, you should already understand most of the code for this class.

However, this class has a few new statements that work with the Application object. These statements are highlighted. In the onResume method, the second statement gets the milliseconds for the publication date of the feed from the Application object. Then, an if statement checks whether the milliseconds for the publication date of the feed are equal to -1. If so, the app is being started for the first time. As a result, this code downloads the feed, reads the feed, and displays it. Otherwise, this code reads and displays the feed, or just displays it.

In the onPostExecute method of the ReadFeed class, the second statement sets the milliseconds for the publication date of the feed in the Application object. In other words, every time the app reads the feed, it sets the milliseconds in the Application object. As a result, the milliseconds for the feed are only equal to -1 the first time the app starts.

The FileIO class

Figure 11-16 shows the FileIO class. For the most part, this class is the same as the FileIO class described in the previous chapter. As a result, you should already understand most of the code for this class.

However, the downloadFile method of this class uses a system service to check if a network connection is available before it attempts to download the file. These statements are highlighted. Here, the first two statements get a NetworkInfo object. Then, this code uses the NetworkInfo object to check whether a network connection is available. If so, this code downloads the file from the Internet. If not, this code does not attempt to connect to the network, which is good since that conserves the battery of the device.

The ItemsActivity class

```java
package com.murach.newsreader;

import java.util.ArrayList;
import java.util.HashMap;
import java.util.List;

import android.os.AsyncTask;
import android.os.Bundle;
import android.app.Activity;
import android.content.Intent;
import android.util.Log;
import android.view.Menu;
import android.view.MenuItem;
import android.view.View;
import android.widget.AdapterView;
import android.widget.ListView;
import android.widget.SimpleAdapter;
import android.widget.TextView;
import android.widget.Toast;
import android.widget.AdapterView.OnItemClickListener;

public class ItemsActivity extends Activity
implements OnItemClickListener {

    private NewsReaderApp app;
    private RSSFeed feed;
    private long feedPubDateMillis;
    private FileIO io;

    private TextView titleTextView;
    private ListView itemsListView;

    @Override
    protected void onCreate(Bundle savedInstanceState) {
        super.onCreate(savedInstanceState);
        setContentView(R.layout.activity_items);

        titleTextView = (TextView) findViewById(R.id.titleTextView);
        itemsListView = (ListView) findViewById(R.id.itemsListView);

        itemsListView.setOnItemClickListener(this);

        // get references to Application and FileIO objects
        app = (NewsReaderApp) getApplication();
        io = new FileIO(getApplicationContext());
    }

    @Override
    public void onResume() {
        super.onResume();

        // get feed from app object
        feedPubDateMillis = app.getFeedMillis();
```

Figure 11-15 The ItemsActivity class (part 1 of 3)

The ItemsActivity class

```
        if (feedPubDateMillis == -1) {
            new DownloadFeed().execute();   // download, read, and display
        }
        else if (feed == null) {
            new ReadFeed().execute();       // read and display
        }
        else {
            updateDisplay();                // just display
        }
    }

    class DownloadFeed extends AsyncTask<Void, Void, Void> {
        @Override
        protected Void doInBackground(Void... params) {
            io.downloadFile();
            return null;
        }

        @Override
        protected void onPostExecute(Void result) {
            Log.d("News reader", "Feed downloaded");
            new ReadFeed().execute();
        }
    }

    class ReadFeed extends AsyncTask<Void, Void, Void> {
        @Override
        protected Void doInBackground(Void... params) {
            feed = io.readFile();
            return null;
        }

        @Override
        protected void onPostExecute(Void result) {
            Log.d("News reader", "Feed read");
            app.setFeedMillis(feed.getPubDateMillis());
            ItemsActivity.this.updateDisplay();
        }
    }

    public void updateDisplay()
    {
        if (feed == null) {
            titleTextView.setText("Unable to get RSS feed");
            return;
        }

        // set the title for the feed
        titleTextView.setText(feed.getTitle());

        // get the items for the feed
        List<RSSItem> items = feed.getAllItems();
```

Figure 11-15   The ItemsActivity class (part 2 of 3)

The ItemsActivity class

```java
        // create a List of Map<String, ?> objects
        ArrayList<HashMap<String, String>> data =
                new ArrayList<HashMap<String, String>>();
        for (RSSItem item : items) {
            HashMap<String, String> map = new HashMap<String, String>();
            map.put("date", item.getPubDateFormatted());
            map.put("title", item.getTitle());
            data.add(map);
        }

        // create the resource, from, and to variables
        int resource = R.layout.listview_item;
        String[] from = {"date", "title"};
        int[] to = { R.id.pubDateTextView, R.id.titleTextView};

        // create and set the adapter
        SimpleAdapter adapter =
                new SimpleAdapter(this, data, resource, from, to);
        itemsListView.setAdapter(adapter);

        Log.d("News reader", "Feed displayed");
    }

    @Override
    public void onItemClick(AdapterView<?> parent, View v,
            int position, long id) {
        RSSItem item = feed.getItem(position);
        Intent intent = new Intent(this, ItemActivity.class);
        intent.putExtra("pubdate", item.getPubDate());
        intent.putExtra("title", item.getTitle());
        intent.putExtra("description", item.getDescription());
        intent.putExtra("link", item.getLink());
        this.startActivity(intent);
    }

    @Override
    public boolean onCreateOptionsMenu(Menu menu) {
        getMenuInflater().inflate(R.menu.activity_items, menu);
        return true;
    }

    @Override
    public boolean onOptionsItemSelected(MenuItem item) {
        switch (item.getItemId()) {
            case R.id.menu_refresh:
                new DownloadFeed().execute();
                Toast.makeText(this, "Feed refreshed!",
                        Toast.LENGTH_SHORT).show();
                return true;
            default:
                return super.onOptionsItemSelected(item);
        }
    }
}
```

Figure 11-15   The ItemsActivity class (part 3 of 3)

The FileIO class

```java
package com.murach.newsreader;

import java.io.FileInputStream;
import java.io.FileOutputStream;
import java.io.IOException;
import java.io.InputStream;
import java.net.URL;

import javax.xml.parsers.SAXParser;
import javax.xml.parsers.SAXParserFactory;
import org.xml.sax.InputSource;
import org.xml.sax.XMLReader;

import android.content.Context;
import android.net.ConnectivityManager;
import android.net.NetworkInfo;
import android.util.Log;

public class FileIO {

    private final String URL_STRING = "http://rss.cnn.com/rss/cnn_tech.rss";
    private final String FILENAME = "news_feed.xml";
    private Context context = null;

    public FileIO (Context context) {
        this.context = context;
    }

    public void downloadFile() {
        // get NetworkInfo object
        ConnectivityManager cm = (ConnectivityManager)
            context.getSystemService(Context.CONNECTIVITY_SERVICE);
        NetworkInfo networkInfo = cm.getActiveNetworkInfo();

        // if network is connected, download feed
        if (networkInfo != null &&  networkInfo.isConnected()) {

            try{
                // get the URL
                URL url = new URL(URL_STRING);

                // get the input stream
                InputStream in = url.openStream();

                // get the output stream
                FileOutputStream out =
                    context.openFileOutput(FILENAME, Context.MODE_PRIVATE);
```

Figure 11-16   The FileIO class (part 1 of 2)

The FileIO class

```java
                // read input and write output
                byte[] buffer = new byte[1024];
                int bytesRead = in.read(buffer);
                while (bytesRead != -1)
                {
                    out.write(buffer, 0, bytesRead);
                    bytesRead = in.read(buffer);
                }
                out.close();
                in.close();
            }
            catch (IOException e) {
                Log.e("News reader", e.toString());
            }
        }
    }

    public RSSFeed readFile() {
        try {
            // get the XML reader
            SAXParserFactory factory = SAXParserFactory.newInstance();
            SAXParser parser = factory.newSAXParser();
            XMLReader xmlreader = parser.getXMLReader();

            // set content handler
            RSSFeedHandler theRssHandler = new RSSFeedHandler();
            xmlreader.setContentHandler(theRssHandler);

            // read the file from internal storage
            FileInputStream in = context.openFileInput(FILENAME);

            // parse the data
            InputSource is = new InputSource(in);
            xmlreader.parse(is);

            // set the feed in the activity
            RSSFeed feed = theRssHandler.getFeed();
            return feed;
        }
        catch (Exception e) {
            Log.e("News reader", e.toString());
            return null;
        }
    }
}
```

Figure 11-16   The FileIO class (part 2 of 2)

Perspective

The skills presented in this chapter should be enough to get you started with services and notifications. However, this chapter doesn't show how to work with bound services. As a result, if you need to use bound services, you can search the Internet for more information about working with them.

In this chapter, the service is started when the app starts, which is appropriate for some apps. However, you may also want to start a service when the device boots or when the network becomes available. In the next chapter, you'll learn how to do that.

Terms

service
unbound service
bound service
notification
notification area
notification drawer

pending intent
back stack
task
task history
system service

Summary

- To store data and methods that apply to the entire application, you can extend the Application class and add instance variables and methods.

- A *service* performs tasks in the background, does not provide a user interface, and continues to run even if the user switches to another app.

- An *unbound service* does not interact with other components such as activities, and runs until it's stopped by another component or by itself.

- A *bound service* can interact with components such as activities. This type of service runs only as long as another component is bound to it. Multiple components can bind to the service at once, but when all of them unbind, the service is destroyed.

- To test a service, you can print messages to the LogCat view.

- You can use the Settings app to view all of the apps and services that are running on a device or emulator.

- Before you can use an Application object or a service, you must register it.

- A *notification* provides a way for a service to display a message even when another app is running.

- When it's first displayed, a notification appears as an icon in the *notification area* at the top of the screen.

- To view a notification, the user can pull down on the notification area to open the *notification drawer*.

- A *pending intent* is an intent that can be passed to other apps so that they can execute the intent at a later time.

- The *back stack* is a stack of all recently used activities. These activities are sorted in the order in which they were used.

- A *task* is a cohesive unit that can contain multiple activities. Every activity belongs to a task.

- The *task history* is the back stack for a particular task.

- A *system service* is a service that's provided by the Android operating system.

- You can use the NotificationManager object to display or remove a notification, and you can use the ConnectivityManager object to check if a network connection is available to the device.

Exercise 11-1   Work with a service

In this exercise, you'll create, register, and test a service. Then, you'll modify the service so it stops itself.

Test the app

1. Open the project named ch11_ex1_Tester that's in the ex_starts directory.

2. Run the app. This should launch an activity that displays a message that indicates that you should start this exercise by adding a service.

Create a service that runs indefinitely

3. Add a class named TesterService to the product. Modify this class so it extends the Service class.

4. Add code that implements all four methods as described in figure 11-4. Within each method, add a statement that displays a message in the LogCat view. For the onCreate method, for example, add a statement like this:

   ```
   Log.d("Test service", "Service created");
   ```

5. Open the AndroidManifest.xml file for the project and register this service as described in figure 11-5.

6. Open the TesterActivity class, and add a statement to its onCreate method that starts the service.

7. Run the app. This should launch the activity described earlier in this exercise, and it should display these messages in the LogCat view:

   ```
   Service created
   Service started
   ```

8. Change the orientation for the app one or more times. This should display additional "Service started" messages in the LogCat view.

9. Navigate to another app. Then, navigate back to the Tester app. Note that this does not display the "Service destroyed" message. As a result, this approach is useful for services that you want to continue running.

10. Use the Settings app to view the service for the Tester app. This app should be using 1 process and 1 service.

Create a service that stops itself

11. Open the TesterService class, and add two statements to the onStartCommand method. The first statement should display a message in the LogCat view that says, "Task completed". The second statement should call the stopSelf method.

12. Run the app. This should display these messages in the LogCat view:

    ```
    Service created
    Service started
    Task completed
    Service destroyed
    ```

    This approach is useful for a service that you want to stop after it completes a task. For long-running tasks, it's generally considered a best practice to use the AsyncTask class to execute the task in a separate thread.

13. Comment out the two statements that you added to the onStartCommand method in step 11.

Add a timer that performs a task at specified intervals

14. Open the TesterService class, and add the startTimer and stopTimer methods shown in figure 11-6.

15. Modify the startTimer method so it displays the first message after a delay of 1 second. Then, it should display the message every 10 seconds.

16. Modify the onCreate method so it calls the startTimer method.

17. Modify the onDestroy method so it calls the stopTimer method.

18. Run the app. This should display these messages in the LogCat view:

    ```
    Service created
    Service started
    Timer task executed
    Timer task executed
    Timer task executed
    ...
    ```

19. Comment out the statements that call the startTimer and stopTimer methods.

20. Run the app. This should stop the timer from displaying a new message in the LogCat view every 10 seconds.

Exercise 11-2   Modify the notification for the News Reader app

In this exercise, you'll modify the notification used by the News Reader app presented in this chapter.

1. Open the project named ch11_ex2_NewsReader that's in the ex_starts directory.

2. Open the NewsReaderService class. Then, review the code, and note how the various methods display messages in the LogCat view.

3. In the startTimer method, modify the code so it only delays for 5 seconds (instead of 1 hour) before it runs the timer task. Then, in the if/else statement, add a statement that displays a notification even when no update is available. The text for this notification should be: "Updated feed NOT available."

4. Run the app. After a delay of 5 seconds, it should display a notification in the notification area. Pull down on this area to display the drawer and select the notification. This should display the News Reader app.

5. Press the Back button to remove the current News Reader activity from the stack and return to the previous activity such as the Home screen.

6. Modify the sendNotification method so it uses the "new task" flag instead of the "clear top" flag.

7. Repeat step 4.

8. Press the Back button to remove the current activity from the stack. This should exit the News Reader activity displayed by the notification, but it should leave the original News Reader activity on the stack.

9. Press the Back button again to remove that activity from the stack. This should remove the original News Reader activity from the stack and display the previous activity such as the Home screen.

12

How to work with broadcast receivers

Some Android apps need to respond to actions that are broadcast by the system. For example, an app may need to start a service after the user turns the device on. To do that, you can use an Android component known as a broadcast receiver.

Other apps need to broadcast an action. For example, the service for the News Reader app may want to broadcast an action when a new RSS feed is available. Then, other components within the News Reader app, such as the Items activity, can use a broadcast receiver to respond to that action. Or, other apps can use a broadcast receiver to respond to that action.

How to work with system broadcasts

As you use your Android device, the operating system can *broadcast* an *action* that has occurred. For example, when the device finishes starting up, Android broadcasts an action that indicates that the device has completed booting. To receive these broadcasts, an app can include a component known as a *broadcast receiver*. A broadcast receiver provides a way for the app to execute code when the broadcast is received.

A summary of the system broadcasts

Every broadcast has an *action string* that uniquely identifies the action. Figure 12-1 begins by showing five constants that define the action string for five broadcasts that are commonly made by the Android operating system. The first four constants are stored in the Intent class. These constants are general-purpose constants that apply to many different types of apps. However, the fifth constant is stored in the ConnectivityManager class. This constant is stored here because it only applies to apps that use a network connection.

This figure only shows the action strings for five system broadcasts. However, Android 4.0 (API 15) provides 121 of these actions, and each new version of Android seems to add a few more. To view a list of these action strings, you can use a text editor to open the broadcast_actions.txt file for one of the Android platforms that have been installed on your system. This figure, for example, shows the path for this text file for Android 6.0 (API 23).

Two classes used to work with broadcasts

```
android.content.Intent
android.net.ConnectivityManager
```

Some action constants of the Intent class

Constant	Action string / Description
ACTION_BOOT_COMPLETED	android.intent.action.BOOT_COMPLETED
	Boot completed. This broadcast is typically used to start services that should always run.
ACTION_BATTERY_LOW	android.intent.action.BATTERY_LOW
	Battery is low. This broadcast is typically used to pause activities that consume power.
ACTION_BATTERY_OKAY	android.intent.action.BATTERY_OKAY
	Battery is OK. This broadcast can be used to resume activities that consume power
ACTION_POWER_CONNECTED	android.intent.action.ACTION_POWER_CONNECTED
	Power source connected to the device.

An action constant of the ConnectivityManager class

Constant	Action string / Description
CONNECTIVITY_ACTION	android.net.conn.CONNECTIVITY_CHANGE
	A change in network connectivity has occurred.

A text file that contains the broadcast actions for the system

```
\sdk\platforms\android-23\data\broadcast_actions.txt
```

Description

- The Android operating system *broadcasts* certain *actions* that occur as a device is being used.

- Every broadcast has an *action string* that uniquely identifies the action.

- A *broadcast receiver* is an application component that listens for a broadcast and executes when it receives that broadcast.

Figure 12-1   A summary of system broadcasts

How to code a receiver
for the boot completed broadcast

Figure 12-2 shows how to code a receiver for the boot completed broadcast. To do that, you can code a class that extends the BroadcastReceiver class. In this figure, for example, the BootReceiver class extends the BroadcastReceiver class.

Within the class for the receiver, you can override the onReceive method. This method is executed when the broadcast is received. In this figure, this method starts the service for the News Reader app described in the previous chapter.

After you create the class for a broadcast receiver, you need to register the receiver, so Android knows what broadcasts it should receive. The easiest way to register a receiver is to add a receiver element to the project's AndroidManifest.xml file. This receiver element should specify the name of the class for the receiver, and it should specify the action or actions to receive. To do that, you can use the name attribute of the action element to specify the action string. In this figure, the second example shows how to register the BootReceiver class so it listens for the BOOT_COMPLETED action.

Some broadcast receivers require additional permissions. For example, the BOOT_COMPLETED action in this figure requires the RECEIVE_BOOT_COMPLETED permission. To request this permission, you can add a uses-permission element to the AndroidManifest.xml file. Then, when users install the app, they can decide whether or not to grant this permission.

The BootReceiver class

```
package com.murach.newsreader;

import android.content.BroadcastReceiver;
import android.content.Context;
import android.content.Intent;
import android.util.Log;

public class BootReceiver extends BroadcastReceiver {

    @Override
    public void onReceive(Context context, Intent intent) {
        Log.d("News reader", "Boot completed");

        // start service
        Intent service = new Intent(context, NewsReaderService.class);
        context.startService(service);
    }
}
```

The receiver element in the AndroidManifest.xml file

```
<receiver android:name="BootReceiver" >
    <intent-filter>
        <action android:name="android.intent.action.BOOT_COMPLETED" />
    </intent-filter>
</receiver>
```

The permission in the AndroidManifest.xml file

```
<uses-permission android:name="android.permission.RECEIVE_BOOT_COMPLETED" />
```

Description

- You can create a broadcast receiver that executes code when the device finishes starting, which is also known as booting.

- To define a broadcast receiver, code a class that inherits the BroadcastReceiver class and override its onReceive method. This method is executed when the broadcast is received.

- To register a broadcast receiver, open the project's AndroidManifest.xml file and add a receiver element. This element should specify the name of the class for the receiver, and it should specify the action or actions to receive. To do that, you can add an action element and use its name attribute to specify the action string for the action.

- If a broadcast receiver requires permissions, you can request those permissions by adding a uses-permission element to the AndroidManifest.xml file.

Figure 12-2   How to code a receiver for the boot completed broadcast

How to code a receiver
for the connectivity changed broadcast

Figure 12-3 shows how to code a receiver for the connectivity changed broadcast. This works similarly to coding a receiver for the boot completed broadcast as shown in the previous figure. As a result, you shouldn't have much trouble understanding how this works.

The ConnectivityReceiver class shown in this figure extends the BroadcastReceiver class and overrides its onReceive method. Within this method, the second and third statements get a NetworkInfo object as described in the previous chapter. Then, the fourth statement creates an intent that can be used to start or stop the service for the News Reader class. After that, this code checks whether a network connection is available. If so, this code starts the service. Otherwise, it stops the service.

Since this receiver makes sure that the service doesn't attempt to connect to the network when a connection isn't available, it conserves the battery of the device. In the previous chapter, you learned another technique for conserving battery life. Although these techniques work differently, both accomplish the same goal of conserving battery life.

Like the BootReceiver class, you can register the ConnectivityReceiver class by adding a receiver element to the AndroidManifest.xml file. Similarly, you can add a uses-permissions element to request permissions for the ACCESS_NETWORK_STATE privilege that's required to allow this class to access information about the network. This permission isn't required by the CONNECTIVITY_CHANGE action. However, it is required by the getActiveNetworkInfo method that's used by the onReceive method.

The ConnectivityReceiver class

```java
package com.murach.newsreader;

import android.content.BroadcastReceiver;
import android.content.Context;
import android.content.Intent;
import android.net.ConnectivityManager;
import android.net.NetworkInfo;
import android.util.Log;

public class ConnectivityReceiver extends BroadcastReceiver {

    @Override
    public void onReceive(Context context, Intent intent) {
        Log.d("News reader", "Connectivity changed");

        ConnectivityManager connectivityManager = (ConnectivityManager)
                context.getSystemService(Context.CONNECTIVITY_SERVICE);
        NetworkInfo networkInfo =
                connectivityManager.getActiveNetworkInfo();

        Intent service = new Intent(context, NewsReaderService.class);
        if (networkInfo != null && networkInfo.isConnected()){
            Log.d("News reader", "Connected");
            context.startService(service);
        }
        else {
            Log.d("News reader", "NOT connected");
            context.stopService(service);
        }
    }
}
```

The receiver element in the AndroidManifest.xml file

```xml
<receiver android:name="ConnectivityReceiver" >
    <intent-filter>
        <action android:name="android.net.conn.CONNECTIVITY_CHANGE" />
    </intent-filter>
</receiver>
```

The permission in the AndroidManifest.xml file

```xml
<uses-permission android:name="android.permission.ACCESS_NETWORK_STATE" />
```

Description

- You can code a broadcast receiver to execute code when the connectivity for the device changes.

Figure 12-3   How to code a receiver for the connectivity change broadcast

How to work with custom broadcasts

In some cases, you may want to send a custom broadcast that other components within your app can receive. Or, you may want to send a custom broadcast that other apps can receive. Then, you can code a receiver for that broadcast.

In the News Reader app, for example, you may want to automatically update the display for the Items activity if a new feed becomes available. That way, if a user is viewing the Items activity and a new update becomes available, the News Reader app will automatically update the display. To accomplish this, the service for this app can send a custom broadcast that indicates that a new feed is available. Then, the Items activity can receive that broadcast and update the display.

How to create and send a custom broadcast

Figure 12-4 shows how to create and send a custom broadcast. To start, you can define a constant for the action string for the broadcast. By convention, this string should begin with the name of the package for your app, followed by the name of the constant. This should create an action string that's globally unique. In addition, you should typically code this constant in a class that's related to the action. In this figure, for example, the first step defines a constant named NEW_FEED in the RSSFeed class. This constant indicates that a new feed is available.

Once you create a constant for a broadcast, you can add code that creates the intent for the broadcast and sends the broadcast. To start, you can create an Intent object for the broadcast by passing the constant for the broadcast to the constructor of this class. Then, if necessary, you can use the putExtra method to store data in the Intent object. In this figure, the example puts a test string in the Intent object. However, you can use the putExtra method to store most types of data that you want to pass to the receiver. Finally, you can use the sendBroadcast method to send the broadcast.

Two classes used to work with broadcasts

```
android.content.Intent
android.content.Context
```

How to send a broadcast

Step 1: Define a constant for the broadcast (in the RSSFeed class)

```
public final static String NEW_FEED = "com.murach.newsreader.NEW_FEED";
```

Step 2: Create and send the broadcast (in the NewsReaderService class)

```
Intent intent = new Intent(RSSFeed.NEW_FEED);
intent.putExtra("test", "test1 test2");
sendBroadcast(intent);
```

A constructor of the Intent class

Constructor	Description
`Intent(actionString)`	Create an intent with the specified action string.

A method of the Context class

Method	Description
`sendBroadcast(intent)`	Broadcast the specified intent to all registered receivers.

Description

- The constant for a custom broadcast should store an action string that's globally unique. By convention, this string should begin with the name of the package for your app, followed by the name of the constant.

- To send a custom broadcast, create an Intent object for the broadcast by passing the constant for the action string to the constructor of the Intent class. Then, use the sendBroadcast method to send the broadcast.

- If you want to pass data to the receiver, you can use the putExtra method to store extra data in the Intent object.

Figure 12-4   How to create and send a custom broadcast

How to code a receiver for a custom broadcast

Figure 12-5 shows how to code a receiver for a custom broadcast. The steps for coding a receiver for a custom broadcast are similar to the steps for coding a receiver for a system broadcast. As a result, you shouldn't have much trouble understanding how to code the receiver shown in this figure.

However, if you want to update the user interface when a broadcast is received, you can code the receiver as an inner class of the activity. That way, you can easily access elements of the user interface, including any methods that update the user interface. In this figure, for example, the NewFeedReceiver class is coded as an inner class of the ItemsActivity class. As a result, it can access the updateDisplay method that's available from the ItemsActivity class.

If you want to get extra data from the Intent for the action, you can use the getXxxExtra methods of the Intent class. In this figure, the second statement in the onReceive method uses the getStringExtra method to get the test string that was set by the code that broadcast the action. Then, the third statement sends this data to the LogCat view. Although this doesn't do anything useful for the News Reader app, it illustrates how a broadcast can be used to pass data from one component to another.

When you code a receiver as an inner class of an activity, you typically use Java code to register and unregister the receiver. To do that, you create a receiver object and a filter object for the intent that has the specified action string. To create an IntentFilter object, you can pass the constant for the action string to the constructor of the IntentFilter class.

After you create these objects, you can use the methods of the Context class to register and unregister the receiver. Typically, you add these methods to the onResume and onPause methods of the activity. In this figure, the code examples register and unregister the receiver defined by the NewFeedReceiver class.

Since the NEW_FEED action that's broadcast and received by the News Reader app doesn't contain any sensitive information, I decided to not require privileges to broadcast or receive this action. As a result, it may be possible for another app to broadcast or receive this action. However, it's unlikely that another app (even a malicious one) would want to do this, and it wouldn't be too harmful if it did.

However, if your app broadcasts or receives an action that should be secure, you can use permissions to secure the broadcast. To do that, you can define the send and receive permissions in the manifest file for your project. Then, you can modify the sendBroadcast and registerReceiver methods so they enforce these permissions. For more information, you can search the Internet.

A class used to work with receivers

```
android.content.Context
```

How to receive a broadcast in the ItemsActivity class

Step 1: Define an inner class for the broadcast receiver

```
class NewFeedReceiver extends BroadcastReceiver {

    @Override
    public void onReceive(Context context, Intent intent) {
        Log.d("News reader", "New items broadcast received");

        // if necessary, get data from intent
        String test = intent.getStringExtra("test");
        Log.d("News reader", "test: " + test);

        // update the display
        updateDisplay();
    }
}
```

Step 2: Define instance variables for the broadcast receiver and intent filter

```
private NewFeedReceiver newFeedReceiver;
private IntentFilter newFeedFilter;
```

Step 3: Create the broadcast receiver and intent filter (onCreate)

```
newFeedFilter = new IntentFilter(RSSFeed.NEW_FEED);
newFeedReceiver = new NewFeedReceiver();
```

Step 4: Register the receiver (onResume)

```
registerReceiver(newFeedReceiver, newFeedFilter);
```

Step 5: Unregister the receiver (onPause)

```
unregisterReceiver(newFeedReceiver);
```

Two methods of the Context class

Methods	Description
`registerReceiver(receiver, filter)`	Register the specified broadcast receiver with the specified intent filter.
`unregisterReceiver(receiver)`	Unregister the specified broadcast receiver.

Description

- To code a receiver for a custom broadcast, you can define a class for the receiver. If you want to update the user interface when a broadcast is received, you can code the class for the receiver as an inner class of the activity.

- If you want to get extra data from the Intent for the action, you can use the getXxxExtra methods of the Intent class.

- You can use Java code to register and unregister a receiver. To do that, you can create a receiver object and a filter object for the intent with the specified action.

Figure 12-5   How to code a receiver for a custom broadcast

Perspective

The skills presented in this chapter should be enough to get you started with broadcast receivers. However, Android broadcasts actions that you can use to execute code due to changes in the battery, text messages, phone state, external hardware devices, and so on.

If you are developing an app that needs to receive and work with these types of broadcasts, you can search the Internet for more information. To do that, you can begin by using the broadcast_actions.txt described in figure 12-1 to find the action string for the broadcast. Then, you can search the Internet to find the API documentation for that broadcast action. This documentation should include a description of the action, a description of any extra data that's stored in the intent for the action, and a description of any permissions that are required by the action.

Terms

broadcast
action
broadcast receiver
action string

Summary

- The Android operating system *broadcasts* certain *actions* that occur as a device is being used.

- Every broadcast has an *action string* that uniquely identifies the action.

- A *broadcast receiver* is an application component that listens for a broadcast and executes code when it receives that broadcast.

- You can code a broadcast receiver to execute code when the connectivity for the device changes.

- To code a receiver for a custom broadcast, you can define a class for the receiver.

- If you want to get extra data from the Intent for the action, you can use the getXxxExtra methods of the Intent class.

- You can use Java code to register and unregister a receiver.

Exercise 12-1   Work with broadcast receivers

In this exercise, you'll experiment with the broadcast receivers that are available from the News Reader app.

Test the existing broadcast receivers

1. Open the project named ch12_ex1_NewsReader.

2. Open the BootReceiver and ConnectivityReceiver classes. Then, review the code for these classes.

3. Open the AndroidManifest.xml file. Note that it registers both of the broadcast receivers described in the previous step and that it requests all permissions required by those receivers.

4. Run the app. If you're using an actual device, test the Connectivity receiver by turning on airplane mode. This should display messages in the LogCat view that say:

    ```
    Connectivity changed
    NOT connected
    ```

    NOTE: This might not work correctly on an emulator.

5. Turn off airplane mode. This should display messages in the LogCat view that say:

    ```
    Connectivity changed
    Connected
    ```

    NOTE: Again, this might not work correctly on an emulator.

6. Open the broadcast_actions.txt file for one of the Android SDKs that's installed on your system. Then, review the various broadcasts that your application can receive.

Add a new broadcast receiver

7. Add a broadcast receiver that displays a message on the LogCat view that says "Battery low!". This receiver should also stop the service for the app.

8. Use the AndroidManifest.xml file to register this receiver. Unfortunately, there's no easy way to test this receiver. As a result, don't test this receiver yet. Instead, review the code and make sure that it seems right.

Modify a broadcast

9. Open the RSSFeed class. Then, change the name of the constant for the broadcast from NEW_FEED to ACTION_UPDATE_AVAILABLE. Also, change the name of the action string so it matches the name of the constant.

10. Open the ItemsActivity and NewsReaderService classes. Then, make sure the statement that creates the IntentFilter object uses the new name for the constant.

Section 4

The Task List app

This section shows how to develop the Task List app. As you learn to develop this app, you'll also learn many important Android skills that can be applied to other apps. To start, chapter 13 shows how to use SQLite to create and work with the database for an app. Then, chapter 14 shows how to use tabs and custom adapters to display data on the user interface for the app. Next, chapter 15 shows how to use a content provider to make the data for an app available to other apps. Finally, chapter 16 shows how to create an app widget than can display the app's data on a device's Home screen.

13

How to work
with SQLite databases

Some types of Android apps need to store data in a local database. For example, a Contacts app needs to store contact information such as names and phone numbers in a database. Fortunately, Android includes libraries that make it easy to create and use local databases. In this chapter, you'll learn how to create and use a database for a Task List app.

An introduction to databases

The skills presented in this chapter are necessary to create the Task List app. To put these skills in context, this chapter begins by showing the user interface for this app.

The user interface for the Task List app

Figure 13-1 shows the interface for the Task List app. When the app starts, the Task List activity displays a list of tasks where each task has a check box that indicates whether it has been completed, a name, and optional notes.

In addition, the Task List activity provides tabs for two lists: Personal and Business. To display lists, the user can click on the tabs that are available from the Task List activity. In this figure, for example, the Personal tab is selected and the tasks for that tab are displayed. However, the user could click on the Business tab to display the tasks for that list.

From the Task List activity, the user can mark a task as completed by selecting a check box. Then, the user can remove all tasks that have been marked as completed by clicking on the Delete Tasks icon (the small trash can). This icon is available from the task bar. However, older versions of Android don't provide for the task bar. As a result, on older devices, the user can display the options menu and select the Delete Tasks item.

The user can add a task by clicking on the Add Task icon (the + sign) that's available from the task bar. Or, if the device is too old to display a task bar, the user can display the options menu and select the Add Task item. This displays the Add/Edit activity in "add mode". At this point, the user can use the spinner to select the list. Then, the user can use the editable text views to add or edit the text for the task.

Similarly, the user can edit a task by clicking on the task in the Task List activity. This displays the Add/Edit activity in "edit mode" so the user can edit the data for the task. In this figure for example, I clicked on the "Get hair cut" task in the Task List activity. As a result, this task is displayed in the Add/Edit activity and is ready to be edited.

Take a moment to reflect on the data this app needs. First, it needs the names of the lists. Second, it needs at least four pieces of data about a task: its list, its name, its notes, and whether it has been completed. In this chapter, you'll learn how to store this type of data in a local database.

The Task List activity

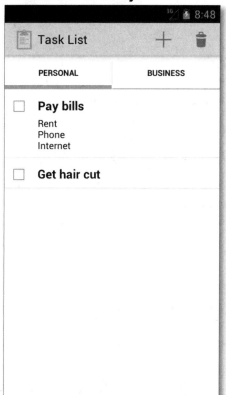

The Add/Edit activity

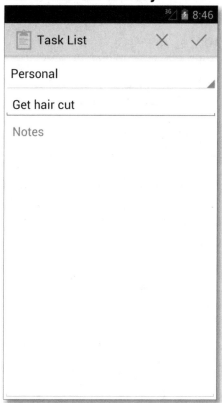

Description

- When the app starts, the Task List activity displays a list of tasks where each task has a check box that indicates whether it has been completed, a name, and optional notes.

- The app provides for two lists: Personal and Business.

- To display lists, the user can click on the tabs that are available in the Task List activity.

- To edit a task, the user can click on the task in the Task List activity. This displays the Add/Edit activity in "edit mode" so the user can edit the data for the task.

- To mark a task as completed, the user can select the check box in the Task List activity.

- To remove all completed tasks, the user can click the Delete Tasks button that's available from the task bar.

- To add a new task, the user can navigate to the Task List activity and click on the Add Task icon. This displays the Add/Edit activity in "add mode" so the user can add a new task.

- In the Add/Edit activity, the user can use the spinner to select the list. Then, the user can use the editable text views to add or edit the text for the task.

Figure 13-1   The user interface for the Task List app

An introduction to SQLite

Figure 13-2 describes the SQLite database that's available from Android. What is SQLite? To start, SQLite is a *relational database management system* (*RDBMS*) that implements most, but not all, of the SQL standard. SQLite is a lightweight (approximately 350KB) programming library. This library is available on every Android device and does not require any configuration or database administration.

When you create a SQLite database, that database is stored in a single file on a device. In this figure, for example, you can see where the file for the Task List app is stored on a device. Since that file is only accessible by the app that created the file, a SQLite database is relatively secure.

SQLite is a popular choice as an embedded database. It is used today by several browsers and operating systems, including the Android operating system. One of the reasons that SQLite is used so widely is because it is open-source. As a result, it's free for most purposes.

If you're familiar with other database management systems such as MySQL or Oracle, you shouldn't have much trouble learning to work with SQLite. However, SQLite is different than other database systems in a few ways.

First, SQLite does not run in a server process that's accessed by client apps, which run in separate processes. Instead, it's embedded as part of the client app and runs in the same process.

Second, SQLite only supports three data types (TEXT, INTEGER, and REAL). These data types correspond with these three Java data types: String, long, and double. As a result, other Java data types such as date/time and Boolean values must be converted into one of the SQLite types before saving them in the database. For example, a Date type can be converted to a TEXT type before it's stored in the database. Then, when it's retrieved from the database, it can be converted from a TEXT type to a Date type.

Third, SQLite is weakly-typed. In other words, a column in a SQLite database does not reject a value of an incorrect data type. For example, if you define a database column to accept the INTEGER type, you could still store a string value in this column and it would not be rejected by the database. Usually, this isn't a problem as the code for the app shouldn't attempt to store a string in an INTEGER column. However, you should be aware of this as it means that the code for your app must make sure to insert values of the correct type in each column.

SQLite is...

- **A RDBMS.** SQLite is a *relational database management system* (*DBMS*).

- **Standards-compliant.** SQLite implements most of the SQL standard.

- **Embedded.** Unlike most database management systems, SQLite does not run in a server process that's accessed by client apps, which run in separate processes. Instead, it's embedded as part of the client app and runs in the same process.

- **Zero-configuration.** SQLite is available on every Android device and does not require any database administration.

- **Lightweight**. The SQLite programming library is approximately 350 KB.

- **Secure**. A SQLite database is only accessible by the app that created the file for the database.

- **Popular.** SQLite is a popular choice as an embedded database.

- **Open-source.** The source code for SQLite is in the public domain.

Data types supported by SQLite

SQLite data type	Java data type
TEXT	String
INTEGER	long
REAL	double

The location of the SQLite database file for the Task List app

`/data/data/com.murach.tasklist/databases/tasklist.db`

Description

- SQLite only supports three data types (String, long, and double). As a result, the code for the app must convert other Java data types such as date/time and Boolean values into one of the SQLite types before saving them in the database.

- SQLite is weakly-typed. In other words, a column in a SQLite database does not reject a value of an incorrect data type. As a result, the code for the app must make sure to insert values of the correct type.

Figure 13-2   An introduction to SQLite

An introduction to the Task List database

Figure 13-3 begins by showing a diagram for the Task List database. This diagram shows that the data for the Task List database is stored in two tables: the List table and the Task table. These tables are related to each other in a one-to-many relationship. As a result, a single list can have many tasks, but a task can only have a single list.

The List table only has two columns. The _id column uniquely identifies each row, and the list_name column provides the name of the list.

The Task table has six columns. The _id column uniquely identifies each row, and the list_id column identifies the list for this task. The task_name and notes columns store the name and notes for the task.

The date_completed column stores the date and time that the task was completed. As a result, this column can be used to determine whether the task has been completed. Alternately, this column could store a Boolean value. However, for this database, I choose to store the date/time value since this allows the database to store more potentially useful information about the task.

The hidden column stores a Boolean value that indicates whether the task should be shown on a list. I choose to use this column because I'd prefer to hide tasks instead of deleting them. In other words, when a user marks a task as complete and removes it from a list, the Task List app doesn't delete it from the database. Instead, it marks it as hidden, and doesn't display it on the user interface anymore. That way, a history of completed tasks remains in the database.

When naming columns, it's a common coding convention to use _id as the name of the primary key column. This isn't required, but it makes it easier to create a content provider for a database as described in chapter 15.

The CREATE TABLE statements provide the names of the tables as well as the column names, data types, and attributes of those tables. Here, the first column in each table uses the PRIMARY KEY attribute to define a primary key column. This column requires that each row must have a value, and that value must be unique. Then, the first column uses the AUTO_INCREMENT attribute to automatically increment the value of the column when a new row is inserted.

In the List table, the list_name column uses the NOT NULL attribute to specify that this column must store a not-null value. In addition, it uses the UNIQUE attribute to specify that each value stored in the column must be unique. As a result, for a new row, the database requires the user to specify a non-null value for the list name, and it doesn't allow a list name that already exists in the database. This is usually what you want, since it doesn't make sense to use a null value for a list name or to have two lists with the same name.

In the Task table, the list_id and task_name columns use the NOT NULL attribute to specify that these columns must store a non-null value. This is usually what you want since a task should have a list and a name.

The DROP TABLE statements delete the specified table. This deletes the structure for the table as well as any data that's stored in the table. Since these statements include the IF EXISTS keywords, they don't cause an error if the table doesn't already exist.

The diagram for the Task List database

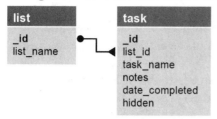

The SQL statement that creates the List table

```
CREATE TABLE list (
    _id         INTEGER PRIMARY KEY AUTOINCREMENT,
    list_name   TEXT    NOT NULL UNIQUE
)
```

The SQL statement that creates the Task table

```
CREATE TABLE task (
    _id             INTEGER PRIMARY KEY AUTOINCREMENT,
    list_id         INTEGER NOT NULL,
    task_name       TEXT    NOT NULL,
    notes           TEXT,
    date_completed  TEXT,
    hidden          TEXT
)
```

The SQL statement that drops the List table

```
DROP TABLE IF EXISTS list
```

The SQL statement that drops the Task table

```
DROP TABLE IF EXISTS task
```

Common column attributes

Attribute	Description
PRIMARY KEY	Defines a primary key column. A *primary key* column requires that each row must have a value, and that value must be unique.
AUTO_INCREMENT	Identifies a column whose value is automatically incremented by SQLite when a new row is inserted. An auto increment column is typically defined with the INTEGER data type.
NOT NULL	Specifies that the column can't store NULL values.
UNIQUE	Specifies that each value stored in the column must be unique.

Description

- The CREATE TABLE statement creates a table based on the column names, data types, and attributes that you specify.

- The DROP TABLE statement deletes the specified table.

Figure 13-3   An introduction to the Task List database

The business objects for the Task List app

Figure 13-4 shows the classes that define the *business objects* for the Task List app. These business objects map to the tables of the database. As a result, the data for a row can be stored in its corresponding business object. For example, a row from the List table can be stored in a List object. Similarly, a row from the Task table can be stored in a Task object. Then, if necessary, the business object can provide additional methods for working with the data.

When working with databases, it's usually considered a good practice to map rows to Java objects like this. This allows you to store all database code in a class or a series of classes that's known as the database layer. Since this layer provides access to *plain old Java objects* (sometimes called *POJOs*), it separates the database layer from the user interface.

The List class defines an instance variable for each column in the List table. Then, it defines three contructors that can be used to create a List object. Next, it defines get and set methods for the instance variables. Finally, it defines a toString method that provides a way to convert a List object to a string. In this case, it does that by returning the name of the list.

The Task class works similarly. However, it provides TRUE and FALSE constants that define two strings that can be used to represent Boolean values. Here, a string of "0" represents a false value, and a string of "1" represents a true value. Within this class, the first constructor uses the FALSE constant to set default values for the completed and hidden variables.

Of the get and set methods, the getCompletedDate method returns the string for the completed_date column. This string stores the number of milliseconds for the date. However, the getCompletedDateMillis method converts the string for the date_completed column from its string value to a long value and it returns that long value. Similarly, the setCompletedDate method is overridden so it can accept a string value or a long value. If the user passes a long value to this method, it converts the long value to a string value.

The List class

```java
package com.murach.tasklist;

public class List {

    private int id;
    private String name;

    public List() {}

    public List(String name) {
        this.name = name;
    }

    public List(int id, String name) {
        this.id = id;
        this.name = name;
    }

    public void setId(int id) {
        this.id = id;
    }

    public int getId() {
        return id;
    }

    public void setName(String name) {
        this.name = name;
    }

    public String getName() {
        return name;
    }

    @Override
    public String toString() {
        return name;    // used for add/edit spinner
    }
}
```

Description

- The List class defines an object that can store the data from the List table.

Figure 13-4   The business objects for the Task List app (part 1 of 3)

The Task class

```java
package com.murach.tasklist;

public class Task {

    private int taskId;
    private int listId;
    private String name;
    private String notes;
    private String completed;
    private String hidden;

    public static final String TRUE = "1";
    public static final String FALSE = "0";

    public Task() {
        name = "";
        notes = "";
        completed = FALSE;
        hidden = FALSE;
    }

    public Task(int listId, String name, String notes,
            String completed, String hidden) {
        this.listId = listId;
        this.name = name;
        this.notes = notes;
        this.completed = completed;
        this.hidden = hidden;
    }

    public Task(int taskId, int listId, String name, String notes,
            String completed, String hidden) {
        this.taskId = taskId;
        this.listId = listId;
        this.name = name;
        this.notes = notes;
        this.completed = completed;
        this.hidden = hidden;
    }

    public int getId() {
        return taskId;
    }

    public void setId(int taskId) {
        this.taskId = taskId;
    }

    public int getListId() {
        return listId;
    }

    public void setListId(int listId) {
        this.listId = listId;
    }
```

Figure 13-4   The business objects for the Task List app (part 2 of 3)

The Task class

```java
    public String getName() {
        return name;
    }

    public void setName(String name) {
        this.name = name;
    }

    public String getNotes() {
        return notes;
    }

    public void setNotes(String notes) {
        this.notes = notes;
    }

    public String getCompleted() {
        return completed;
    }

    public long getCompletedMillis() {
        return Long.valueOf(completed);
    }

    public void setCompletedDate(String completed) {
        this.completed = completed;
    }

    public void setCompletedDate(long millis) {
        this.completedDate = Long.toString(millis);
    }

    public String getHidden(){
        return hidden;
    }

    public void setHidden(String hidden) {
        this.hidden = hidden;
    }
}
```

Description

- The Task class defines an object that can store the data from the Task table.

Figure 13-4   The business objects for the Task List app (part 3 of 3)

How to create a database class

Now that you have some background about SQLite and the Task List database, you're ready to learn how to code a class that you can use to work with a database. This approach allows you to store all of the database code for the app in a single class. This is generally considered a best practice. However, if the class becomes too long to easily maintain, you can split it into multiple database classes.

How to define the constants for a database

Figure 13-5 shows the beginning of the TaskListDB class. This class contains all of the database code for the Task List app.

The TaskListDB class begins by defining the constants for the name and version number of the database. To start, the DB_NAME constant specifies a name of "tasklist.db". As a result, Android stores the file for the database in a file named tasklist.db. Although you can use any name you want, it's a common convention to use ".db" as the filename extension.

The DB_VERSION constant specifies the database version. To start, you can create version 1 of the database. Then, if you make changes to the structure of the database, you can increment this version number. When you do that, Android updates the database as specified later in this class.

This class also defines the constants for the List and Task tables. These constants include the names of the tables, their columns, and their column numbers. For example, the LIST_TABLE constant specifies the name of the List table ("list"), the LIST_ID constant specifies the name of the _id column ("_id"), and the LIST_ID_COL constant specifies the column number of the _id column (0). When specifying the column numbers, you should begin the numbering with 0. As a result, the first column is 0, the second column is 1, and so on.

The TaskListDB class   **Page 1**

```java
package com.murach.tasklist;

import java.util.ArrayList;

import android.content.ContentValues;
import android.content.Context;
import android.database.Cursor;
import android.database.sqlite.SQLiteDatabase;
import android.database.sqlite.SQLiteDatabase.CursorFactory;
import android.database.sqlite.SQLiteOpenHelper;
import android.util.Log;

public class TaskListDB {

    // database constants
    public static final String DB_NAME = "tasklist.db";
    public static final int    DB_VERSION = 1;

    // list table constants
    public static final String LIST_TABLE = "list";

    public static final String LIST_ID = "_id";
    public static final int    LIST_ID_COL = 0;

    public static final String LIST_NAME = "list_name";
    public static final int    LIST_NAME_COL = 1;

    // task table constants
    public static final String TASK_TABLE = "task";

    public static final String TASK_ID = "_id";
    public static final int    TASK_ID_COL = 0;

    public static final String TASK_LIST_ID = "list_id";
    public static final int    TASK_LIST_ID_COL = 1;

    public static final String TASK_NAME = "task_name";
    public static final int    TASK_NAME_COL = 2;

    public static final String TASK_NOTES = "notes";
    public static final int    TASK_NOTES_COL = 3;

    public static final String TASK_COMPLETED = "date_completed";
    public static final int    TASK_COMPLETED_COL = 4;

    public static final String TASK_HIDDEN = "hidden";
    public static final int    TASK_HIDDEN_COL = 5;
```

Description

- It's generally considered a good practice to define constants for the name and version number of the database.

- It's generally considered a good practice to define constants for the table names, column names, and column numbers.

Figure 13-5   How to define the constants for the database

How to define the SQL statements
that create a database

Now that the class has defined the constants for the table and column names, it can build the CREATE TABLE statements needed to create the List and Task tables. In figure 13-6, for example, the first constant contains the CREATE TABLE statement needed to create the List table. This statement uses the LIST_TABLE constant to specify the name of the table, and it uses the LIST_ID and LIST_NAME constants to specify the names of the columns.

Similarly, the second constant contains the CREATE TABLE statement needed to create the Task table. These constants create the same CREATE TABLE statements described earlier in this chapter.

This code also defines the constants for the DROP TABLE statements that are needed to delete the List and Task tables. These constants are used to upgrade a database as shown in the next figure.

The TaskListDB class   Page 2

```java
// CREATE and DROP TABLE statements
public static final String CREATE_LIST_TABLE =
        "CREATE TABLE " + LIST_TABLE + " (" +
        LIST_ID   + " INTEGER PRIMARY KEY AUTOINCREMENT, " +
        LIST_NAME + " TEXT    NOT NULL UNIQUE);";

public static final String CREATE_TASK_TABLE =
        "CREATE TABLE " + TASK_TABLE + " (" +
        TASK_ID         + " INTEGER PRIMARY KEY AUTOINCREMENT, " +
        TASK_LIST_ID    + " INTEGER NOT NULL, " +
        TASK_NAME       + " TEXT    NOT NULL, " +
        TASK_NOTES      + " TEXT, " +
        TASK_COMPLETED  + " TEXT, " +
        TASK_HIDDEN     + " TEXT);";

public static final String DROP_LIST_TABLE =
        "DROP TABLE IF EXISTS " + LIST_TABLE;

public static final String DROP_TASK_TABLE =
        "DROP TABLE IF EXISTS " + TASK_TABLE;
```

Description

- You can use the constants defined in the previous figure to build the CREATE
  TABLE and DROP TABLE statements needed to create and upgrade a database.

Figure 13-6   How to define the SQL statements that create a database

How to create or upgrade a database

Figure 13-7 shows the code for a static inner class named DBHelper that extends the SQLiteOpenHelper class. This class begins by defining a constructor. This constructor contains a single statement that passes four arguments to the superclass (the SQLiteOpenHelper class).

After the constructor, this class overrides the onCreate method of the superclass. If Android doesn't find the specified database on the device, it executes this method. Within this method, the first two statements call the execSQL method of the SQLiteDatabase parameter to execute the CREATE TABLE statements that create the List and Task tables. Then, the next four statements execute INSERT statements to insert rows into the List and Task tables. These rows provide some default and sample data.

After the onCreate method, this class overrides the onUpgrade method of the superclass. If Android finds the database on the device and a newer version of the database is available, it executes this method. The first statement in this method displays a message in the LogCat view. This message indicates that the database is being upgraded, and it includes the old and new version numbers. Then, the second and third statements call the execSQL method of the SQLiteDatabase parameter to execute the DROP TABLE statements that drop the List and Task tables. This deletes the structure of the database as well as all of its data. Finally, the fourth statement calls the onCreate method to recreate the database and to reinsert the starting data.

During development, it's a common practice for the onUpgrade method to drop a database and recreate it. However, this deletes all data in the database. Although this is often acceptable during testing, it's not usually what you want for a production app. In that case, the onUpgrade method typically uses an ALTER TABLE statement to alter the structure of the table without deleting all of its data. For example, you could use an ALTER TABLE statement to add a new column to a table without deleting the data that's already in the other columns of the table.

The execSQL method that's available from the SQLiteDatabase class is useful for executing SQL statements such as the CREATE TABLE and DROP TABLE statements shown earlier in this chapter. In addition, it's useful for executing INSERT statements such as the ones shown in this figure. However, if a SQL statement allows the user to insert a parameter, the execSQL method is susceptible to a security vulnerability known as a SQL injection attack. In that case, you should use the techniques specified later in this chapter. For example, you can use the insert method to allow the user to insert data into the database.

The TaskListDB class   **Page 3**

```java
        private static class DBHelper extends SQLiteOpenHelper {

            public DBHelper(Context context, String name,
                    CursorFactory factory, int version) {
                super(context, name, factory, version);
            }

            @Override
            public void onCreate(SQLiteDatabase db) {
                db.execSQL(CREATE_LIST_TABLE);
                db.execSQL(CREATE_TASK_TABLE);

                // insert default lists
                db.execSQL("INSERT INTO list VALUES (1, 'Personal')");
                db.execSQL("INSERT INTO list VALUES (2, 'Business')");

                // insert sample tasks
                db.execSQL("INSERT INTO task VALUES (1, 1, 'Pay bills', " +
                        "'Rent\nPhone\nInternet', '0', '0')");
                db.execSQL("INSERT INTO task VALUES (2, 1, 'Get hair cut', " +
                        "'', '0', '0')");
            }

            @Override
            public void onUpgrade(SQLiteDatabase db,
                    int oldVersion, int newVersion) {

                Log.d("Task list", "Upgrading db from version "
                        + oldVersion + "to " + newVersion);

                db.execSQL(TaskListDB.DROP_LIST_TABLE);
                db.execSQL(TaskListDB.DROP_TASK_TABLE);
                onCreate(db);
            }
        }
    }
```

A method of the SQLiteDatabase class

Method	Description
execSQL(sqlString)	Executes the specified SQL string.

Description

- If Android doesn't find the database on the device, it executes the onCreate method.
- If Android finds the database on the device and a newer version of the database is available, it executes the onUpgrade method.
- During development, it's a common practice for the onUpgrade method to use a DROP TABLE statement to drop a database and to call the onCreate method to recreate the database. This deletes all data in the database.
- For a production app, the onUpgrade method typically uses an ALTER TABLE statement to alter a table without deleting all of its data.

Figure 13-7   How to create or upgrade a database

How to open and close a database connection

Figure 13-8 shows how to code the private methods that make it easy to open and close a connection to the database. These private methods are used by the public methods shown in the next few figures.

The code in this figure begins by declaring the SQLiteDatabase and DBHelper objects as instance variables. The SQLiteDatabase object defines a connection to the database, and the DBHelper object contains the methods that can open that connection.

The constructor of the TaskListDB class creates an instance of the DBHelper object. To do that, this code passes the constants for the database name and version to the DBHelper class.

The openReadableDB method calls the getReadableDatabase method from the DBHelper object to return a SQLiteDatabase object for a read-only connection to the database. As a result, you can't use this connection to write data to the database.

The openWriteableDB method works like the openReadableDB method. However, the openWriteableDB method returns a SQLiteDatabase object for a read-write connection to the database. As a result, you can use this connection to read or write data.

The closeDB method begins by checking to make sure the SQLiteDatabase object exists. If so, it calls the close method of this object to close the connection. In general, it's considered a best practice to close the database connection after you're done using it.

The TaskListDB class   **Page 4**

```
    // database object and database helper object
    private SQLiteDatabase db;
    private DBHelper dbHelper;

    // constructor
    public TaskListDB(Context context) {
        dbHelper = new DBHelper(context, DB_NAME, null, DB_VERSION);
    }

    // private methods
    private void openReadableDB() {
        db = dbHelper.getReadableDatabase();
    }

    private void openWriteableDB() {
        db = dbHelper.getWritableDatabase();
    }

    private void closeDB() {
        if (db != null)
            db.close();
    }
```

Two methods of the SQLiteOpenHelper class

Method	Description
`getReadableDatabase()`	Opens a read-only connection to the database.
`getWriteableDatabase()`	Opens a read-write connection to the database.

A method of the SQLiteDatabase class

Method	Description
`close()`	Closes a connection.

Description

- Within a database class, you can use private methods that make it easy to open and close a connection to the database.

Figure 13-8   How to open and close a database connection

How to add public methods to a database class

Now that you understand how to code the private methods for a database class, you're ready to learn how to code its public methods. These methods provide a way for the rest of the app to work with the database.

How to retrieve multiple rows from a table

The first method in figure 13-9 shows how to code a public method named getTasks that retrieves multiple rows from a table. To start, the declaration for this method indicates that the method accepts the name of the list as a parameter and that it returns an ArrayList of Task objects.

Within the method, the first statement defines a string that specifies which rows to retrieve. If you're familiar with SQL, you should recognize that this works like the condition in a WHERE clause. Here, the question mark (?) identifies a parameter that will be supplied later. As a result, the WHERE clause retrieves all rows where the list_id column is equal to the supplied listID value and where the hidden column is not equal to a true value. Since the hidden column is stored with the TEXT data type, this code uses a string of '1' for a true value.

The second statement calls the getList method that's defined later in this class to return a List object that corresponds with the specified list name. Although this method isn't explained in this chapter, it works much like the getTask method. Then, it calls the getID method to get the ID for that list.

The third statement defines an array of strings that contains the arguments for the WHERE clause. In this case, the WHERE clause contains a single parameter. As a result, this code supplies a single string for the array. However, if the WHERE clause contained multiple parameters, you could supply multiple strings, separating each string with a comma.

The fourth statement uses one of the private methods defined earlier in this class to open a read-only connection to the database.

The fifth statement uses the query method of the database object to return a Cursor object that contains the rows that are retrieved. In this case, the first argument specifies the name of the table, the third argument specifies the string for the WHERE clause, and the fourth argument specifies the array that contains the arguments for the WHERE clause.

This code specifies null values for the other arguments. As a result, the query method retrieves all columns and doesn't include GROUP BY, HAVING, or ORDER BY clauses. However, if you want to specify the columns to retrieve, you can code an array of strings that contains the names of the columns that you want to retrieve. Then, you can supply this array as the second argument. Or, if you want to group or sort the rows that are retrieved, you can supply a string for the appropriate clause. For example, to sort the rows, you can specify a string for the ORDER BY clause.

The TaskListDB class

```java
public ArrayList<Task> getTasks(String listName) {
    String where =
            TASK_LIST_ID + "= ? AND " +
            TASK_HIDDEN + "!='1'";
    int listID = getList(listName).getId();
    String[] whereArgs = { Integer.toString(listID) };

    this.openReadableDB();
    Cursor cursor = db.query(TASK_TABLE, null,
            where, whereArgs,
            null, null, null);
    ArrayList<Task> tasks = new ArrayList<Task>();
    while (cursor.moveToNext()) {
        tasks.add(getTaskFromCursor(cursor));
    }
    if (cursor != null)
        cursor.close();
    this.closeDB();

    return tasks;
}

public Task getTask(int id) {
    String where = TASK_ID + "= ?";
    String[] whereArgs = { Integer.toString(id) };

    this.openReadableDB();
    Cursor cursor = db.query(TASK_TABLE,
            null, where, whereArgs, null, null, null);
    cursor.moveToFirst();
    Task task = getTaskFromCursor(cursor);
    if (cursor != null)
        cursor.close();
    this.closeDB();

    return task;
}
```

A method of the SQLiteDatabase class

Method	Description
query(table, columns, where, whereArgs, groupBy, having, orderBy)	Queries the specified table and returns a cursor for the result set. All arguments are a string or an array of strings. If you don't want to use an argument, you can specify a null value.

Some methods of the Cursor class

Method	Description
moveToFirst()	Moves to the first row in the cursor.
moveToNext()	Moves to the next row in the cursor.
close()	Closes the cursor.

Figure 13-9 How to retrieve rows from a table

The sixth statement declares an ArrayList of Task objects. Then a while loop uses the moveToNext method of the Cursor object to loop through all rows in the cursor. Within the while loop, the statement uses the getTaskFromCursor method shown in figure 13-10 to create a Task object from the current row and to add that task to the array of tasks.

After the while loop, the last statement in the method cleans up resources by closing the cursor as well as the connection to the database. Then, it returns the array of tasks.

How to retrieve a single row from a table

The second method in figure 13-9 shows how to code the public getTask method. This method works similarly to the getTasks method, but it retrieves a single row. To start, the declaration for the getTask method indicates that it returns a Task object for the task with the specified ID.

Within the method, the first two statements define the condition for the WHERE clause and its arguments. Once again, the WHERE clause only contains a single parameter. This time, the parameter is for the task ID instead of the list ID.

After defining the arguments for the WHERE clause, this method opens a read-only connection and uses the query method to retrieve a cursor for a single row. Then, it uses the moveToFirst method to move the cursor to the first (and only) row in the cursor. Next, it uses the getTaskFromCursor method to get the Task object from the row. Finally, it closes the cursor and the database connection and returns the Task object.

How to get data from a cursor

Figure 13-10 shows the getTaskFromCursor method that's used by the getTasks and getTask methods shown earlier. This method accepts a Cursor object and returns a Task object.

The getTaskFromCursor method begins by checking whether the cursor is null or doesn't contain any rows. If so, it returns a null value and exits the method. Otherwise, it attempts to create a Task object from the current row.

To do that, this code creates a new Task object and uses the getXxx methods to get the data from the current row and passes that data to the constructor of the Task class. For example, this code uses the getInt method to get the value for the task ID, and it uses the getString method to get the value for the task name. Both of these methods accept a parameter for the column index, not the column name. As a result, this code uses the constants for column indexes, not the column names.

If the method successfully creates the Task object, the try clause returns that object and exits. Otherwise, the catch clause returns a null value.

The TaskListDB class   **Page 6**

```
private static Task getTaskFromCursor(Cursor cursor) {
    if (cursor == null || cursor.getCount() == 0){
        return null;
    }
    else {
        try {
            Task task = new Task(
                cursor.getInt(TASK_ID_COL),
                cursor.getInt(TASK_LIST_ID_COL),
                cursor.getString(TASK_NAME_COL),
                cursor.getString(TASK_NOTES_COL),
                cursor.getString(TASK_COMPLETED_COL),
                cursor.getString(TASK_HIDDEN_COL));
            return task;
        }
        catch(Exception e) {
            return null;
        }
    }
}
```

More methods of the Cursor class

Method	Description
getInt(columnIndex)	Gets an int value from the cursor.
getDouble(columnIndex)	Gets a double value from the cursor.
getString(columnIndex)	Gets a string from the cursor.

Description

- Once you have moved to a row in a cursor, you can use the getXxx methods to get data from that row.

Figure 13-10   How to get data from a cursor

How to insert, update, and delete rows

Figure 13-11 shows how to code the public methods that insert, update, and delete rows. Although these methods work with the Task table, you can use similar techniques to work with the List table and other tables.

The insertTask method provides a way to insert a row into the database. This method accepts a Task paramter and returns a long value for the ID of the newly inserted row.

Within the method, the first statement creates a ContentValues object that's used to store the column names and their corresponding values. Then, the next five statements use the column constants to specify the column names, and they use the methods of the Task object to get the corresponding values. Since the task_id column is defined with the AUTO_INCREMENT attribute, the database automatically generates a value for this column. As a result, you don't have to supply one.

After creating the ContentValues object, this code calls the openWriteableDB method defined earlier in this class to open a read-write connection to the database. Then, it calls the insert method from the database object to insert the row and to store the ID of the newly inserted row. Here, the first argument specifies the name of the table, and the third argument specifies the column names and values. Since the second argument isn't needed, this code passes a null value for this argument.

After inserting the row, this method calls the closeDB method defined earlier in this class to close the connection to the database. Finally, it returns the ID of the newly inserted row.

The updateTask method provides a way to update an existing row in the database. This method accepts a Task parameter and returns an int value for the count of updated rows. If the update is successful, this should return a value of 1. Otherwise, it should return a value of 0.

Like the insertTask method, the updateTask method begins by creating a ContentValues object to store the column names and their corresponding values. Then, it creates a string for the WHERE clause and an array of strings that contains its argument. This works similarly to the WHERE clause for the getTask method.

After creating the WHERE clause and its arguments, this code opens a read-write connection to the database. Then, it calls the update method from the database connection to update the row and return the count of updated rows. Here, the first argument specifies the name of the table, the second argument specifies columns and their values, the third argument specifies the condition for the WHERE clause, and the fourth argument specifies the argument for the WHERE clause. Finally, this code closes the database connection and returns the count of updated rows.

The TaskListDB class

```
    public long insertTask(Task task) {
        ContentValues cv = new ContentValues();
        cv.put(TASK_LIST_ID, task.getListId());
        cv.put(TASK_NAME, task.getName());
        cv.put(TASK_NOTES, task.getDescription());
        cv.put(TASK_COMPLETED, task.getCompleted());
        cv.put(TASK_HIDDEN, task.getHidden());

        this.openWriteableDB();
        long rowID = db.insert(TASK_TABLE, null, cv);
        this.closeDB();

        return rowID;
    }

    public int updateTask(Task task) {
        ContentValues cv = new ContentValues();
        cv.put(TASK_LIST_ID, task.getListId());
        cv.put(TASK_NAME, task.getName());
        cv.put(TASK_NOTES, task.getDescription());
        cv.put(TASK_COMPLETED, task.getCompleted());
        cv.put(TASK_HIDDEN, task.getHidden());

        String where = TASK_ID + "= ?";
        String[] whereArgs = { String.valueOf(task.getId()) };

        this.openWriteableDB();
        int rowCount = db.update(TASK_TABLE, cv, where, whereArgs);
        this.closeDB();

        return rowCount;
    }
```

Two methods of the SQLiteDatabase class

Method	Description
insert(table, columns, values)	Inserts a row into the table and returns a long value for the ID of the newly inserted row.
update(table, values, where, whereArgs)	Updates one or more rows in a table and returns an int value for the number of rows that were successfully updated.

One method of the ContentValues class

Method	Description
put(column, value)	Specifies the column name and value.

Description

- You can use the methods of a SQLiteDatabase object to insert or update a row. For these operations, you can store the column names and their corresponding values in a ContentValues object.

Figure 13-11   How to insert, update, and delete rows (part 1 of 2)

The deleteTask method provides a way to delete a row from the database. This method accepts the ID of the task to delete and returns an int value for the count of deleted rows. If the deletion is successful, this should return a value of 1. Otherwise, it should return a value of 0.

Within the method, the first two statements define the condition and arguments for the WHERE clause. Then, the third statement opens a read-write connection to the database.

The fourth statement calls the delete method from the database connection to delete the row and return the count of deleted rows. Here, the first argument specifies the name of the table, the second argument specifies the condition for the WHERE clause, and the third argument specifies the argument for the WHERE clause. Finally, this code closes the database connection and returns the count of deleted rows.

The TaskListDB class   **Page 8**

```
    public int deleteTask(long id) {
        String where = TASK_ID + "= ?";
        String[] whereArgs = { String.valueOf(id) };

        this.openWriteableDB();
        int rowCount = db.delete(TASK_TABLE, where, whereArgs);
        this.closeDB();

        return rowCount;
    }
}
```

Another method of the SQLiteDatabase class

Method	Description
delete(table, where, whereArgs)	Deletes one or more rows in a table and returns an int value for the number of rows that were deleted.

Description

- You can use the delete method of a SQLiteDatabase object to delete one or more rows.

Figure 13-11   How to insert, update, and delete rows (part 2 of 2)

How to test the database

Once you write a database class, you need to test it to make sure it works correctly. To do that, you can write code that creates an object from the database class. Then, you can call public methods from that object to make sure they work correctly. As you test and modify the database class, you may want to delete the database files from a device, or you may want to use a graphical tool such as DB Browser for SQLite to view a database.

How to test the database class

Figure 13-12 shows how to test the TaskListDB class described in the previous figures. You can add code like this to an activity. Then, you can run that activity to display the results on the user interface. Or, if you prefer, you can add code that uses the Log class to display similar messages in the LogCat view.

The code in this figure begins by creating a TaskListDB object named db. Then, it creates a StringBuilder object named sb. This object is used to build the string that's displayed on the user interface.

The code that tests the insertTask method begins by creating a Task object that has some test data in it. This test data includes a list ID of 1. Then, this code calls the insertTask method from the db object and stores the long value that's returned in a variable named insertID. Next, this code checks whether the value of the insert ID is greater than 0. If so, the insert was successful, and the code appends an appropriate message to the StringBuilder object.

The code that tests the updateTask method begins by setting the task ID in the Task object. That way, the Task object contains all of the columns necessary for an update operation. Next, this code changes the name that's stored in the Task object, calls the updateTask method to update the row, and stores the count of updated rows in a variable named updateCount. Finally, this code checks whether the count of updated rows is equal to 1. If so, the update was successful, and the code appends an appropriate message to the StringBuilder object.

The code that tests the deleteTask method begins by calling the deleteTask method from the db object. This code passes the ID of the newly inserted row to this method and store the count of deleted rows in a variable named deleteCount. Then, this code checks whether the count of deleted rows is equal to 1. If so, the deletion was successful, and the code appends an appropriate message to the StringBuilder object.

The code that tests the getTasks method begins by getting all tasks for the list with a name of "Personal". Then, this code loops through all of the tasks and appends the task ID and name for each task to the StringBuilder object.

After this code finishes building the string, it displays that string on a TextView widget. To do that, this code gets a reference to the widget. Then, it calls the setText method of that widget to display the string that's stored in the StringBuilder object.

Code that tests the database class

```
// get db and StringBuilder objects
TaskListDB db = new TaskListDB(this);
StringBuilder sb = new StringBuilder();

// insert a task
Task task = new Task(1, "Make dentist appointment", "", "0", "0");
long insertId = db.insertTask(task);
if (insertId > 0) {
    sb.append("Row inserted! Insert Id: " + insertId + "\n");
}

// update a task
task.setId((int) insertId);
task.setName("Update test");
int updateCount = db.updateTask(task);
if (updateCount == 1) {
    sb.append("Task updated! Update count: " + updateCount + "\n");
}

// delete a task
int deleteCount = db.deleteTask(insertId);
if (deleteCount == 1) {
    sb.append("Task deleted! Delete count: " + deleteCount + "\n\n");
}

// display all tasks (id + name)
ArrayList<Task> tasks = db.getTasks("Personal");
for (Task t : tasks) {
    sb.append(t.getId() + "|" + t.getName() + "\n");
}

// display string on UI
TextView taskListTextView = (TextView)
        findViewById (R.id.taskListTextView);
taskListTextView.setText(sb.toString());
```

Sample text displayed on the user interface

```
Task inserted! Insert Id: 3
Task updated! Update count: 1
Task deleted! Delete count: 1

1|Pay bills
2|Get hair cut
```

Description

- To test the database class, you can write code that creates an instance of the database class and uses its public methods. Then, you can display data on the user interface. Alternately, you can display the data in the LogCat view.

- If you make changes to the structure of the database, you can increment the version number for the database.

Figure 13-12   How to test the database class

As you test an app, you may decide that you need to change the structure of the database. To do that, you can modify the code in the onUpgrade method of the DBHelper class. However, to get Android to execute this method on each device, you need to increment the version number for the database. Then, the next time you run the app, Android calls the onUpgrade method of the helper class and upgrades the database to the most current version.

How to clear test data from a device

If you don't want to continue to increment the version number for a database, you can delete all data for the app from your test devices. The easiest way to do this is to uninstall the app on your device. To do that, you can long-click on the app and drag it onto the Uninstall icon that appears. This should uninstall the app and delete its data. Then, when you run the app, Android executes the onCreate method of the DBHelper class since no database exists on the device. This allows you to keep the database version at 1 or to reset the database version to 1.

When you use this technique, keep in mind that it also deletes all other data for the app including any shared preferences. As a result, if you want to keep shared preferences or any other test data that has been stored on the device, you shouldn't use this technique.

How to work with the database file

As you test a database, you may want to view the database file that's created on a device. To do that, you can use the standalone Android Device Monitor tool shown figure 13-13 to select an emulator or a physical device that has been rooted. Then, you can use the File Explorer tab shown in this figure to navigate to the databases directory for your app. In this figure, for example, the File Explorer tab shows the databases directory for the Task List app. This directory shows the tasklist.db file that's used by the TaskListDB class. For security reasons, this technique doesn't work for a physical device that hasn't been rooted. In that case, the database files are stored in a private area that the Android Device Monitor can't view.

If you want to copy the database file from the device to your computer, you can select the file and click the "Pull a file" button. Then, you can use the resulting dialog box to specify the directory on your computer where you'd like to store the file. This can be useful if you want to use an app like the one shown in the next figure to work with a database.

If you want to copy a file from your computer to a device, you can select the directory on the device and click the "Push a file" button. Then, you can use the resulting dialog box to specify the file that you want to put on the device. This can be useful if you have a SQLite database on your computer that you want to use for testing and you need to copy it onto a device.

The file for the Task List database

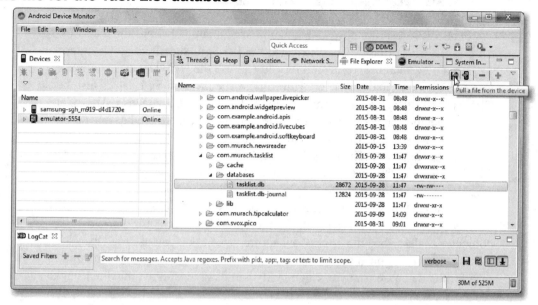

How to view the file for a database

1. Start Android Studio.
2. Open the standalone Android Device Monitor tool by selecting the Tools→Android→Android Device Monitor item.
3. In the Devices view, select the device.
4. In the main window, click on the File Explorer tab.
5. Navigate to the databases directory for the app.

The directory and file for the Task List database

```
/data/data/com.murach.tasklist/databases/tasklist.db
```

Description

- To copy a file from the device to your computer, select the file and click on the "Pull a file" button. Then, use the resulting dialog box to locate the file that you want to get from the device.

- To copy a file from your computer to a device, select the directory on the device and click the "Push a file" button. Then, use the resulting dialog box to specify the file that you want to put on the device.

Figure 13-13   How to work with the database file

How to use DB Browser for SQLite

Figure 13-14 shows how to install and use an app known as the DB Browser for SQLite. This often makes it easier to test the SQL statements for your app.

For a new database, for example, you can use the New Database button to create the new database. Then, you can use the Execute SQL tab to test the CREATE TABLE statements that create the tables for the database. After that, you can use the Database Structure tab to verify that each table has been created.

Similarly, you can use this tab to test the INSERT statements that insert test data into the database. Then, you can use the Browse Data tab to view this data. Once you're sure that your SQL statements are working correctly, you can use Java statements to build the SQL statements in the database class.

If you want to work with an existing database that's on a device or emulator, you need to start by pulling the file from the device or emulator onto your computer as described in the previous figure. Then, you can use the Open Database button to open that database. After that, you can view the structure or data for the database, and you can quickly and easily test SQL statements against that database.

The DB Browser for SQLite app

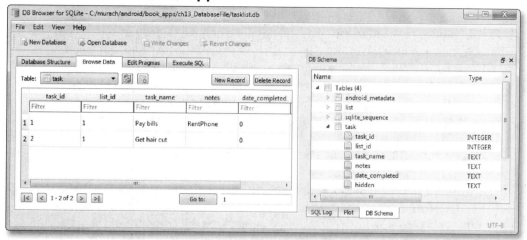

The URL for downloading DB Browser for SQLite

```
http://sqlitebrowser.org/
```

Description

- You can use a tool like the DB Browser for SQLite to work with a database that's on your computer (not on a device or emulator).

- To install the DB Browser for SQLite, you can search the Internet and follow the instructions to download and install it.

- To create a new database, click the New Database button in the toolbar and respond to the resulting dialog boxes.

- To open an existing database, click the Open Database button in the toolbar and respond to the resulting dialog boxes.

- To view the structure of the database, use the Database Structure tab.

- To view or edit the data for a table, use the Browse Data tab.

- To execute SQL statements, use the Execute SQL tab.

Figure 13-14   How to use DB Browser for SQLite

Perspective

The skills presented in this chapter should be enough to get you started with SQLite databases. However, there is much more to learn. For example, the SQLiteDatabase class provides many more methods for querying a database and for inserting, updating, and deleting data. In addition, it provides methods for other features such as working with transactions. To get an idea of the methods that are available from this class, you can review the documentation for the SQLiteDatabase class.

Another issue to consider is that it often makes sense to store the data for a mobile app in the cloud. That way, if you have multiple devices, they can all work with the same data and that data should always be current. If you store the data for the Tasks List application in the cloud, for example, you should be able to update tasks from any of your devices (computer, tablet, phone, and so on) and the most current tasks should always be available to these devices.

To get this to work correctly, you usually need to store data on the local device. That way, the data is available even if a connection to the cloud isn't available. However, when you are able to connect to the cloud, the app should be able to synch the data on your devices with the data that's stored in the cloud. There are several Task List apps available from the Android market, including GTasks, that use this approach.

Although you learned how to work with a SQLite database in this chapter, you haven't yet learned how to display that data on a user interface. In the next chapter, you'll learn how to use tabs and a custom adapter to do that.

Terms

relational database management system (RDBMS)
business objects
plain old Java objects (POJOs)

Summary

- SQLite only supports three data types (String, long, and double), and is weakly-typed, meaning a column in a SQLite database does not reject a value of an incorrect data type

- It's generally considered a good practice to define constants for the name and version number of the database, and to define constants for the table names, column names, and column numbers.

- The SQLiteOpenHelper class provides the onCreate and onUpgrade methods that are used to create and upgrade the database.

- Within a database class, you can use private methods that make it easy to open and close a connection to the database.

- You can use the methods of a SQLiteDatabase object to insert, update, and delete rows.

- To test the database class, you can write code that creates an instance of the database class and uses its public methods. Then, you can display data on the user interface. Alternately, you can display the data in the LogCat view.

- If you make changes to the structure of the database, you can increment the version number for the database.

- You can uninstall an app from a device to delete its database. This also deletes all other data for the app including any shared preferences.

- You can use a tool like the DB Browser for SQLite to work with a database file that's on your computer (not on a device or emulator).

Exercise 13-1   Review the Task List app and use its database class

In this exercise, you'll review the code that creates and uses the Task List database. Then, you'll modify the code that uses the Task List app.

1. Open the project named ch13_ex1_TaskList.

2. Open the Task, List, TaskListDB, and TaskListActivity classes and review their code.

3. Run the app. This should display a series of messages on the user interface for the activity.

4. Switch to the TaskListActivity class and add code that inserts a second task with a list ID of 1 and a name of "Take car in for oil change".

5. Run the app. This should display a similar series of messages as before, but it should also display the task that you inserted in step 4.

6. Change the orientation of the activity one or more times. This should run the test code again, which should insert the task from step 4 again. This shows that the database doesn't use the UNIQUE attribute to require that all names are unique.

7. Add code that deletes the rows that you added in steps 4-6. To do that, you can get the IDs for the tasks by viewing the messages on the Task List activity. Then, you can pass these IDs to the deleteTask method.

8. Run the app. This time the tasks inserted in steps 4-6 should be deleted, but the task inserted by this step should still be in the database.

Exercise 13-2   Use the DB Browser for SQLite

In this exercise, you'll use the DB Browser for SQLite app to work with an existing Task List database.

1.   Open the project named ch13_ex2_TaskList.

2.   Run the app in an emulator. This should display a series of messages on the user interface for the activity.

3.   If necessary, install DB Browser for SQLite on your computer.

4.   Start the DB Browser for SQLite app.

5.   Use the Android Device Monitor tool to pull the tasklist.db file for the database from the emulator to the ch13_ex2_TaskList directory on your computer.

6.   Use the DB Browser for SQLite to open this tasklist.db file.

7.   View the database structure for the List and Task tables. You should be able to view all of the column names and their data types.

8.   Browse the data that's stored in the List and Task tables. This should help you understand how these tables work.

9.   Execute the following SQL statement:
```
INSERT INTO task VALUES (1, 'Test!', '', '0', '0')
```
This should display an error that indicates that the Task table requires 6 columns.

10.   Execute the following SQL statement:
```
INSERT INTO task VALUES (100, 1, 'Test!', '', '0', '0')
```
This should not display any errors.

11.   View the data for the Task table to make sure the "Test!" row was inserted.

Exercise 13-3   Modify the database class for the Task List app

In this exercise, you'll modify the database class for the Task List app.

Modify the data types that are used by the database

1.   Open the project named ch13_ex3_TaskList.

2.   Run the app. This should display a series of messages on the user interface for the activity.

3.   Open the TaskListDB class.

4.   Modify the CREATE TABLE statement for the Task table so it stores the date_completed and hidden columns using the INTEGER types instead of using the TEXT type.

5. Modify the INSERT statements for the Task table so they don't include single quotes around the last two columns. (You don't need to use quotes for INTEGER values, only for TEXT values.)

6. In the getTasks method, modify the variable named where so it doesn't include quotes around the 1. (They are no longer needed.)

7. Open the Task class and modify it so it uses the long type for the completedDate instance variable and the int type for the hidden variable. To get this to work, you can change the data type for the TRUE and FALSE constants to the int type. In addition, you need to modify the constructors, and some of the methods of this class, so they use the correct data types.

8. Switch to the TaskListDB class. Then, modify the getTaskFromCursor method so it uses the getInt method to get values from the date_completed and hidden columns.

9. Open the TaskListActivity class. Then, modify the code that creates that Task object so that it doesn't include quotes around the completedDate and hidden values.

10. Increment the database version number to 2.

11. Run the app. It should work the same as before. However, the app is now storing its data with different types. To verify this, you can check the LogCat view. It should display a message that indicates that the database has upgraded from version 1 to 2.

12. Set the database version number to 1 and run the app again. The app should crash because the database can't be downgraded from 2 to 1.

13. Uninstall the app on the device. This should also delete the database file for the app.

14. Run the app again. This time, the app should run, and the database version number should be reset to 1.

Add a private method to the TaskListDB class

15. Switch to the TaskListDB class and scroll down to the closeDB method. After this method, add a private method named closeCursor that accepts a Cursor object. This method should check whether the Cursor object exists (is not null). If so, this method should close the Cursor object.

16. Modify the getLists, getList, getTasks, and getTask methods so they use the closeCursor method.

17. Run the app again. It should work the same as before, but you have made a minor improvement to the database class.

14

How to work with tabs and custom adapters

In the last chapter, you learned how to write the database and business classes for the Task List app. Now, you'll learn how to use those classes to display data on the user interface for the Task List app. To do that, you'll learn how to use tabs and custom adapters.

How to use tabs

The Task List app displays the name of each list on a tab. To do that, this app uses the TabManager class that was originally made available by Google. This class provides a way to use a fragment to display the content for each tab.

How to add the TabManager class to your project

The TabManager class isn't available as part of the Android API. As a result, before you can use this class, you must add its library to your project. To do that, you can get the TabManager.jar file for the library by downloading the source code for this book. Then, you can copy the TabManager.jar file into the libs directory for your project. Once you have copied the TabManager.jar file into the libs directory, you can open the Gradle script for the app and add a dependency to it like the first one shown in figure 14-1.

In addition, the TabManager class uses classes from the Android v4 support library. This library provides support for the ActivityFragment and Fragment classes used in this chapter. As a result, you also need to add a dependency for the android-support-v4.jar file like the second one show in this figure.

The layout for an activity that displays tabs

Figure 14-1 also shows the layout for an activity that displays tabs. If you want to display tabs across the top of an activity, you can add code like the code shown in this figure. This code is boilerplate code that you can copy and paste from the downloadable source code for this book into your activity.

The TabHost widget defines a container for tabs. This container has two sets of children: (1) a TabWidget object that can contain one or more tabs (2) a FrameLayout object that can display the content for the selected tab. The code in this figure provides a TabHost object named tabhost, a TabWidget object named tabs, and two FrameLayout objects named tabcontent and realtabcontent. Here, the first FrameLayout object is a placeholder that's required by the TabHost object. However, it isn't used by the app. That's why both its height and width have been set to 0dp. The second FrameLayout object, on the other hand, has a height of 0dp, but a weight of 1. As a result, it takes all of the vertical space between the TabWidget object and the bottom of the screen.

Since the tabs aren't added to the TabWidget object until runtime, the graphical layout editor doesn't show the tabs. Instead, it shows an area for the tabs with a label that says "TAB LABEL". Below that, the graphical editor displays a white area for the content of the selected tab.

The dependencies needed to work with a tab widget

```
compile files('libs/TabManager.jar')
compile 'com.android.support:support-v4:23.0.0'
```

A tab widget displayed in the graphical editor

The XML for the layout

```xml
<TabHost xmlns:android="http://schemas.android.com/apk/res/android"
    android:id="@android:id/tabhost"
    android:layout_width="match_parent"
    android:layout_height="match_parent">

    <LinearLayout android:orientation="vertical"
        android:layout_width="match_parent"
        android:layout_height="match_parent">

        <TabWidget android:id="@android:id/tabs"
            android:orientation="horizontal"
            android:layout_width="match_parent"
            android:layout_height="wrap_content"
            android:layout_weight="0"/>

        <FrameLayout android:id="@android:id/tabcontent"
            android:layout_width="0dp"
            android:layout_height="0dp"
            android:layout_weight="0"/>

        <FrameLayout android:id="@+android:id/realtabcontent"
            android:layout_width="match_parent"
            android:layout_height="0dp"
            android:layout_weight="1"/>
    </LinearLayout>
</TabHost>
```

Description

- To get the TabManager.jar file, you can download the source code for this book. This .jar file is stored in the book_apps\ch14_TaskList\libs directory.

- To add the TabManager library to your project, copy the TabManager.jar file into the libs directory for your project.

Figure 14-1   The layout for an activity that displays tabs

The class for an activity that displays tabs

Figure 14-2 shows the class for an activity that displays tabs. This class begins by importing the TabManager class that's stored in the TabManager library. Then, it imports the FragmentActivity class that's included with the Android v4 support library. In addition, it imports the TabHost and TabSpec classes that are used to work with tabs.

The TaskListActivity class begins by extending the FragmentActivity class. As mentioned earlier, this is generally considered the best way to display tabs since it allows you to display the content for a tab as a fragment.

Within the TaskListActivity class, the onCreate method gets a reference to the TabHost object. Then, it calls the setup method from that object. When you load the TabHost object by calling the findViewById method, you must call this method before you add tabs to the TabHost object.

After setting up the TabHost object, this code creates a new TabManager object. To do that, this code passes three arguments to the constructor of the TabManager class: (1) the context, (2) the TabHost object, and (3) the ID of the FrameLayout widget that's used to display the content for the current tab. This ID corresponds with the second FrameLayout widget shown in the previous figure.

After creating the TabManager object, this code creates a new object from the TaskListDB class described in the previous chapter. Then, it calls the getLists method from this object to get an ArrayList of List objects.

If this ArrayList object contains one or more List objects, this code loops through all of the List objects. Then, it displays the name of each list on a tab. To do that, this code creates a TabSpec object that represents the specification for the tab, and it adds that tab to the tabs.

Within the loop, the first statement uses the newTabSpec method of the TabHost class to create a TabSpec object and set its tag to the name of the list. Then, the second statement uses the setIndicator method to set the label that's used for the tab to the name of the list. Next, the third statement uses the addTab method of the TabManager class to add the tab and its content to the TabHost widget. To do that, this statement passes the TabSpec object and the name of the class for the fragment that displays the content of the tab. In the next figure, you'll see how this class works.

After the while loop, the onCreate method finishes by checking whether the Bundle object that's passed to it is not null. If so, this code gets the name of the current tab's tag from the Bundle object. Then, it sets the current tab to that tag.

The onSaveInstanceState method begins by passing the Bundle object to the onSaveInstanceState superclass. Then, it saves the name of the current tab's tag in the Bundle object. This makes the tag for the current tab available from the Bundle object of the onCreate method.

The class for an activity that displays tabs

```
package com.murach.tasklist;

import java.util.ArrayList;

import com.google.tabmanager.TabManager;

import android.os.Bundle;
import android.support.v4.app.FragmentActivity;
import android.widget.TabHost;
import android.widget.TabHost.TabSpec;

public class TaskListActivity extends FragmentActivity {
    TabHost tabHost;
    TabManager tabManager;

    @Override
    protected void onCreate(Bundle savedInstanceState) {
        super.onCreate(savedInstanceState);
        setContentView(R.layout.activity_task_list);

        // get tab manager
        tabHost = (TabHost) findViewById(android.R.id.tabhost);
        tabHost.setup();
        tabManager = new TabManager(this, tabHost, R.id.realtabcontent);

        // get the lists from the database
        TaskListDB db = new TaskListDB(this);
        ArrayList<List> lists = db.getLists();

        // add a tab for each list
        if (lists != null && ! lists.isEmpty()) {
            for (List list : lists) {
                TabSpec tabSpec = tabHost.newTabSpec(list.getName());
                tabSpec.setIndicator(list.getName());
                tabManager.addTab(tabSpec, TaskListFragment.class, null);
            }
        }

        // set current tab to the last tab opened
        if (savedInstanceState != null) {
            tabHost.setCurrentTabByTag(savedInstanceState.getString("tab"));
        }
    }

    @Override
    protected void onSaveInstanceState(Bundle outState) {
        super.onSaveInstanceState(outState);
        outState.putString("tab", tabHost.getCurrentTabTag());
    }
}
```

Description

- Within the class for an activity, you can use a TabManager object to add one or more tabs. For example, you can add one tab for each list stored in a database.

Figure 14-2   The class for an activity that displays tabs

The class for a fragment
that displays tab content

Figure 14-3 shows the class for a fragment that displays the content for a tab. This class begins by extending the Fragment class that's available from the Android v4 support library. For the most part, this Fragment class works like the Fragment class described in chapter 9.

The onCreate method begins by inflating the layout for the fragment. Although it isn't shown here, the layout for this class uses a vertical linear layout that contains a single TextView widget named taskTextView. The second statement gets a reference to this TextView widget.

The third statement gets a reference the TabHost object. To do that, this statement calls the getParent method of the container object twice. The first call gets the LinearLayout widget that's used by the fragment, and the second call gets the TabHost widget that's used by the activity.

The fourth statement uses the getCurrentTabTag method of the TabHost object to get the name of the current tab. Then, the fifth statement calls the refreshTaskList method. This method displays a message on the TextView widget that includes the name of the current tab. For example, if the user clicks on the Personal tab, the content for that tab displays a message that says, "This is the Personal tab."

The onResume method begins by calling the onResume method of the superclass. Then, it calls the refreshTaskList method to display a message on the TextView widget.

Now that you can display a simple message as the content for a tab, you're just a step or two away from displaying more complex data. For example, you might want to look up all of the tasks for the current tab and display those tasks as the content for that tab. In the next few figures, you'll learn how to do that.

The class for a fragment that displays tab content

```
package com.murach.tasklist;

import android.os.Bundle;
import android.support.v4.app.Fragment;
import android.view.LayoutInflater;
import android.view.View;
import android.view.ViewGroup;
import android.widget.TextView;
import android.widget.TabHost;

public class TaskListFragment extends Fragment {

    private TextView taskTextView;
    private String currentTabTag;

    @Override
    public View onCreateView(LayoutInflater inflater, ViewGroup container,
            Bundle savedInstanceState) {

        // inflate the layout for this fragment
        View view = inflater.inflate(R.layout.fragment_task_list,
            container, false);

        // get references to widgets
        taskTextView = (TextView) view.findViewById (R.id.taskTextView);

        // get the current tab
        TabHost tabHost = (TabHost) container.getParent().getParent();
        currentTabTag = tabHost.getCurrentTabTag();

        // refresh the task list view
        refreshTaskList();

        // return the view
        return view;
    }

    public void refreshTaskList() {
        String text = "This is the " + currentTabTag + " list.";
        taskTextView.setText(text);
    }

    @Override
    public void onResume() {
        super.onResume();
        refreshTaskList();
    }
}
```

Description

- Within the class for a fragment, you can use a TabHost class to get the tag for the current tab. You can use this tag to display the appropriate data for the current tab.

Figure 14-3   The class for a fragment that displays tab content

How to use a custom adapter

After the Task List app reads a list of tasks from the database, it needs to display those tasks on the user interface. To do that, you could use a simple adapter to display those tasks in a ListView widget as shown in the previous chapters. However, the Task List app includes a check box in the ListView widget. To get this check box to work correctly, you can use a custom adapter as shown in the next few figures.

A layout for a list view item

Figure 14-4 shows the layout for an item in the ListView widget that's used to display a list of tasks. This layout displays a CheckBox widget that indicates whether the task has been completed, a TextView widget for the name of the task, and a TextView widget for any notes for the task. If you've read chapter 5, you shouldn't have any trouble understanding how this layout works. However, there are a couple of attributes that haven't been presented yet.

For the CheckBox widget, the button attribute specifies a custom layout for the button. This layout is stored in a file named btn_check.xml in the res\drawable directory of the project, and it specifies the .png files for the checked and unchecked versions of the check box. Although this code isn't necessary, it provides a way to control the appearance of the check box. By default, the CheckBox widget was too light to see on some devices, so I specified some slightly darker .png files for these check boxes. These .png files are stored in the appropriate res\drawable-xxx directories for the project.

For the TextView widgets, the textColor attribute specifies black for the color of the text. Again, this code isn't necessary. However, the default color was light gray on some devices, and I think black is easier to read.

The listview_task layout

The XML for the layout

```xml
<?xml version="1.0" encoding="utf-8"?>
<RelativeLayout xmlns:android="http://schemas.android.com/apk/res/android"
    android:id="@+id/relativeLayoutTaskItem"
    android:layout_width="match_parent"
    android:layout_height="match_parent" >

    <CheckBox
        android:id="@+id/completedCheckBox"
        android:layout_width="wrap_content"
        android:layout_height="wrap_content"
        android:layout_alignParentLeft="true"
        android:layout_alignParentTop="true"
        android:layout_margin="5dp"
        android:button="@drawable/btn_check" />

    <TextView
        android:id="@+id/nameTextView"
        android:layout_width="wrap_content"
        android:layout_height="wrap_content"
        android:layout_alignBaseline="@+id/completedCheckBox"
        android:layout_alignBottom="@+id/completedCheckBox"
        android:layout_toRightOf="@+id/completedCheckBox"
        android:text="@string/task_name"
        android:textColor="@android:color/black"
        android:textSize="18sp"
        android:textStyle="bold" />

    <TextView
        android:id="@+id/notesTextView"
        android:layout_width="wrap_content"
        android:layout_height="wrap_content"
        android:layout_below="@+id/nameTextView"
        android:layout_marginTop="-5dp"
        android:layout_toRightOf="@+id/completedCheckBox"
        android:paddingBottom="7dp"
        android:text="@string/task_notes"
        android:textColor="@android:color/black"/>

</RelativeLayout>
```

Figure 14-4   The layout for a list view item

A class that extends the layout for a list view item

When the layout for a ListView widget contains complex widgets such as a check box, you code a class that extends the layout as shown in figure 14-5. This allows you to display the widgets correctly and to handle events that occur upon them.

The TaskLayout class begins by extending the RelativeLayout class and implementing the OnClickListener interface. The first constructor of this class supports some Android tools. It isn't required for this app to work, but it's considered a good practice to include it.

The second constructor accepts two parameters: a Context object and a Task object. Within this constructor, the first statement passes the Context object to the constructor of the superclass. Then, the next two statements set the context for the current object and create a TaskListDB object.

The fourth statement uses the getSystemService method of the Context object to get a LayoutInflater object. Then, the fifth statement uses that object to inflate the layout defined in the previous figure. After that, the next three statements get references to the widgets on the layout.

The eighth and ninth statements set the listeners for the widgets. Here, the eighth statement sets the listener for the check box, and the ninth statement sets the listener for the entire layout.

A class that extends the layout   **Page 1**

```
package com.murach.tasklist;

import android.content.Context;
import android.content.Intent;
import android.view.LayoutInflater;
import android.view.View;
import android.view.View.OnClickListener;
import android.widget.CheckBox;
import android.widget.RelativeLayout;
import android.widget.TextView;

public class TaskLayout extends RelativeLayout implements OnClickListener {

    private CheckBox completedCheckBox;
    private TextView nameTextView;
    private TextView notesTextView;

    private Task task;
    private TaskListDB db;
    private Context context;

    public TaskLayout(Context context) {    // used by Android tools
        super(context);
    }

    public TaskLayout(Context context, Task t) {
        super(context);

        // set context and get db object
        this.context = context;
        db = new TaskListDB(context);

        // inflate the layout
        LayoutInflater inflater = (LayoutInflater)
                context.getSystemService(Context.LAYOUT_INFLATER_SERVICE);
        inflater.inflate(R.layout.listview_task, this, true);

        // get references to widgets
        completedCheckBox = (CheckBox) findViewById(R.id.completedCheckBox);
        nameTextView = (TextView) findViewById(R.id.nameTextView);
        notesTextView = (TextView) findViewById(R.id.notesTextView);

        // set listeners
        completedCheckBox.setOnClickListener(this);
        this.setOnClickListener(this);
```

Figure 14-5   A class that extends the layout (part 1 of 2)

The last statement in the constructor calls the setTask method that's defined later in this class. This method sets the Task object as an instance variable, and it displays the data in the Task object on the user interface.

The setTask method begins by setting the instance variable for the Task object equal to the Task parameter. Then, it updates the user interface. First, it sets the name of the task on the appropriate TextView widget. Then, it checks whether the notes for the task are equal to an empty string. If so, it hides the TextView widget for the notes. Otherwise, it sets the notes for the task on the appropriate TextView widget.

This code continues by checking whether the milliseconds for the completed date are greater than zero. If so, the task has been marked as completed. As a result, this code checks the CheckBox widget. Otherwise, this code removes the check from the CheckBox widget.

The onClick method begins by using a switch statement to check whether the CheckBox widget was clicked. If so, this code checks whether the CheckBox widget was checked. If so, this code sets the completed date for the task to the number of milliseconds for the current date. Otherwise, this code sets the completed date for the task to zero. This indicates that the task has not been completed. Either way, this code uses the database object to update the current task in the database.

If the CheckBox widget wasn't clicked, some other area of the layout for the item was clicked. For example, the user may have clicked the name or notes for the task. If so, this code executes the code for the default case. This code begins by creating an intent for the Add/Edit activity. Then, it adds a "new task" flag to the intent and stores two pieces of extra data for the intent: (1) the ID of the task and (2) a Boolean variable that indicates that the layout is in edit mode. Finally, this code uses the intent to start the Add/Edit activity.

A class that extends the layout   **Page 2**

```
            // set task data on widgets
            setTask(t);
        }

    public void setTask(Task t) {
        task = t;
        nameTextView.setText(task.getName());

        // Remove the notes if empty
        if (task.getNotes().equalsIgnoreCase("")) {
            notesTextView.setVisibility(GONE);
        }
        else {
            notesTextView.setText(task.getNotes());
        }

        if (task.getCompletedDateMillis() > 0) {
            completedCheckBox.setChecked(true);
        }
        else{
            completedCheckBox.setChecked(false);
        }
    }

    @Override
    public void onClick(View v) {
        switch (v.getId()) {
            case R.id.completedCheckBox:
                if (completedCheckBox.isChecked()) {
                    task.setCompletedDate(System.currentTimeMillis());
                }
                else {
                    task.setCompletedDate(0);
                }
                db.updateTask(task);
                break;
            default:
                Intent intent = new Intent(context, AddEditActivity.class);
                intent.addFlags(Intent.FLAG_ACTIVITY_NEW_TASK);
                intent.putExtra("taskId", task.getId());
                intent.putExtra("editMode", true);
                context.startActivity(intent);
                break;
        }
    }
}
```

Description

- When the layout for a ListView widget contains complex widgets such as a check box, you code a class that extends the layout. This allows you to display the widgets correctly and to handle events that occur upon them.

Figure 14-5   A class that extends the layout (part 2 of 2)

A class for a custom adapter

Figure 14-6 shows a class for a custom adapter. This class begins by extending the BaseAdapter class. Then, it declares two instance variables: (1) a Content object and (2) an ArrayList of Task objects. Both of these instance variables are assigned values in the constructor for this class.

After the constructor, the TaskListAdapter class overrides four methods of the Adapter class: (1) getCount, (2) getItem, (3) getItemId, and (4) getView. These methods are needed to display the tasks in the adapter as items on a ListView widget.

The first three methods are easy to understand. The getCount method returns a count of the number of items in the ArrayList object. The getItem method returns the Task object that's at the specified position in the ArrayList object. And the getItemId method returns the position parameter. As a result, this method uses the same value for position and item ID.

The fourth method, the getView method, is more complicated. This method returns a TaskLayout object like the one defined by the class shown in the previous figure. This object corresponds with the data for the task at the specified position. To start, the getView method defines a variable for a TaskLayout object. Then, it gets the Task object for the specified position parameter. Next, it checks whether the View parameter is null. If so, it creates a new TaskLayout object for the current task. Otherwise, it converts the View parameter to a TaskLayout object and uses the setTask method of that object to set the current task for the layout. Either way, this method returns the TaskLayout object.

A class for a custom adapter

```
package com.murach.tasklist;

import java.util.ArrayList;

import android.content.Context;
import android.view.View;
import android.view.ViewGroup;
import android.widget.BaseAdapter;

public class TaskListAdapter extends BaseAdapter {

    private Context context;
    private ArrayList<Task> tasks;

    public TaskListAdapter(Context context, ArrayList<Task> tasks){
        this.context = context;
        this.tasks = tasks;
    }

    @Override
    public int getCount() {
        return tasks.size();
    }

    @Override
    public Object getItem(int position) {
        return tasks.get(position);
    }

    @Override
    public long getItemId(int position) {
        return position;
    }

    @Override
    public View getView(int position, View convertView, ViewGroup parent) {
        TaskLayout taskLayout = null;
        Task task = tasks.get(position);

        if (convertView == null) {
            taskLayout = new TaskLayout(context, task);
        }
        else {
            taskLayout = (TaskLayout) convertView;
            taskLayout.setTask(task);
        }
        return taskLayout;
    }
}
```

Description

- The class for a custom adapter that extends the BaseAdapter class can supply the data for ListView and Spinner widgets.

Figure 14-6   A class for a custom adapter

A class for a fragment that uses a custom adapter

Figure 14-7 shows a class for a fragment that uses a custom adapter like the one shown in the previous figure. This class works much like the class shown in figure 14-3. However, the refreshTaskList method in figure 14-7 gets the tasks for a list from the database and uses a custom adapter to display them.

The refreshTaskList method begins by getting the context for the application. Then, it creates an object from the TaskListDB class and calls the getTasks method from that object to get the tasks for the current tab.

The refreshTaskList method finishes by creating a TaskListAdapter object and passing it the application context as well as the tasks. Then, it sets that object as the adapter for the ListView control for the fragment.

A class for a fragment that uses a custom adapter

```java
package com.murach.tasklist;

import java.util.ArrayList;

import android.content.Context;
import android.os.Bundle;
import android.support.v4.app.Fragment;
import android.view.LayoutInflater;
import android.view.View;
import android.view.ViewGroup;
import android.widget.ListView;
import android.widget.TabHost;

public class TaskListFragment extends Fragment {
    private ListView taskListView;
    private String currentTabTag;

    @Override
    public View onCreateView(LayoutInflater inflater, ViewGroup container,
            Bundle savedInstanceState) {
        View view = inflater.inflate(R.layout.fragment_task_list,
            container, false);
        taskListView = (ListView) view.findViewById (R.id.taskListView);
        TabHost tabHost = (TabHost) container.getParent().getParent();
        currentTabTag = tabHost.getCurrentTabTag();
        refreshTaskList();
        return view;
    }

    public void refreshTaskList() {
        // get task list for current tab from database
        Context context = getActivity().getApplicationContext();
        TaskListDB db = new TaskListDB(context);
        ArrayList<Task> tasks = db.getTasks(currentTabTag);

        // create adapter and set it in the ListView widget
        TaskListAdapter adapter = new TaskListAdapter(context, tasks);
        taskListView.setAdapter(adapter);
    }

    @Override
    public void onResume() {
        super.onResume();
        refreshTaskList();
    }
}
```

Description

- Within the class for a fragment, you can use a custom adapter to display appropriate data for the current tab.

Figure 14-7   A class for a fragment that uses a custom adapter

The Task List app

So far, this chapter has presented enough of the code for the Task List app that you should have a general idea of how it works. Now, this chapter presents some more of the code to fill in the gaps about how this app works. But first, to put this code in context, this chapter reviews the user interface for the app, which was originally presented at the beginning of the previous chapter.

The user interface

Figure 14-8 reviews the interface for the Task List app. When the app starts, the Task List activity displays a list of tasks where each task has a check box that indicates whether it has been completed, a name, and optional notes.

In addition, the Task List activity provides tabs for two lists: Personal and Business. To display lists, the user can click on the tabs that are available from the Task List activity. In this figure, for example, the Personal tab is selected and the tasks for that tab are displayed. However, the user could click on the Business tab to display the tasks for that list.

From the Task List activity, the user can mark a task as completed by selecting a check box. Then, the user can remove all tasks that have been marked as completed by clicking on the Delete Tasks icon (the small trash can). This icon is available from the task bar. However, older versions of Android don't provide for the task bar. As a result, on older devices, the user can display the options menu and select the Delete Tasks item.

The user can add a task by clicking on the Add Task icon (the + sign) that's available from the task bar. Or, if the device is too old to display a task bar, the user can display the options menu and select the Add Task item. This displays the Add/Edit activity in add mode. At this point, the user can use the spinner to select the list. Then, the user can use the editable text views to add or edit the text for the task.

Similarly, the user can edit a task by clicking on the task in the Task List activity. This displays the Add/Edit activity in edit mode so the user can edit the data for the task. In this figure for example, I clicked on the "Get hair cut" task in the Task List activity. As a result, the Add/Edit activity displays this task in edit mode.

The Task List activity

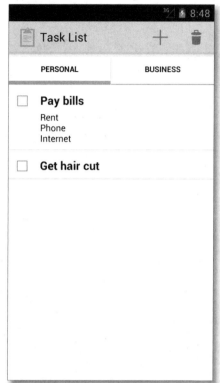

The Add/Edit activity

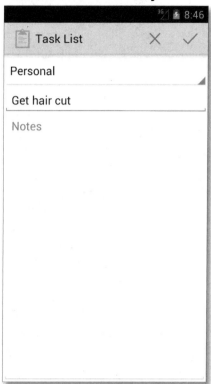

Description

- When the app starts, the Task List activity displays a list of tasks where each task has a check box that indicates whether it has been completed, a name, and optional notes.

- The app provides for two lists: Personal and Business.

- To display lists, the user can click on the tabs that are available in the Task List activity.

- To edit a task, the user can click on the task in the Task List activity. This displays the Add/Edit activity in edit mode so the user can edit the data for the task.

- To mark a task as completed, the user can select the check box in the Task List activity.

- To remove all completed tasks, the user can click the Delete Tasks button that's available from the task bar.

- To add a new task, the user can navigate to the Task List activity and click on the Add Task icon. This displays the Add/Edit activity in add mode so the user can add a new task.

- In the Add/Edit activity, the user can use the spinner to select the list. Then, the user can use the editable text views to add or edit the text for the task.

Figure 14-8   The user interface for the Task List app

The activity_task_list menu

Figure 14-9 shows the code that defines the menu for the TaskListActivity class. This code displays two menu items: (1) an Add Task item and (2) a Delete item.

Both of these menu items display icons in the task bar if there is enough room. These icons are stored in the res\drawable-xxx directory that's appropriate for the resolution of the icon. Otherwise, they display text in the options menu. Also, if the device is old and doesn't support the task bar, these menu items are available from the options menu.

The TaskListActivity class

Figure 14-10 shows the complete code for the TaskListActivity class. Part 1 of this figure works like the class described in figure 14-2. As a result, you shouldn't have much trouble understanding how it works. However, part 2 of this figure includes code that handles the events that occur when the user selects an item from the options menu.

To start, the onOptionsItemSelected method checks which menu item was selected. If the Add Task item was selected, the first statement creates an intent for the Add/Edit activity. Then, the second statement stores the current tab in the extra data for the intent. Finally, the third statement uses the intent to start the Add/Edit activity. Since this intent doesn't include extra data that specifies edit mode, this displays the Add/Edit activity in add mode.

If the Delete item was selected, this code hides all tasks that have been marked as complete. To do that, the first statement gets all tasks for the current tab. Then, the code loops through all of these tasks. If the milliseconds for the completed date are greater than 0, the task is marked as hidden, and the database is updated.

After updating the database, this code refreshes the task list. To do that, this code gets a reference to the fragment that corresponds with the current tab. Then, it calls the refreshTaskList method from that fragment to refresh the list of tasks. As a result, any tasks marked as hidden will no longer be displayed on the list.

The activity_task_list menu

```xml
<menu xmlns:android="http://schemas.android.com/apk/res/android" >
    <item
        android:id="@+id/menuAddTask"
        android:icon="@drawable/ic_action_add"
        android:orderInCategory="1"
        android:showAsAction="ifRoom"
        android:title="@string/add_task"/>
    <item
        android:id="@+id/menuDelete"
        android:icon="@drawable/ic_action_delete"
        android:orderInCategory="2"
        android:showAsAction="ifRoom"
        android:title="@string/delete"/>
</menu>
```

Figure 14-9   The activity_task_list menu

The TaskListActivity class

```java
package com.murach.tasklist;

import java.util.ArrayList;

import com.google.tabmanager.TabManager;

import android.content.Intent;
import android.os.Bundle;
import android.support.v4.app.FragmentActivity;
import android.view.Menu;
import android.view.MenuItem;
import android.widget.TabHost;
import android.widget.TabHost.TabSpec;

public class TaskListActivity extends FragmentActivity {
    TabHost tabHost;
    TabManager tabManager;
    TaskListDB db;

    @Override
    protected void onCreate(Bundle savedInstanceState) {
        super.onCreate(savedInstanceState);
        setContentView(R.layout.activity_task_list);

        // get tab manager
        tabHost = (TabHost) findViewById(android.R.id.tabhost);
        tabHost.setup();
        tabManager = new TabManager(this, tabHost, R.id.realtabcontent);

        // get database
        db = new TaskListDB(getApplicationContext());

        // add a tab for each list in the database
        ArrayList<List> lists = db.getLists();
        if (lists != null && lists.size() > 0) {
            for (List list : lists) {
                TabSpec tabSpec = tabHost.newTabSpec(list.getName());
                tabSpec.setIndicator(list.getName());
                tabManager.addTab(tabSpec, TaskListFragment.class, null);
            }
        }

        // sets current tab to the last tab opened
        if (savedInstanceState != null) {
            tabHost.setCurrentTabByTag(savedInstanceState.getString("tab"));
        }

    }

    @Override
    protected void onSaveInstanceState(Bundle outState) {
        super.onSaveInstanceState(outState);
        outState.putString("tab", tabHost.getCurrentTabTag());
    }
```

Figure 14-10   The TaskListActivity class (part 1 of 2)

The TaskListActivity class   Page 2

```java
@Override
public boolean onCreateOptionsMenu(Menu menu) {
    getMenuInflater().inflate(R.menu.activity_task_list, menu);
    return true;
}

@Override
public boolean onOptionsItemSelected(MenuItem item) {
    switch (item.getItemId()){
        case R.id.menuAddTask:
            Intent intent = new Intent(this, AddEditActivity.class);
            intent.putExtra("tab", tabHost.getCurrentTabTag());
            startActivity(intent);
            break;
        case R.id.menuDelete:
            // Hide all tasks marked as complete
            ArrayList<Task> tasks =
                db.getTasks(tabHost.getCurrentTabTag());
            for (Task task : tasks){
                if (task.getCompletedDateMillis() > 0){
                    task.setHidden(Task.TRUE);
                    db.updateTask(task);
                }
            }

            // Refresh list
            TaskListFragment currentFragment = (TaskListFragment)
                getSupportFragmentManager().
                findFragmentByTag(tabHost.getCurrentTabTag());
            currentFragment.refreshTaskList();

            break;
    }
    return super.onOptionsItemSelected(item);
}
}
```

Figure 14-10   The TaskListActivity class (part 2 of 2)

The activity_add_edit and spinner_list layout

Figure 14-11 begins by showing the layout for the Add/Edit activity. This activity allows the user to use a spinner to select the name of the list. Then, the user can use the two editable text views to edit the name and notes of the task. Both of the editable text views use the hint attribute to specify a hint to the user. For example, the editable text view for notes specifies a hint of "Notes". That way, the user knows to enter notes in this editable text view.

Since the layout for the activity uses a custom layout for the spinner, this figure also shows the spinner_list layout. This layout is used to display the names of the lists in the spinner. To do that, it uses a TextView widget with black text color and a text size of 18sp.

The activity_add_edit menu

Figure 14-11 also shows the code that defines the menus for the Add/Edit activity. This displays two items: (1) the Save item and (2) the Cancel item. Like the menu items for the Task List activity, both of these menu items display icons in the task bar if there is enough room. Otherwise, they display text in the options menu.

The activity_add_edit layout

```xml
<?xml version="1.0" encoding="utf-8"?>
<LinearLayout xmlns:android="http://schemas.android.com/apk/res/android"
    android:layout_width="match_parent"
    android:layout_height="match_parent"
    android:orientation="vertical" >

    <Spinner
        android:id="@+id/listSpinner"
        android:layout_width="match_parent"
        android:layout_height="wrap_content" />

    <EditText
        android:id="@+id/nameEditText"
        android:layout_width="match_parent"
        android:layout_height="wrap_content"
        android:hint="@string/task_name_hint" >

        <requestFocus />
    </EditText>

    <EditText
        android:id="@+id/notesEditText"
        android:layout_width="match_parent"
        android:layout_height="0dp"
        android:layout_weight="1"
        android:gravity="top"
        android:hint="@string/task_notes_hint"
        android:inputType="textMultiLine" />

</LinearLayout>
```

The spinner_list layout

```xml
<?xml version="1.0" encoding="utf-8"?>
<TextView xmlns:android="http://schemas.android.com/apk/res/android"
    android:layout_width="fill_parent"
    android:layout_height="wrap_content"
    android:padding="5dp"
    android:textColor="@android:color/black"
    android:textSize="18sp" />
```

The activity_add_edit menu

```xml
<?xml version="1.0" encoding="utf-8"?>
<menu xmlns:android="http://schemas.android.com/apk/res/android" >
    <item android:id="@+id/menuSave"
        android:showAsAction="ifRoom"
        android:title="@string/save"
        android:icon="@drawable/ic_action_save"
        android:orderInCategory="2"></item>
    <item android:id="@+id/menuCancel"
        android:showAsAction="ifRoom"
        android:title="@string/cancel"
        android:icon="@drawable/ic_action_cancel"
        android:orderInCategory="1"></item>
</menu>
```

Figure 14-11   The activity_add_edit layout and menu

The AddEditActivity class

Figure 14-12 shows the code for the AddEditActivity class. In part 1, the code works like most of the other activities you've seen throughout this book. However, it also creates a TaskListDB object. Then, it uses that database object to get an array of lists that are used as the data source for the adapter of the spinner.

Part 2 begins by getting the edit mode from the intent. Then, this code checks if the activity is in edit mode. If so, it uses the ID of the task to retrieve that task from the database. Then, it updates the task name and notes on the user interface. To do that, this code uses the setText methods of the editable text views for the name and notes of the task.

After setting the name and notes for the task, this code sets the appropriate list in the spinner. To do that, it checks if the activity is in edit mode. If so, it gets the ID for the same list as the selected task. Otherwise, it gets the ID for the list that was current when the user clicked the Add Task item. Either way, this code subtracts 1 from the database ID to get the correct position. This is necessary because the database numbering starts with 1, and the list item positions start with 0. Then, this code uses the current list item position to select the correct list from the spinner.

The onOptionsItemSelected method checks which item was selected. If the Save item was selected, this code saves all data on the activity to the database. To do that, it calls the saveToDB method shown in part 3 of this figure. Then, it calls the finish method of the activity to remove the Add/Edit activity from the stack. This displays the Task List activity.

If the Cancel item was selected, this code calls the finish method of the activity. This removes the Add/Edit activity from the stack without saving any of its data to the database.

The saveToDB method shown in part 3 begins by getting data from the three widgets displayed on the Add/Edit activity. Then, this code checks to make sure that the user has entered a task name. If the user hasn't entered a task name, this code exits the method without saving the task. That's because it doesn't make sense to save a task that doesn't have a name.

After making sure the user has entered a name, this code checks if the activity is in add mode. If so, this code creates a new Task object. This isn't necessary if the activity is in edit mode as the Task object should already exist.

After creating the Task object, this code stores the data in the Task object. Then, it checks if the activity is in edit mode. If so, this updates the task in the database. Otherwise, the activity is in add mode. As a result, this code adds the task to the database.

The onKey method hides the soft keyboard if the user presses the Center key on a DPad. Also, if the user presses the Back key, the onKey method saves the data to the database before it allows the system to process this key, which causes the Add/Edit activity to be removed from the stack. That's usually what you want since it's usually nice to save the user's data even if he or she presses the Back key.

The AddEditActivity class

```
package com.murach.tasklist;

import java.util.ArrayList;

import android.app.Activity;
import android.content.Context;
import android.content.Intent;
import android.os.Bundle;
import android.view.KeyEvent;
import android.view.Menu;
import android.view.MenuItem;
import android.view.View;
import android.view.View.OnKeyListener;
import android.view.inputmethod.InputMethodManager;
import android.widget.ArrayAdapter;
import android.widget.EditText;
import android.widget.Spinner;

public class AddEditActivity extends Activity
implements OnKeyListener {

    private EditText nameEditText;
    private EditText notesEditText;
    private Spinner listSpinner;

    private TaskListDB db;
    private boolean editMode;
    private String currentTabName = "";
    private Task task;

    public void onCreate(Bundle savedInstanceState) {
        super.onCreate(savedInstanceState);
        setContentView(R.layout.activity_add_edit);

        // get references to widgets
        listSpinner = (Spinner) findViewById(R.id.listSpinner);
        nameEditText = (EditText) findViewById(R.id.nameEditText);
        notesEditText = (EditText) findViewById(R.id.notesEditText);

        // set listeners
        nameEditText.setOnKeyListener(this);
        notesEditText.setOnKeyListener(this);

        // get the database object
        db = new TaskListDB(this);

        // set the adapter for the spinner
        ArrayList<List> lists = db.getLists();
        ArrayAdapter<List> adapter = new ArrayAdapter<List>(
                this, R.layout.spinner_list, lists);
        listSpinner.setAdapter(adapter);
```

Figure 14-12   The AddEditActivity class (part 1 of 3)

The AddEditActivity class

```java
        // get edit mode from intent
        Intent intent = getIntent();
        editMode = intent.getBooleanExtra("editMode", false);

        // if editing
        if (editMode) {
            // get task
            long taskId = intent.getLongExtra("taskId", -1);
            task = db.getTask(taskId);

            // update UI with task
            nameEditText.setText(task.getName());
            notesEditText.setText(task.getNotes());
        }

        // set the correct list for the spinner
        long listID;
        if (editMode) {   // edit mode - use same list as selected task
            listID = (int) task.getListId();
        }
        else {            // add mode - use the list for the current tab
            currentTabName = intent.getStringExtra("tab");
            listID = (int) db.getList(currentTabName).getId();
        }
        // subtract 1 from database ID to get correct list position
        int listPosition = (int) listID - 1;
        listSpinner.setSelection(listPosition);
    }

    @Override
    public boolean onCreateOptionsMenu(Menu menu) {
        getMenuInflater().inflate(R.menu.activity_add_edit, menu);
        return true;
    }

    @Override
    public boolean onOptionsItemSelected(MenuItem item) {
        switch (item.getItemId()){
            case R.id.menuSave:
                saveToDB();
                this.finish();
                break;
            case R.id.menuCancel:
                this.finish();
                break;
        }
        return super.onOptionsItemSelected(item);
    }
```

Figure 14-12   The AddEditActivity class (part 2 of 3)

The AddEditActivity class   **Page 3**

```java
        private void saveToDB() {
            // get data from widgets
            int listID = listSpinner.getSelectedItemPosition() + 1;
            String name = nameEditText.getText().toString();
            String notes = notesEditText.getText().toString();

            // if no task name, exit method
            if (name == null || name.equals("")) {
                return;
            }

            // if add mode, create new task
            if (!editMode) {
                task = new Task();
            }

            // put data in task
            task.setListId(listID);
            task.setName(name);
            task.setNotes(notes);

            // update or insert task
            if (editMode) {
                db.updateTask(task);
            }
            else {
                db.insertTask(task);
            }
        }

        @Override
        public boolean onKey(View view, int keyCode, KeyEvent event) {
            if (keyCode == KeyEvent.KEYCODE_DPAD_CENTER) {
                // hide the soft Keyboard
                InputMethodManager imm = (InputMethodManager)
                        getSystemService(Context.INPUT_METHOD_SERVICE);
                imm.hideSoftInputFromWindow(view.getWindowToken(), 0);
                return true;
            }
            else if (keyCode == KeyEvent.KEYCODE_BACK) {
                saveToDB();
                return false;
            }
            return false;
        }
    }
```

Figure 14-12   The AddEditActivity class (part 3 of 3)

Perspective

The skills presented in this chapter should be enough to get you started with tabs and custom adapters. This is a complicated subject. Fortunately, you can begin by using the Task List app presented in this chapter as a starting point. Then, you can copy and paste the necessary code into your app and modify it as necessary to get your tabs and custom adapters to work correctly.

So far, the database that's used by the Task List app is only available to the Task List app. In the next chapter, you'll learn how make the data in this database available to other apps. In addition, you'll learn how to work with other system databases that are available to your apps.

Summary

- To add the TabManager library to your project, copy the TabManager.jar file into the libs directory for your project. To get the TabManager.jar file, you can download the source code for this book.

- Within the class for an activity, you can use a TabManager object to add one or more tabs. For example, you can add one tab for each list stored in a database.

- Within the class for a fragment, you can use a TabHost class to get the tag for the current tab. You can use this tag to display the appropriate data for the current tab.

- When the layout for a ListView widget contains complex widgets such as a check box, you code a class that extends the layout, which allows you to display the widgets correctly and to handle events that occur upon them.

- The class for a custom adapter that extends the BaseAdapter class can supply the data for ListView and Spinner widgets.

- Within the class for a fragment, you can use a custom adapter to display appropriate data for the current tab.

Exercise 14-1 Work with tabs

In this exercise, you'll work with the tabs that are available from a simplified version of the TaskList app.

Review the app

1. Open the project named ch14_ex1_TaskList.

2. Open the TaskListActivity and TaskListFragment classes and review the code.

3. Run the app. It should display a Task List app that displays two tabs. The Personal tab should be selected, and it should display a message that says, "This is the Personal list."

4. Click on the Business tab. It should display a message that says, "This is the Business list".

Use a string to display the content for a tab

5. Switch to the TaskListFragment class.

6. Modify the refreshTaskList method so it displays a string that contains all task names for the current list. To do that, create a TaskListDB object and use it to get the tasks for the current list from the database. Then, loop through the tasks and build a string that contains each task on a separate line. Finally, display the string on the fragment.

7. Run the app. This should display the task names for the selected list. These names should reflect the data that's in the current Task List database. Since this data is stored on the device (or emulator), the data that's displayed may vary from device to device.

Use a simple adapter to display the content for a tab

8. Open the fragment_task_list layout and add a ListView widget below the TextView widget.

9. Switch to the TaskListFragment class and modify the refreshTaskList method so it uses a simple adapter to display each task as a ListView item. To get this to work, you can use the listview_item layout that's already in the project. In addition, you may be able to reuse some code from the News Reader app presented earlier in this book.

10. Run the app. This should display the task names for the selected list in the TextView and ListView widgets.

Exercise 14-2   Work with a custom adapter

In this exercise, you'll modify the News Reader app presented earlier in this book so it uses a custom adapter instead of a simple adapter.

Review the app

1. Open the project named ch14_ex2_NewsReader.

2. Open the listview_item layout. This layout contains the TextView widgets for displaying the publication date and title of a news item.

3. Run the app. It should display a list of news items. Then, click on a news item. It should display more information about the news item.

Add a custom adapter

4. Copy the TaskLayout class from the book_apps\ch14_TaskList project to this project.

5. Rename this class from TaskLayout to ItemLayout.

6. Modify the newly renamed ItemLayout class so it works with an RSSItem object and a listview_item layout instead of a Task object and a listview_task layout. This class should map the data from an RSSItem to the two widgets in the listview_item layout.

7. Modify the onClick method of the ItemLayout class so it creates an intent for the ItemActivity class, stores all necessary extra data in that intent, and uses the intent to start the activity. To do that, you can open the ItemsActivity class and copy the code that stores the extra data for the intent into this class.

8. Switch to the ItemsActivity class. Then, delete the onItemClick method and all other code that implements the OnItemClickListener.

9. Copy the TaskListAdapter class from the book_apps\ch14_TaskList project to this project.

10. Rename this class from TaskListAdapter to ItemsAdapter.

11. Modify the newly renamed ItemsAdapter class so it works with a List of RSSItem objects instead of an ArrayList of Task objects.

12. Switch to the ItemsActivity class. Then, delete all code that's necessary to create the SimpleAdapter object and replace it with a single statement that creates the ItemsAdapter object.

13. Run the app. It should work the same as before. However, the code in the ItemsActivity class should be shorter and easier to read since the adapter code has been moved from the ItemsActivity class into the ItemsAdapter and ItemLayout classes.

15

How to work
with content providers

In Android, a content provider allows multiple apps to share the same data. Android uses built-in content providers to make certain types of data available to multiple apps. For example, Android includes a content provider for your contacts. That way, built-in Android apps such as the phone dialer and text messaging apps can share this data. Similarly, third-party apps such as a custom contacts app can share this data.

In this chapter, you'll learn how to add a content provider to the Task List app from the previous two chapters. This makes the data in the Task List database available to all apps. Then, you'll learn how to create a Task History app that uses this content provider. This illustrates most of the concepts you need to work with content providers. In addition, you'll learn a little about using Android's built-in content provider for contacts.

An introduction to content providers

As mentioned earlier, a *content provider* allows multiple apps to share the same data. This data can be stored in a database or in files. Android includes content providers for your contacts, calendar, settings, bookmarks, and media including images, music, and video. That way, built-in and third-party apps can share this data.

URIs for content providers

Figure 15-1 begins by showing the syntax for a *URI* (*Uniform Resource Identifier*) for a content provider. This syntax begins with "content://", followed by the name of the provider (the *authority*), followed by a *path* to the content. Typically, a path begins with a name that corresponds with a table or file. In addition, a path can include an optional ID that points to an individual row in a table. In this figure, the first URI specifies all rows in the Task table of the Task List database. The second URI specifies all rows in the List table of the Task List database. And the third URI specifies the row in the Task table that has an ID of 2. In other words, the row with a primary key value of 2.

When you work with URIs, you should know that each front slash in the path begins a new *segment*. For the third URI, for example, the first segment is "tasks" and the second segment is "2". Typically, the URI for a content provider only has one or two segments. However, a URI for a content provider can have more segments if that helps organize and structure the data.

MIME types for content providers

When you create a URI for a content provider, you need to specify the type of content that's returned by that URI. To do that, you use a *MIME type*, which is an Internet standard for defining types of content.

This figure shows how to specify the MIME type for a URI. Here, the first syntax is for a URI that can return multiple rows, and the second syntax is for a URI that returns a single row. The only difference between these syntaxes is that the first syntax uses "dir" to indicate that it returns a directory of items, and the second syntax uses "item" to indicate that it returns a single item. For example, the first MIME type indicates that the URI can return multiple tasks, and the second MIME type indicates that the URI only returns a single task.

Built-in Android content providers

- Contacts
- Calendar
- Settings
- Bookmarks
- Media (images, music, video)

URIs for content providers

The syntax
```
content://authority/path[/id]
```

A URI for all rows of the Task table
```
content://com.murach.tasklist.provider/tasks
```

A URI for all rows of the List table
```
content://com.murach.tasklist.provider/lists
```

A URI for a single row in the Task table with an ID of 2
```
content://com.murach.tasklist.provider/tasks/2
```

MIME types for content providers

The syntax for multiple rows
```
vnd.android.cursor.dir/vnd.company_name.content_type
```

The syntax for a single row
```
vnd.android.cursor.item/vnd.company_name.content_type
```

A MIME type for multiple rows
```
vnd.android.cursor.dir/vnd.murach.tasklist.tasks
```

A MIME type for a single row
```
vnd.android.cursor.item/vnd.murach.tasklist.tasks
```

Description

- A *content provider* allows multiple apps to share the same data. This data can be stored in a database or in files. Android uses built-in content providers to make certain types of data available to all apps.

- A *URI* (*Uniform Resource Identifier*) for a content provider includes the name of the provider (the *authority*) and a *path* to the content. This path typically begins with a name that corresponds with a table or file, and it can include an optional ID that points to an individual row in a table.

- Within a path, each front slash begins a new *segment*.

- A *MIME type* is an Internet standard for defining types of content.

Figure 15-1   An introduction to content providers

How to add supporting methods to the database class

In chapter 13, you learned how to create a database class named TaskListDB. This class includes methods like getTasks and insertTask that are used by the app to work with the database. Now, figure 15-2 shows some methods that you can add to this database class that can be used by the content provider class. These supporting methods provide basic query, insert, update, and delete operations for the Task table. In the next few figures, you'll see how these methods are used by the content provider.

These supporting methods open connections to the database, but they don't close those connections. That's because the database connections are closed automatically when the process that hosts the content provider is destroyed.

Four supporting methods in the TaskListDB class

```
public Cursor queryTasks(String[] columns, String where,
        String[] whereArgs, String orderBy) {
    this.openReadableDB();
    return db.query(TASK_TABLE, columns, where, whereArgs,
            null, null, orderBy);
}

public long insertTask(ContentValues values) {
    this.openWriteableDB();
    return db.insert(TASK_TABLE, null, values);
}

public int updateTask(ContentValues values, String where,
        String[] whereArgs) {
    this.openWriteableDB();
    return db.update(TASK_TABLE, values, where, whereArgs);
}

public int deleteTask(String where, String[] whereArgs) {
    this.openWriteableDB();
    return db.delete(TASK_TABLE, where, whereArgs);
}
```

Description

- To make it easy to create a content provider, you can include some supporting methods in your database class that provide for the query, insert, update, and delete operations.

- These methods should not close the database connection after performing the operation.

Figure 15-2   How to add supporting methods to the database class

How to create a content provider

Now that you understand some basic concepts about content providers, you're ready to learn how to create a content provider. In particular, you're ready to learn how to add a content provider to the Task List app described in the previous chapter. This allows other apps to read and write data from the database that's used by the Task List app.

How to start a content provider class

Figure 15-3 shows how to start a class for a content provider. To begin, you declare a class like the TaskListProvider class shown in this figure that extends the ContentProvider class.

Within the TaskListProvider class, the first four statements declare the constants for the class. The first statement declares the constant for the authority that's used within the URI. Then, the next three statements declare constants that are used to determine whether the URI matches one of the URIs that are supported by this class.

After declaring the constants, this class declares two instance variables. The first instance variable is a TaskListDB object for working with the Task List database. The second instance variable is a UriMatcher object for checking whether a URI matches one of the supported URIs.

The onCreate method contains the code that initializes these variables. This method overrides the onCreate method of the ContentProvider class. As a result, Android calls this method when it creates the content provider.

Within the onCreate method, the first statement creates the TaskListDB object. Then, the next three statements create the UriMatcher object. Of these statements, the first creates the UriMatcher object and uses the NO_MATCH variable to specify the value that's returned if no matching URI is found. The second statement adds this URI:

```
content://com.murach.tasklist.provider/tasks
```

to the UriMatcher object and uses the ALL_TASKS_URI constant to specify the value that's returned if a URI matches this URI. The third statement adds this URI:

```
content://com.murach.tasklist.provider/tasks/#
```

and uses the SINGLE_TASK_URI constant to specify the value that's returned if a URI matches this URI. Here, the last segment uses the # wildcard to specify any number. As a result, this URI pattern matches a URI that specifies a single row like this:

```
content://com.murach.tasklist.provider/tasks/5
```

The TaskListProvider class   **Page 1**

```
package com.murach.tasklist;

import android.content.ContentProvider;
import android.content.ContentValues;
import android.content.UriMatcher;
import android.database.Cursor;
import android.net.Uri;

public class TaskListProvider extends ContentProvider {

    public static final String AUTHORITY = "com.murach.tasklist.provider";

    public static final int NO_MATCH = -1;
    public static final int ALL_TASKS_URI = 0;
    public static final int SINGLE_TASK_URI = 1;

    private TaskListDB db;
    private UriMatcher uriMatcher;

    @Override
    public boolean onCreate() {
        db = new TaskListDB(getContext());

        uriMatcher = new UriMatcher(NO_MATCH);
        uriMatcher.addURI(AUTHORITY, "tasks", ALL_TASKS_URI);
        uriMatcher.addURI(AUTHORITY, "tasks/#", SINGLE_TASK_URI);

        return true;
    }
```

A constructor and method of the UriMatcher class

Constructor/Method	Description
UriMatcher(code)	Creates a new UriMatcher object and specifies the code to return when no matching URI is found.
addURI(authority, path, code)	Adds a URI to the UriMatcher object and specifies the code to return when a matching URI is found for this URI.

Description

- The class for a content provider must inherit the ContentProvider class.
- To initialize variables, your content provider class can override the onCreate method of the ContentProvider class.

Figure 15-3   How to start a content provider class

How to provide for querying

Figure 15-4 shows how to code the methods of a content provider that provides for querying. To start, you can override the query method of the ContentProvider class. This method accepts five arguments (the URI, an array of column names, a WHERE clause, an array of arguments for the WHERE clause, and an ORDER BY clause), and it returns a Cursor object.

Within the query method, this code uses a switch statement to check the URI argument. Then, you can execute appropriate code depending on the URI argument. In this figure, for example, the switch statement checks whether the URI argument matches the URI that corresponds with the ALL_TASKS_URI constant. If so, this code calls the queryTasks method from the TaskListDB object and passes it the other four arguments of the query method. This returns a Cursor object that contains the columns and rows specified by those arguments.

If the URI argument doesn't match the URI that corresponds with the ALL_TASKS_URI constant, this code throws an exception that indicates that this type of operation isn't supported. In this figure, for example, if you used a URI that corresponded with the SINGLE_TASK_URI constant, this code would throw an exception. In other words, this query method doesn't support the URI for a single task. As a result, if you attempted to use a URI for a single row, it would throw an exception.

This exception sends an error message to the developers of other apps that are trying to use this content provider. This message should help the developer figure out how to use the content provider. In this case, the exception clearly indicates that the operation being requested by the URI isn't supported. If you want, you can use a different type of exception object such as an IllegalArgumentException. However, the goal is to send an error message to other developers that clearly identifies why the URI that they sent didn't work.

For querying to work, the content provider class must override the getType method of the ContentProvider class. This method accepts one argument (the URI) and returns a string that uniquely identifies the content type.

Within the getType method, this code uses a switch statement to check the URI argument. Then, you can execute the appropriate code depending on the URI argument. In this figure, for example, the switch statement returns the appropriate MIME type if the URI argument matches one of the URIs that correspond with the ALL_TASKS_URI or SINGLE_TASK_URI constants. However, if the URI argument doesn't match either of these URIs, this code throws an exception to indicate that the operation is not supported for the specified URI.

The TaskListProvider class   **Page 2**

```java
@Override
public Cursor query(Uri uri, String[] columns, String where,
        String[] whereArgs, String orderBy) {
    switch(uriMatcher.match(uri)) {
        case ALL_TASKS_URI:
            return db.queryTasks(columns, where,
                    whereArgs, orderBy);
        default:
            throw new UnsupportedOperationException (
                    "URI " + uri + " is not supported.");
    }
}

@Override
public String getType(Uri uri) {
    switch(uriMatcher.match(uri)) {
        case ALL_TASKS_URI:
            return "vnd.android.cursor.dir/vnd.murach.tasklist.tasks";
        case SINGLE_TASK_URI:
            return "vnd.android.cursor.item/vnd.murach.tasklist.tasks";
        default:
            throw new UnsupportedOperationException(
                    "URI " + uri + " is not supported.");
    }
}
```

A method of the UriMatcher class

Method	Description
`match(uri)`	If the specified Uri object matches any of the URIs stored in the UriMatcher object, this method returns the code for that URI.

Description

- To allow other apps to query your data, your content provider class can override the query and getType methods of the ContentProvider class.

- Every data access method of the ContentProvider class has a Uri object as its first argument. This allows the method to determine what table, row, or file to access.

- In a URI, you can use the # wildcard to match any number.

Figure 15-4   How to provide for querying

How to provide for inserting rows

Figure 15-5 shows how to code the method of a content provider that provides for inserting rows. To do that, you can override the insert method of the ContentProvider class. This method accepts two arguments (the URI, and a ConventValues object that contains the column names and values for the row). This method returns a URI for the row that's inserted.

Within the insert method, this code uses a switch statement to check the URI argument. This works similarly to the query method described in the previous figure. If the URI argument matches the URI that corresponds with the ALL_TASKS_URI constant, the first statement calls the insertTask method from the TaskListDB object and passes it the second argument of the insert method. This returns the ID for the row that was inserted.

The second statement gets a ContentResolver object and calls the notifyChange method from it to notify registered observers that a row was changed. That way, the content provider notifies registered observers when the data changes so those observers can re-requery the provider to keep their data current. Without this statement, apps that use this provider would have to re-query the provider manually to make sure that their content is current. By default, CursorAdapter objects get this notification. As a result, you don't need to register them.

The third statement uses several methods of the Uri.Builder class to build the Uri object for the row that was inserted. To do that, it appends the ID for the row that was inserted to the URI argument. For example, if you insert a row and the database assigns that row a primary key value of 14, this statement returns this URI:

```
content://com.murach.tasklist.provider/tasks/14
```

The TaskListProvider class   **Page 3**

```
@Override
public Uri insert(Uri uri, ContentValues values) {
    switch(uriMatcher.match(uri)) {
        case ALL_TASKS_URI:
            long insertId = db.insertTask(values);
            getContext().getContentResolver().notifyChange(uri, null);
            return uri.buildUpon().appendPath(
                    Long.toString(insertId)).build();
        default:
            throw new UnsupportedOperationException(
                    "URI: " + uri + " is not supported.");
    }
}
```

A method of the Context class

Method	Description
`getContentResolver()`	Returns a ContentResolver object for your app.

A method of the ContentResolver class

Method	Description
`notifyChange(uri, observer)`	Notify registered observers that a row was updated. By default, CursorAdapter objects get this notification.

Some methods of the Uri.Builder class

Method	Description
`buildUpon()`	Creates a Uri.Builder object for building a new Uri object.
`appendPath(segment)`	Encodes the specified segment and appends it to the path. Returns a Uri.Builder object to provide for method chaining.
`build()`	Returns a Uri object for the current Uri.Builder object.

Description

- To allow other apps to insert rows into your database, you can override the insert method of the ContentProvider class.
- The insert method should return a URI for the inserted row.

Figure 15-5   How to provide for inserting rows

How to provide for updating rows

Figure 15-6 shows how to code the method of a content provider that provides for updating rows. To do that, you can override the update method of the ContentProvider class. This method accepts four arguments (the URI, a ContentValues object, a WHERE clause, and an array that contains the arguments for the WHERE clause). This method returns an int value for the count of rows that were updated.

Within the update method, this code uses a switch statement to check the URI argument. This works similarly to the query method described earlier in this chapter. However, unlike the query method described earlier in this chapter, the update method supports two URIs: (1) the URI that corresponds with the ALL_TASKS_URI constant and (2) the URI that corresponds with the SINGLE_TASK_URI constant.

If the URI argument matches the SINGLE_TASK_URI constant, this code updates the specified row. To do that, the first statement uses the getLastPath-Segment method to get the ID for the row to update. The next two statements use that ID to build a WHERE clause that selects the row for the specified update. The fourth statement calls the updateTask method of the TaskListDB object to update the row and get the count of updated rows. The fifth statement notifies any registered observers that a row was changed. And the sixth statement returns the count of rows that were updated. If the update is successful, this should return a count of 1.

If the URI argument matches the ALL_TASKS_URI constant, this code updates the specified rows. To do that, the first statement passes the second, third, and fourth arguments to the updateTask method of the TaskListDB object. This may update one or more rows. Then, the second statement notifies any registered observers that a row was changed, and the third statement returns the count of rows that were updated.

How to provide for deleting rows

This figure also shows how to code the method of a content provider that provides for deleting rows. Since this method works much like the update method described in this figure, you shouldn't have much trouble understanding how it works.

The TaskListProvider class   **Page 4**

```
@Override
public int update(Uri uri, ContentValues values, String where,
        String[] whereArgs) {
    int updateCount;
    switch(uriMatcher.match(uri)) {
        case SINGLE_TASK_URI:
            String taskId = uri.getLastPathSegment();
            String where2 = "_id = ?";
            String[] whereArgs2 = { taskId };
            updateCount = db.updateTask(values, where2, whereArgs2);
            getContext().getContentResolver().notifyChange(uri, null);
            return updateCount;
        case ALL_TASKS_URI:
            updateCount = db.updateTask(values, where, whereArgs);
            getContext().getContentResolver().notifyChange(uri, null);
            return updateCount;
        default:
            throw new UnsupportedOperationException (
                    "URI " + uri + " is not supported.");
    }
}

@Override
public int delete(Uri uri, String where, String[] whereArgs) {
    int deleteCount;
    switch(uriMatcher.match(uri)) {
        case SINGLE_TASK_URI:
            String taskId = uri.getLastPathSegment();
            String where2 = "_id = ?";
            String[] whereArgs2 = { taskId };
            deleteCount = db.deleteTask(where2, whereArgs2);
            getContext().getContentResolver().notifyChange(uri, null);
            return deleteCount;
        case ALL_TASKS_URI:
            deleteCount = db.deleteTask(where, whereArgs);
            getContext().getContentResolver().notifyChange(uri, null);
            return deleteCount;
        default:
            throw new UnsupportedOperationException (
                    "URI " + uri + " is not supported.");
    }
}
}
```

A method of the Uri class

Method	Description
`getLastPathSegment()`	Returns the value for the last path segment. For a URI that specifies a single row, this gets the ID for the row.

Description

- To allow other apps to update or delete rows in your database, you can override the update and delete methods of the ContentProvider class.

Figure 15-6   How to provide for updating and deleting rows

How to register a content provider

After you create the class for a content provider, you need to register the provider as described in figure 15-7. The easiest way to register a provider is to add a provider element to the project's AndroidManifest.xml file. When you code a provider element, you can code it at the same level as an activity element.

A provider element should use the name attribute to specify the name of the class for the provider, and it should use the authorities attribute to specify the authority for the provider. In this figure, for example, the provider element specifies the name and authority for the content provider for the Task List app.

In addition, a provider element should use the exported attribute to specify whether other apps can use the content provider. Since content providers are designed to be used by other apps, you typically set the exported attribute to a value of "true". However, in some cases, you may only want the current app to be able to use the content provider. For example, you may want to use a content provider to allow an app widget to work with the data for an app. For more information about app widgets, please see the next chapter.

The provider element in the AndroidManifest.xml file

```
<provider
    android:name="com.murach.tasklist.TaskListProvider"
    android:authorities="com.murach.tasklist.provider"
    android:exported="true" >
</provider>
```

The attributes of the provider element

Attribute	Description
name	Specifies the package and name of the content provider class.
authorities	Specifies the authority for the content provider.
exported	Specifies whether the content provider can be used by other apps.

Description

- Before other apps can use your content provider class, you must register it by adding a provider element to the AndroidManifest.xml file.

Figure 15-7   How to register a content provider

How to use a content provider

Now that you know how to create a content provider, you're ready to learn how to use a content provider. To start, you'll learn how to use the content provider for the Task List app that's described in the previous figures. Then, you'll learn how to use one of the built-in content providers that's available from Android, the Contacts Provider. This should give you a good idea of how content providers work.

How to use a custom content provider

Figure 15-8 shows how to use the content provider for the Task List app that's presented earlier in this chapter. To start, the first code example defines the constants that this code needs to refer to the columns names and indexes.

To keep things simple, I copied these constants from the TaskListDB class presented in chapter 13. However, a more elegant approach would be to store these constants in a library to make it easy to share them between apps. For this content provider, for example, you could store the constants in a class named TaskListContract. Then, you could store that class in a library named TaskListContract.jar. That way, both the Task List and Task History apps could import this library and use the same set of constants.

The second code example defines the constants for the authority and the URI for multiple tasks. Again, to keep things simple, I copied the code for the AUTHORITY constant from content provider for the Task History app. However, to reduce code duplication, these constants could be stored in a library as described in the previous paragraph.

Once you're done defining the constants that are needed by the code, you can use the getContentResolver method to get a ContentResolver object. Then, you can call any of the methods that are available from the content provider from that object as shown by the rest of the code examples.

The third, fourth, and fifth examples call the query method and pass the URI for all tasks as the first argument. The third example passes null values for all other arguments. As a result, it returns a Cursor object that contains all columns and rows.

The fourth example defines a WHERE clause that only displays tasks that have been hidden. To do that, the where variable only returns rows where the TASK_HIDDEN column is equal to value of '1', which is how the database defines a true value. Then, this example passes this variable to the query method. As a result, this example only returns the rows that have been marked as hidden. In other words, it returns all old tasks that the Task List app no longer displays.

The fifth example defines an array of three columns, the TASK_ID, TASK_NAME, and TASK_NOTES columns. Then, it passes this variable to the query method. As a result, this example only returns these three columns.

How to define constants for columns

```
public static final String TASK_ID = "_id";
public static final int    TASK_ID_COL = 0;
public static final String TASK_NAME = "task_name";
public static final int    TASK_NAME_COL = 2;
...
```

How to create the base Uri object

```
public static final String AUTHORITY = "com.murach.tasklist.provider";
public static final Uri TASKS_URI =
        Uri.parse("content://" + AUTHORITY + "/tasks");
```

How to query a content provider

```
Cursor cursor = getContentResolver()
        .query(TASKS_URI, null, null, null, null);
```

A second way to query a content provider

```
String where = TASK_HIDDEN + " = '1' ";
String orderBy = TASK_COMPLETED + " DESC";
Cursor cursor = getContentResolver()
        .query(TASKS_URI, null, where, null, orderBy);
```

A third way to query a content provider

```
String[] columns = {TASK_ID, TASK_NAME, TASK_NOTES}
Cursor cursor = getContentResolver()
        .query(TASKS_URI, columns, null, null, null);
```

How to delete data from the content provider

```
String where = TASK_ID + " = ?";
String[] whereArgs = { Integer.toString(taskId) };
int deleteCount = getContentResolver()
        .delete(TASKS_URI, where, whereArgs);
```

Another way to delete data from the content provider

```
Uri taskUri = ContentUris.withAppendedId(TASKS_URI, taskId);
int deleteCount = getContentResolver()
        .delete(taskUri, null, null);
```

A method of the ContentUris class

Method	Description
`withAppendedId(Uri, id)`	Returns the specified Uri object with the specified ID appended to the path.

Description

- You can define constants for the columns and URI for the content provider.
- You can call any methods of the content provider from the ContentResolver object.

Figure 15-8   How to use a custom content provider

The sixth and seventh examples call the delete method of the ContentResolver object. The sixth example defines a variable for a WHERE clause that selects the task with the specified ID. Then, this example passes this variable and the URI for all tasks to the delete method. As a result, this example deletes the row that matches the specified ID.

The seventh example appends the task ID to the URI for all tasks. Then, this example passes this URI followed by two null values to the delete method. This is another way to delete a row for the specified ID.

How to use a built-in content provider

Figure 15-9 shows some code that you can use to work with the Contacts Provider that comes with Android. This content provider makes it possible for you to work with your address book: names, phone numbers, email addresses, and so on.

Since the Contacts Provider is a part of the Android API, you can import the classes that contain the constants for working with this provider. In this figure, the first example imports three classes that contain constants. All three of these classes are inner classes of the ContactsContract class.

Once you import these classes, you can use the constants available from them to work with the provider. For example, the second example shows how to use the CONTENT_URI constant that's available from the ContractsContract. Data class to get the URI for the provider.

The third example shows how to query a content provider. To start, this example defines four columns for the cursor: _ID, DISPLAY_NAME, DATA1, and DATA2. To understand the details of these constants, you should view the documentation for the Contacts Provider. In brief, _ID is the primary key that's needed by the adapter, DISPLAY_NAME is the name of the person, DATA1 is the phone number, and DATA2 is the type of phone number (home, mobile, work, etc.).

This example defines a WHERE clause that selects phone number rows for contacts that have been marked as favorites. In other words, it selects contacts where the STARRED constant is true (equal to a value of '1').

This example defines an ORDER BY clause that sorts the rows in descending order by the number of times that they have been contacted. To do that, this code uses the TIMES_CONTACTED constant. This orders the rows so the most frequently contacted numbers are at the top of the cursor.

This example finishes by calling the query method from the ContentResolver object and passing it the DATA_URI constant as well as the columns, where, and orderBy variables. This returns a Cursor object with the specified columns and rows. Although it isn't shown in this figure, you can also call the insert, update, and delete methods to modify the data that's available from the Contacts Provider.

How to import the classes that contain the constants

```
import android.provider.ContactsContract.Contacts;
import android.provider.ContactsContract.Data;
import android.provider.ContactsContract.CommonDataKinds.Phone;
```

How to get a base Uri object

```
private final Uri DATA_URI = Data.CONTENT_URI;
```

How to query a content provider

```
String[] columns = {
        Data._ID,                  // primary key
        Contacts.DISPLAY_NAME,     // person's name
        Data.DATA1,                // phone number
        Data.DATA2                 // phone type (mobile, home, work, etc.)
};
String where =
        "(" + Data.MIMETYPE + "='" + Phone.CONTENT_ITEM_TYPE + "' AND "
            + Contacts.STARRED + "='1' )";
String orderBy = Contacts.TIMES_CONTACTED + " DESC";

Cursor cursor = getContentResolver().query(
        DATA_URI, columns, where, null, orderBy);
```

Description

- When you work with a content provider that's included with Android, you can import the classes that contain the constants for working with the content provider.

- To learn more about a content provider that's included with Android, you can read the documentation for the content provider.

Figure 15-9   How to use a built-in content provider

How to work with a dialog box

At this point, you have all the skills you need to work with a content provider. However, the Task History app that's presented in the next few figures requires one more skill: how to work with a dialog box.

How to import the dialog class and interface

When creating apps, you often need to display a dialog box like the one shown in figure 15-10. To do that, you can begin by importing the AlertDialog. Builder class and the DialogInterface.OnClickListener interface. Then, you can use this class and interface to build and show a dialog box.

How to build and show the dialog box

To build a dialog box, you can begin by creating a new AlertDialog.Builder object. Since the AlertDialog.Builder object supports method chaining, you don't need to provide a name for this object. Instead, you can call a series of methods from it.

To start, you can call the setMessage method to set the message that's displayed across the top of the dialog box. In this figure, for example, this message says, "Are you sure you want to delete this task?"

Then, you can set the buttons on the dialog by calling the setPositiveButton and setNegativeButton methods. Android displays the positive button on the right side of the dialog, and it displays the negative button on the left. In this figure, for example, the Cancel button is the negative button, and the Yes button is the positive button. If necessary, you can specify one or more neutral buttons. In that case, Android displays the neutral buttons between the positive and negative buttons.

When you set a button, you can also specify the event handler that's executed when the user clicks the button. The easiest way to do that is to create an anonymous inner class for the listener. In this figure, for instance, the code that sets the positive button creates an anonymous inner class that implements the DialogInterface.OnClickListener interface. As a result, if the user clicks the Yes button, this code deletes the task.

The code that sets the negative button also creates an anonymous inner class that implements the DialogInterface.OnClickListener interface. However, this event handler cancels the dialog by calling the cancel method of the first parameter of the onClick method. As a result, if the user clicks the Cancel button, this code cancels the dialog box without deleting the task.

Once this code finishes setting the buttons for this dialog, it calls the show method to display the dialog. This displays a dialog like the one shown at the top of this figure.

A dialog box

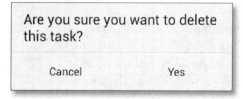

How to use a dialog box

Import the dialog class and interface

```
import android.app.AlertDialog;
import android.content.DialogInterface;
```

Create and show the dialog box

```
new AlertDialog.Builder(this)
.setMessage("Are you sure you want to delete this task?")
.setPositiveButton("Yes", new DialogInterface.OnClickListener() {
    @Override
    public void onClick(DialogInterface dialog, int id) {
        delete(item);
    }
})
.setNegativeButton("Cancel", new DialogInterface.OnClickListener() {
    @Override
    public void onClick(DialogInterface dialog, int id) {
        dialog.cancel();
    }
})
.show();
```

Constructors and methods of the AlertDialog.Builder class

Constructor/Method	Description
AlertDialog.Builder(context)	Creates a new AlertDialog.Builder object.
setMessage(text)	Sets the message that's displayed above the buttons.
setPositiveButton(text, listener)	Sets the text and listener for the positive button.
setNegativeButton(text, listener)	Sets the text and listener for the negative button.
setNeutralButton(text, listener)	Sets the text and listener for a neutral button.
show()	Displays the dialog.

Description

- You can use the AlertDialog.Builder class and the DialogInterface.OnClickListener interface to build and display a dialog box that includes event handlers.

Figure 15-10   How use a dialog box

The Task History app

The Task History app uses the same database that's used by the Task List app described in the previous two chapters. This app only uses the query and delete methods that are available from the content provider, not the insert or update methods. However, this app clearly shows how one app can use a content provider to work with the same data as another app.

The user interface

Figure 15-11 shows the user interface for the Task History app. When the app starts, the Task History activity displays a list of tasks that have been completed and marked as hidden. In other words, it displays all tasks the Task List app no longer displays.

The Task History activity displays the task name, notes, and completion date for each task. If there are no notes for a task, this app displays a message that says, "No notes". In this figure, for example, none of the tasks have notes.

If you want to delete a task permanently, you can click the task. Then, the app displays a dialog box that prompts you to confirm the deletion. If you click the Yes button, this app deletes the task from the database. Otherwise, this app cancels the dialog box and does not delete the task.

The Task History activity

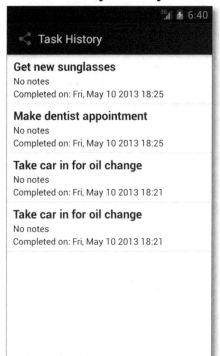

The dialog box for deleting a task

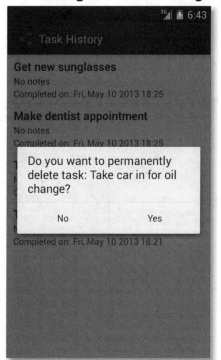

Description

- When the app starts, the Task History activity displays a list of tasks that have been completed and hidden by the Task List app. Each task includes the task name, notes, and completion date.

- To delete a task, the user can click the task. Then, the user can respond to the resulting dialog box to confirm or cancel the deletion.

Figure 15-11   The user interface for the Task History app

The XML for the layouts

Figure 15-12 shows the XML for the layouts that are used by the Task History activity. To start, the layout named activity_task_history uses a LinearLayout container and a ListView widget to display the list of tasks. Then, the listview_task layout uses a LinearLayout and three TextView widgets to format each task. Since you've seen layouts like these in the previous chapter, you shouldn't have much trouble understanding how they work by now.

The activity_task_history layout

```
<?xml version="1.0" encoding="utf-8"?>
<LinearLayout android:layout_width="match_parent"
    android:layout_height="match_parent"
    android:orientation="vertical"
    xmlns:android="http://schemas.android.com/apk/res/android">

    <ListView
        android:id="@+id/taskListView"
        android:layout_width="match_parent"
        android:layout_height="match_parent" >
    </ListView>

</LinearLayout>
```

The listview_task layout

```
<?xml version="1.0" encoding="utf-8"?>
<LinearLayout xmlns:android="http://schemas.android.com/apk/res/android"
    android:layout_width="match_parent"
    android:layout_height="match_parent"
    android:orientation="vertical" >

    <TextView
        android:id="@+id/nameTextView"
        android:layout_width="wrap_content"
        android:layout_height="wrap_content"
        android:layout_marginTop="5dp"
        android:layout_marginLeft="10dp"
        android:textSize="18sp"
        android:textStyle="bold"
        android:text="@string/name" />

    <TextView
        android:id="@+id/notesTextView"
        android:layout_width="wrap_content"
        android:layout_height="wrap_content"
        android:layout_marginLeft="10dp"
        android:text="@string/notes" />

    <TextView
        android:id="@+id/dateTextView"
        android:layout_width="wrap_content"
        android:layout_height="wrap_content"
        android:layout_marginBottom="5dp"
        android:layout_marginLeft="10dp"
        android:text="@string/date" />

</LinearLayout>
```

Figure 15-12   The layouts for the Task History app

The Java code for the activity

Figure 15-13 shows the Java code for the Task List activity. This class begins by defining the constants for the columns that it uses. These constants were copied in from the TaskListDB class presented in chapter 13. Then, this class defines a constant for the authority. This constant was copied in from the TaskListProvider class. Next, this class uses the AUTHORITY constant to define a constant for the URI for all tasks. As mentioned earlier, a more elegant solution would be to create a library for these constants so they can be shared between the Task List and Task History apps.

After defining its constants, this class defines four instance variables that are used throughout the class. The first instance variable is for the list view that stores the tasks. The second is for the cursor adapter. The third is for the cursor. And the fourth is for the application's context.

The onCreate method begins by displaying the activity_task_history layout. Then, it sets the context variable, and gets a reference to the list view.

The TaskHistoryActivity class   **Page 1**

```java
package com.murach.taskhistory;

import java.text.SimpleDateFormat;
import java.util.Date;

import android.app.Activity;
import android.app.AlertDialog;
import android.content.Context;
import android.content.DialogInterface;
import android.database.Cursor;
import android.net.Uri;
import android.os.Bundle;
import android.widget.SimpleCursorAdapter;
import android.widget.SimpleCursorAdapter.ViewBinder;
import android.view.View;
import android.widget.AdapterView;
import android.widget.AdapterView.OnItemClickListener;
import android.widget.ListView;
import android.widget.TextView;
import android.widget.Toast;

public class TaskHistoryActivity extends Activity
implements OnItemClickListener {

    public static final String TASK_ID = "_id";
    public static final int    TASK_ID_COL = 0;
    public static final String TASK_NAME = "task_name";
    public static final int    TASK_NAME_COL = 2;
    public static final String TASK_NOTES = "notes";
    public static final int    TASK_NOTES_COL = 3;
    public static final String TASK_COMPLETED = "date_completed";
    public static final int    TASK_COMPLETED_COL = 4;
    public static final String TASK_HIDDEN = "hidden";
    public static final int    TASK_HIDDEN_COL = 5;

    public static final String AUTHORITY = "com.murach.tasklist.provider";
    public static final Uri TASKS_URI =
            Uri.parse("content://" + AUTHORITY + "/tasks");

    private ListView taskListView;
    private SimpleCursorAdapter adapter;
    private Cursor cursor;
    private Context context;

    @Override
    protected void onCreate(Bundle savedInstanceState) {
        super.onCreate(savedInstanceState);
        setContentView(R.layout.activity_task_history);

        context = this;
        taskListView = (ListView) findViewById(R.id.taskListView);
    }
```

Figure 15-13   The TaskHistoryActivity class (part 1 of 3)

The onResume method begins by using the query method of the ContextResolver object to get a cursor for all tasks that have been marked as hidden. This code also sorts these tasks in descending order by the completed date of the task. As a result, the most recent tasks are at the top of the cursor.

After getting the cursor, the onResume method defines the variables needed to create the adapter. More specifically, it gets the ID for the listview_task layout, it specifies the three columns to get the data from, and it specifies the three TextView widgets to send the data to. Then, this code uses these variables to create the adapter. Finally, this code sets the adapter on the ListView widget, and it sets the listener for the adapter.

After setting the listener for the adapter, this code uses the setViewBinder method to convert column data into readable values. To do that, this code creates an anonymous class that implements the ViewBinder interface and overrides its setViewValue method. Within that method, an if/else statement checks whether the column is the date completed column for the task. If so, it gets the long value for the date, converts it to a readable date value, sets the date on the appropriate TextView widget, and returns a true value to indicate that it has bound data to the view.

Then, the if/else statement checks whether the column is the notes column for the task. If so, it checks whether the string contained in this column is equal to a null value or an empty string. If so, it sets a value of "No notes" on the appropriate TextView widget and returns a true value to indicate that it has bound data to the view. Otherwise, this code returns a false value, which allows the default view binding to occur. As a result, the string contained in the name column is bound to the view.

The TaskHistoryActivity class   **Page 2**

```
    @Override
    public void onResume() {
        super.onResume();

        // get cursor
        String where = TASK_HIDDEN + "= '1' ";
        String orderBy = TASK_COMPLETED + " DESC";
        cursor = getContentResolver()
                .query(TASKS_URI, null, where, null, orderBy);

        // define variables for adapter
        int layout_id = R.layout.listview_task;
        String[] fromColumns = {TASK_NAME, TASK_NOTES,
            TASK_COMPLETED};
        int[] toViews = {R.id.nameTextView, R.id.notesTextView,
            R.id.dateTextView};

        // create and set adapter
        adapter = new SimpleCursorAdapter(this, layout_id,
                cursor, fromColumns, toViews, 0);
        taskListView.setAdapter(adapter);
        taskListView.setOnItemClickListener(this);

        // convert column data to readable values
        adapter.setViewBinder(new ViewBinder() {
            @Override
            public boolean setViewValue(View view, Cursor cursor,
                    int colIndex) {
                if (colIndex == TASK_COMPLETED_COL){
                    long dateMillis = cursor.getLong(colIndex);
                    TextView tv = (TextView) view;
                    SimpleDateFormat date =
                            new SimpleDateFormat("EEE, MMM d yyyy HH:mm");
                    tv.setText("Completed on: " +
                            date.format(new Date(dateMillis)));
                    return true;
                }
                else if (colIndex == TASK_NOTES_COL) {
                    String notes = cursor.getString(colIndex);
                    if (notes == null || notes.equals("")) {
                        TextView tv = (TextView) view;
                        tv.setText("No notes");
                        return true;
                    }
                }
                return false;
            }
        });
    }
```

Figure 15-13   The TaskHistoryActivity class (part 2 of 3)

The onPause method closes the cursor that's held by the adapter. Then, it closes the cursor that's held by the activity. This works because the onResume method always gets a new cursor.

The onItemClick method is executed when the user clicks on a task. This method begins by getting the ID and name for the task that was clicked. Then, it displays a dialog box that confirms the deletion.

If the user clicks the Yes button, this code uses the URI defined earlier in the class to delete the specified task. Then, it uses a toast to display a message that indicates whether the task was successfully deleted. In most cases, the task should be deleted successfully, and the Toast should display a message that says, "Delete operation successful".

If the user clicks the No button, this code calls the cancel method of the DialogInterface object, which is the first argument of the onClick method. This closes the dialog and returns to the Task History activity without deleting the task.

The TaskHistoryActivity class   Page 3

```java
@Override
protected void onPause() {
    adapter.changeCursor(null);    // close cursor for the adapter
    cursor.close();                // close cursor for the activity
    super.onPause();
}

@Override
public void onItemClick(AdapterView<?> adapter, View view,
        int position, long id) {

    // get data from cursor
    cursor.moveToPosition(position);
    final int taskId = cursor.getInt(TASK_ID_COL);
    final String taskName = cursor.getString(TASK_NAME_COL);

    // display a dialog to confirm the delete
    new AlertDialog.Builder(this)
    .setMessage("Do you want to permanently delete task: " +
            taskName + "?")
    .setPositiveButton("Yes", new DialogInterface.OnClickListener() {
        @Override
        public void onClick(DialogInterface dialog, int id) {
            // delete the specified task
            String where = TASK_ID + " = ?";
            String[] whereArgs = { Integer.toString(taskId) };
            int deleteCount = getContentResolver()
                    .delete(TASKS_URI, where, whereArgs);
            if (deleteCount <= 0) {
                Toast.makeText(context, "Delete operation failed",
                    Toast.LENGTH_SHORT).show();
            }
            else {
                Toast.makeText(context, "Delete operation successful",
                    Toast.LENGTH_SHORT).show();
                onResume();
            }
        }
    })
    .setNegativeButton("No", new DialogInterface.OnClickListener() {
        @Override
        public void onClick(DialogInterface dialog, int id) {
            dialog.cancel();
        }
    })
    .show();
}
}
```

Figure 15-13   The TaskHistoryActivity class (part 3 of 3)

Perspective

The skills presented in this chapter should be enough to get you started with content providers. However, there is much more to learn about working with content providers. For example, you may want to provide access to data that's stored in files instead of data that's stored in a database. If so, you can search the Android documentation for more information about how to do that.

Or, you may want to create an app that uses an existing content provider. To do that, you can begin by searching the Internet or the Android documentation for more information about the content provider that you want to use. For example, to learn more about the Contacts Provider, you can search for "android contacts provider". Then, you can use the documentation for that content provider to find the URIs and column names that you need to use the content provider.

Terms

content provider
URI (Uniform Resource Identifier)
authority
path
segment
MIME type

Summary

- A *content provider* allows multiple apps to share the same data. This data can be stored in a database or in files. Android uses built-in content providers to make certain types of data available to all apps.

- A *URI* (*Uniform Resource Identifier*) for a content provider includes the name of the provider (the *authority*) and a *path* to the content. Within a path, each front slash begins a new *segment*.

- A *MIME type* is an Internet standard for defining types of content.

- Every data access method of the ContentProvider class has a Uri object as its first argument. This allows the method to determine what table, row, or file to access.

- To allow other apps to query your database, you can override the query and getType methods of the ContentProvider class.

- To allow other apps to modify the data in your database, you can override the insert, update, and delete methods of the ContentProvider class.

- Before other apps can use your content provider class, you must register it by adding a provider element to the AndroidManifest.xml file.

- You can call any methods of a content provider from the ContentResolver object.

- When you work with a content provider that's included with Android, you can import the classes that contain the constants for working with the content provider.

- You can use the AlertDialog.Builder class and the DialogInterface. OnClickListener interface to build and display a dialog box that includes event handlers.

Exercise 15-1   Review and modify the Task List and Task History apps

In this exercise, you'll review the Task List and Task History apps to see how they share the same data. Then, you'll modify the code for the Task History app.

Review the Task List app

1. Open the project named ch15_TaskList that's in the book_apps directory (not the ex_starts directory).

2. Open the TaskListDB class and scroll down to the end of it to view the last four methods that support the content provider.

3. Open the TaskListProvider class and review its code. Note that the delete method supports a URI for all tasks and a URI for a single task.

4. Run the app. Add at least 4 tasks to the list and then mark at least 3 of these tasks as completed, and delete them from the list. This may look like you are deleting these tasks. However, this code only sets the completed date for these tasks and hides them for this app. That way, these tasks are stored for future reference and can be used by other apps such as the Task History app.

Review the Task History app

5. Open the project named ch15_ex1_TaskHistory that's in the ex_starts directory.

6. Open the TaskHistoryActivity class and review its code.

7. Run the app. This should display all tasks that you removed from the Task List app earlier in this exercise.

8. Click one of the tasks to delete it.

Modify the Task History app

9. Modify the TASKS_URI constant so the path is "task" instead of "tasks".

10. Run the app. The app should crash, and the LogCat view should display an error that indicates that the URI is not supported.

11. Fix the TASKS_URI constant so the path is "tasks" again.

12. Run the app again. It should run correctly this time.

13. Modify the code that deletes the task so it uses a URI for a single row.

14. Test this change to make sure it works correctly by deleting another task.

Exercise 15-2   Review and modify the Favorites app

In this exercise, you'll review and modify the Favorites app. This app uses the Contacts Provider that's introduced in this chapter.

Review the Favorites app

1.   Open the project named ch15_ex2_Favorites.

2.   Open the FavoritesListActivity class and review its code.

3.   Run this app. This should display all phone numbers for all contacts that have been marked as a favorite. It should look something like this:

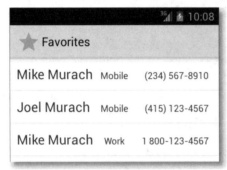

4.   If the app doesn't display any names and numbers, you may need to use the built-in Contacts app to mark at least one contact as a favorite by clicking the star icon. Before you can do that on an emulator, you may need to add a contact.

5.   Click a phone number for a contact. Then, test the resulting dialog box to make sure it works to start a phone call or a text. Of course, you can only complete a text message successfully for a mobile number.

Modify the Favorites app

6.   Modify the app so it only displays mobile phone numbers. To do that, you can modify the WHERE clause so it only returns rows where the DATA2 column is equal to 2.

7.   Run this app to make sure it works correctly.

Learn more about the Contacts Provider

8.   Search the Internet to find the official documentation for the Android Contacts Provider and read more about it.

16

How to work with app widgets

An app widget can display an app's most important information on a device's Home screen. This can allow a user to work with an app without even starting it. Android typically includes several built-in widgets such as the Analog Clock, Power Control, and Music widgets. In this chapter, you'll learn how to build an app widget for the Task List app. This app widget will display the three oldest tasks from the Task List app.

An introduction to app widgets

Before you learn how to create an app widget for your app, you need to learn some background terms and basic skills for working with app widgets.

A Home screen that has app widgets

Figure 16-1 shows a Home screen that has three *app widgets* and seven *app icons*. Here, each icon occupies a single cell. However, each app widget can occupy more than one cell. For example, the Search widget is 4 cells wide by 1 cell tall. As a result, it occupies the entire first row of the screen. Similarly, the second app widget, the Task List widget, is 4 cells wide by 1 cell tall. As a result, it occupies the entire third row of the screen.

The third app widget is the Power Control widget that comes with most versions of Android. This widget makes it easy to extend your phone's battery life by managing power consuming features such as Wi-Fi, bluetooth, GPS, synching, and screen brightness. Of course, the user can also use the Settings app to perform the same functions, but the Power Control widget makes it easier to access these functions.

The user can click any of the app icons to start the related app. For example, the user can click the Task List icon to start the Task List app. However, these icons don't display any data or provide any other functionality.

The user can also click the Task List widget to start the Task List app. However, the user can also get the most relevant data for the app without even starting it. This can potentially save the user time and effort.

How to remove app widgets

If you want to remove an app widget from your Home screen, you can do that by long-clicking on it. Then, you can drag it to the Remove icon that appears at the top of the screen. This doesn't uninstall the app, but it does remove it from your Home screen.

How to add app widgets

When you install an app that includes app widgets, those app widgets are installed on your device, but they aren't usually displayed on the Home screen. However, you can add app widgets to the Home screen. To do that, display all apps. Then, click the Widgets tab to display all app widgets, swipe right or left to find the app widget you want, long-click the app widget, and drag it to the Home screen.

A Home screen with app icons and widgets

Search widget

App icons

Task List widget

Power Control widget

Description

- *App icons* occupy a single cell on the Home screen and provide a way for users to start an app.
- *App widgets* occupy one or more cells on the Home screen. They can display data and receive periodic updates.
- Android typically includes several app widgets such as the Search and Power Control widgets.
- You can create custom app widgets for an app.
- From the Home screen, you can typically start the related app by clicking the app widget.
- To remove an app widget from the Home screen, long-click it and drag it to the Remove icon that appears at the top of the screen.

Figure 16-1   An introduction to app widgets

How to create app widgets

Now that you know how to work with app widgets, you're ready to create one. In particular, you're ready to create a Task List app widget like the one shown in the previous two figures.

How to create the layout

Figure 16-2 shows how to create the layout for an app widget. To start, you can use the graphical editor just as you would use it for any other type of layout. However, an app widget can only use widgets that are supported by the RemoteViews class. This class supports many common widgets such as the TextView and Button widgets. However, it doesn't support other common widgets such as the EditText or CheckBox widgets. For more information about this class, see the next figure.

The layout for the Task List app widget uses a vertical linear layout to display four TextView widgets. Here, the first TextView widget displays the title of the app and the next three TextView widgets display the three oldest tasks.

The layout for the Task List app widget in the graphical editor

The app_widget_top3 layout

```xml
<?xml version="1.0" encoding="utf-8"?>
<LinearLayout xmlns:android="http://schemas.android.com/apk/res/android"
    android:id="@+id/appwidget_top3"
    android:layout_width="match_parent"
    android:layout_height="match_parent"
    android:orientation="vertical"
    android:background="@android:color/background_light" >

    <TextView
        android:id="@+id/titleTextView"
        android:layout_width="wrap_content"
        android:layout_height="wrap_content"
        android:layout_marginLeft="5dp"
        android:layout_marginTop="5dp"
        android:text="@string/app_name"
        android:textColor="@android:color/black"
        android:textStyle="bold" />

    <TextView
        android:id="@+id/task1TextView"
        android:layout_width="wrap_content"
        android:layout_height="wrap_content"
        android:layout_marginLeft="5dp"
        android:text="@string/sample_task"
        android:textColor="@android:color/black" />

    <TextView
        android:id="@+id/task2TextView"
        android:layout_width="wrap_content"
        android:layout_height="wrap_content"
        android:layout_marginLeft="5dp"
        android:text="@string/sample_task"
        android:textColor="@android:color/black" />
```

Figure 16-2   The layout for the app widget (part 1 of 2)

When creating the layout for a widget, it's generally considered a good design guideline to add approximately 8dp of padding to an app widget. That way, there is some padding between the widget, other widgets, and the edge of the screen. However, Android 4.0 (API 14) and later automatically adds padding to app widgets. As a result, you don't typically need to add padding to most app widgets.

The app_widget_top3 layout (continued)

```
    <TextView
        android:id="@+id/task3TextView"
        android:layout_width="wrap_content"
        android:layout_height="wrap_content"
        android:layout_marginLeft="5dp"
        android:text="@string/sample_task"
        android:textColor="@android:color/black" />
</LinearLayout>
```

Description

- You can use the graphical editor to work with the layout for an app widget.

- The layout for an app widget can only use widgets that are supported by the RemoteViews class. For a list of supported layouts and widgets, view the Android documentation for "App Widgets".

- It's generally considered a good practice to add padding to an app widget so there is some padding between the widget, other widgets, and the edge of the screen.

- Android 4.0 (API 14) and later automatically adds padding to app widgets, but if your app supports older APIs, you may need to manually add this padding.

Figure 16-2   The layout for the app widget (part 2 of 2)

How to modify the database class

Figure 16-3 shows how to modify the database class (the TaskListDB class) to support the app widget. To do that, the database class sends a broadcast whenever it modifies the data in the Task table. That way, the class for the app widget can receive the broadcast and update the widget accordingly.

To do that, the database class begins by defining the TASK_MODIFIED constant that uniquely identifies the broadcast action. Then, this class defines the broadcastTaskModified method that sends the broadcast. Finally, this class calls the broadcastTaskModified method from each method that modifies data in the Task table. For example, this figure shows a deleteTask method that deletes a task and then calls the broadcastTaskModified method. However, all methods that insert, update, or delete a task should call the broadcastTaskModified method.

The database class also contains a method named getTopTaskNames that makes it easy to get the task names that are displayed on the widget. This method accepts an argument that specifies the number of names to get. Then, the code within this method gets those names, stores them in an array of strings, and returns the array.

The constant for the action

```
public static final String TASK_MODIFIED =
        "com.murach.tasklist.TASK_MODIFIED";
```

A helper method for broadcasting the action

```
private void broadcastTaskModified() {
    Intent intent = new Intent(TASK_MODIFIED);
    context.sendBroadcast(intent);
}
```

A method that modifies a task and broadcasts the action

```
public int deleteTask(long id) {
    String where = TASK_ID + "= ?";
    String[] whereArgs = { String.valueOf(id) };

    this.openWriteableDB();
    int rowCount = db.delete(TASK_TABLE, where, whereArgs);
    this.closeDB();

    broadcastTaskModified();

    return rowCount;
}
```

A method that gets the data for an app widget

```
public String[] getTopTaskNames(int taskCount) {
    String where = TASK_COMPLETED + "= '0'";
    String orderBy = TASK_COMPLETED + " DESC";
    this.openReadableDB();
    Cursor cursor = db.query(TASK_TABLE, null,
            where, null, null, null, orderBy);

    String[] taskNames = new String[taskCount];
    for (int i = 0; i < taskCount; i++) {
        if (cursor.moveToNext()) {
            Task task = getTaskFromCursor(cursor);
            taskNames[i] = task.getName();
        }
    }

    if (cursor != null)
        cursor.close();
    db.close();

    return taskNames;
}
```

Description

- To keep the data on an app widget current, you can broadcast an action when the app modifies its data. Then, the app widget can receive the broadcast and update itself.

- To get data for an app widget, you can add a method to a database class for the app.

Figure 16-3   How to modify the database class

How to create the provider class

Figure 16-4 shows how to code a provider class for an app widget. This class extends the AppWidgetProvider class.

Within this class, the onUpdate method contains the code that sets up the app widget and displays data on its user interface. Android calls this method when the user adds the app widget to the Home screen. In addition, Android calls this method at the interval specified by the widget's info file, which you'll learn about later in this chapter.

The onUpdate method accepts three parameters: a Context object, an AppWidgetManager object, and an array of IDs for all app widgets for this provider. Within this method, the code begins by looping through all app widgets for this provider. This is necessary because a user can add multiple instances of a widget to the Home screen.

Within this loop, the first two statements create a pending intent for the Task List activity. The third statement creates the RemoteViews object that stores the layout. The fourth statement sets the listener for the widget so clicking on it starts the Task List activity. The fifth and sixth statements use the TaskListDB class to get an array of task names. The next three statements display these names on the corresponding TextView widgets of the layout. Or, if the array contains a null value for one or more of these names, these statements set an empty string on the corresponding TextView widget. Finally, the last two statements use the RemoteViews object to update the app widget with the specified ID. Up until these statements, the changes have only been made to the RemoteViews object but haven't been applied to the app widget.

The RemoteViews object is necessary because it allows Android to display the layout for the app widget in another process. More specifically, it allows Android to display the layout for the app widget in the process for the Home screen. With app widgets, this is the typical use of the RemoteViews object. However, it's possible to use a RemoteViews object to display the layout for an app widget in any process that's designed as a container for an app widget. As a result, other uses of this object are possible.

The AppWidgetTop3 class

```java
package com.murach.tasklist;

import android.app.PendingIntent;
import android.appwidget.AppWidgetManager;
import android.appwidget.AppWidgetProvider;
import android.content.ComponentName;
import android.content.Context;
import android.content.Intent;
import android.widget.RemoteViews;

public class AppWidgetTop3 extends AppWidgetProvider {

    @Override
    public void onUpdate(Context context,
            AppWidgetManager appWidgetManager, int[] appWidgetIds) {

        // loop through all app widgets for this provider
        for (int i = 0; i < appWidgetIds.length; i++) {

            // create a pending intent for the Task List activity
            Intent intent = new Intent(context, TaskListActivity.class);
            PendingIntent pendingIntent =
                    PendingIntent.getActivity(context, 0, intent, 0);

            // get the layout and set the listener for the app widget
            RemoteViews views = new RemoteViews(
                    context.getPackageName(), R.layout.app_widget_top3);
            views.setOnClickPendingIntent(
                    R.id.app-widget_top3, pendingIntent);

            // get the names to display on the app widget
            TaskListDB db = new TaskListDB(context);
            String[] names = db.getTopTaskNames(3);

            // update the user interface
            views.setTextViewText(R.id.task1TextView,
                    names[0] == null ? "" : names[0]);
            views.setTextViewText(R.id.task2TextView,
                    names[1] == null ? "" : names[1]);
            views.setTextViewText(R.id.task3TextView,
                    names[2] == null ? "" : names[2]);

            // update the current app widget
            int appWidgetId = appWidgetIds[i];
            appWidgetManager.updateAppWidget(appWidgetId, views);
        }
    }
```

Figure 16-4   The class for the app widget provider (part 1 of 2)

The onReceive method contains code that's executed when the app widget receives a broadcast. This method accepts two parameters: the application context and the intent that was broadcast.

Within this method, the code begins by checking whether the intent is the TASK_MODIFIED intent that's broadcast by the database class. If so, this code updates the app widget. To do that, the first statement creates an AppWidgetManager object. The second statement creates a ComponentName object for the current class, which is the app widget provider class. The third statement uses these objects to get an array of IDs for all app widgets for this provider. And the fourth statement passes these objects to the onUpdate method shown in part 1 of this figure. This updates the data that's displayed on the app widget. That way, the data that's displayed on the app widget is always synchronized with the data that's displayed by the app.

For most app widgets, you only need to override the onUpdate and onReceive methods of the AppWidgetProvider class to get the widget to work the way you want. However, if you want to execute code when a widget is added to or removed from the Home screen, you may want to override the onEnabled, onDeleted, or onDisabled method. Android calls the onEnabled method when an app widget is added to the Home screen, it calls the onDeleted method when an instance of an app widget is removed from the Home screen, and it calls the onDisabled method when the last instance of an app widget is removed from the Home screen.

The AppWidgetTop3 class

```
@Override
public void onReceive(Context context, Intent intent) {
    super.onReceive(context, intent);

    if (intent.getAction().equals(TaskListDB.TASK_MODIFIED)) {
        AppWidgetManager manager =
                AppWidgetManager.getInstance(context);
        ComponentName provider =
                new ComponentName(context, AppWidgetTop3.class);
        int[] appWidgetIds = manager.getAppWidgetIds(provider);
        onUpdate(context, manager, appWidgetIds);
    }
}
}
```

A constructor and some methods of the RemoteViews class

Constructor/Method	Description
RemoteViews(package, layoutId)	Create a new RemoteViews object that provides access to the views in the specified layout file.
setOnClickPendingIntent(appWidgetId, intent)	Sets the pending intent that's executed when a user clicks on the layout.
setTextViewText(textViewId, text)	Sets the specified text on the specified TextView widget in the layout.

Some methods of the AppWidgetManager class

Method	Description
updateAppWidget(widgetId, views)	Updates the specified widget with the specified RemoteViews object.
getInstance()	Gets an AppWidgetManager object.
getAppWidgetIds(provider)	Gets the IDs of the installed app widgets for the specified provider.

Description

- The class for an app widget must extend the AppWidgetProvider class.
- Android calls the onUpdate method when the user adds the app widget to the Home screen and at the interval specified by the widget's info file.
- Android calls the onReceive method when an action for the app widget is broadcast.
- Android uses a RemoteViews object to display a layout in another process such as the process for the Home screen. The RemoteViews class provides some methods that you can use to modify the layout that's stored in the RemoteViews object.
- The AppWidgetProvider class also provides the onEnabled, onDeleted, and onDisabled methods. If necessary, you can override these methods to execute code when an app widget is added to or removed from the Home screen.

Figure 16-4   The class for the app widget provider (part 2 of 2)

How to configure an app widget

Figure 16-5 shows how to configure an app widget by adding an info file to the res\xml directory. The initialLayout attribute specifies the layout for the app widget.

The minHeight and minWidth attributes specify the minimum height and width for the app widget. In this figure, the minimum height has been set to 40dp, which is approximately 1 cell, and the minimum width has been set to 250dp, which is approximately 4 cells. As a result, the app widget should be 4 cells wide by 1 cell tall.

The updatePeriodMillis attribute specifies the number of milliseconds for the update interval. However, if the device is asleep, these types of updates wake the device to perform the update. As a result, frequent updates of this type can cause significant problems for the battery life. To minimize this issue, Android won't perform this type of update more than once every 30 minutes (1800000 milliseconds), and it's generally recommended to not perform this type of update more than once per hour.

So, if you want to update an app widget more often than once per hour, you need to use a different approach. On one hand, if you want to update the app widget when a particular action occurs in the app, you can broadcast an action from the app that the app widget can receive. Then, you can delete the update-PeriodMillis attribute or set it to a value of 0. This approach is described in the previous figures.

On the other hand, if you need to update an app widget at a specified time interval, you can use an alarm of the ELAPSED_REALTIME or RTC type to broadcast an update action at a specified time interval. Then, the app widget can receive the broadcast and update itself. This approach works similarly to the approach described in this chapter. However, this approach does not wake the device if the device is sleeping. As a result, it conserves battery life. For more details about working with alarms, you can view the documentation for Android's AlarmManager class.

The res\xml\app_widget_top3_info.xml

```
<appwidget-provider
    xmlns:android="http://schemas.android.com/apk/res/android"
    android:initialLayout="@layout/app_widget_top3"
    android:minHeight="40dp"
    android:minWidth="250dp"
    android:updatePeriodMillis="21600000" > <!-- every 6 hours -->
</appwidget-provider>
```

The appwidget-provider element

Attribute	Description
initialLayout	Specifies the layout file for the app widget.
minHeight	Specifies the minimum height for the app widget.
minWidth	Specifies the minimum width for the app widget.
updatePeriodMillis	Specifies the number of milliseconds for the update interval. These updates are not delivered more than once every 30 minutes (1800000 milliseconds). In addition, the actual update is not guaranteed to occur exactly on time.

Approximate sizes for minimum height and width

Cell count	Approximate size
1	40dp
2	110dp
3	180dp
4	250dp

Description

- To configure an app widget, you can add an info file for the app widget to the res\xml directory.

Figure 16-5   How to configure an app widget

How to register an app widget

Before you can use an app widget, you must register it. To do that, you can add a receiver element to the AndroidManifest.xml file as shown in figure 16-6.

When you register an app widget, you need to define a label attribute that specifies the name that Android uses for the app widget in its list of app widgets. For example, this specifies a value of "Task List (4 x 1)" that's stored in the strings.xml file for the app.

Since the AppWidgetProvider class extends the BroadcastReceiver class, registering an app widget is similar to registering a broadcast receiver. In particular, you must add an intent-filter element for every broadcast action that your app listens for. Here, the APPWIDGET_UPDATE action is necessary to update the app widget when it's added to the Home screen and at the interval specified in the info file. Similarly, the TASK_MODIFIED action is necessary to update the app widget when the user inserts, updates, or deletes a task.

To finish registering an app widget, you need to use the meta-data element to specify the type of broadcast receiver and its location. Here, the broadcast receiver is an app widget provider, and its info file is stored in the directory shown in the previous figure.

How to test an app widget

After you register an app widget, you should test it. To do that, you can run the app to install the app widget on a device. Then, you can add the app widget to your Home screen as described in figure 16-2. At this point, you can test the app widget to make sure it works correctly. For the Task List app widget, for example, you can use the Task List app to delete a task. Then, you can make sure the app widget is updated correctly.

The AndroidManifest.xml file

```
<receiver
    android:name="com.murach.tasklist.AppWidgetTop3"
    android:label="@string/appwidget_top3_label" >
    <intent-filter>
        <action
            android:name="android.appwidget.action.APPWIDGET_UPDATE" />
    </intent-filter>
    <intent-filter>
        <action android:name="com.murach.tasklist.TASK_MODIFIED" />
    </intent-filter>
    <meta-data
        android:name="android.appwidget.provider"
        android:resource="@xml/app_widget_top3_info" />
</receiver>
```

The receiver element

Attribute	Description
name	The class for the app widget provider.
label	The name that Android uses for the app widget in its list of app widgets. For example, "Task List (4X1)".

The meta-data element

Attribute	Description
name	The type of broadcast receiver.
resource	The name of the info file for the app widget.

Description

- The AppWidgetProvider class extends the BroadcastReceiver class. As a result, an app widget is also a broadcast receiver.

- To register an app widget, add a receiver element to the AndroidManifest.xml file. For all app widgets, you should add an intent-filter element that listens for the APPWIDGET_UPDATE action. In addition, you may want to add intent-filter elements to listen for other broadcast actions.

- To test an app widget, you can run the app and add the app widget to your Home screen as described in figure 16-2.

Figure 16-6   How to register and test an app widget

Perspective

The skills presented in this chapter should be enough to get you started with app widgets. However, Android provides many more features for working with app widgets. For example, you can create app widgets that can be added to the Lock screen. This provides a way for the user to use your app widget without even unlocking the device. You can add an activity that provides a way for users to configure an app widget. You can create app widgets that the user can resize. And so on. To learn more about these features, you can begin by searching the Android documentation for more information about app widgets.

Terms

app icons
app widgets

Summary

- *App icons* occupy a single cell on the Home screen and provide a way for users to start an app.

- *App widgets* occupy one or more cells on the Home screen and can display data and receive periodic updates. You can create custom app widgets for an app.

- Android calls the onUpdate method of an app widget when the user adds the widget to the Home screen and at the interval specified by the widget's info file.

- Android calls the onReceive method of an app widget when an action for the widget is broadcast.

- Android uses a RemoteViews object to display a layout in another process such as the process for the Home screen.

- To register an app widget, add a receiver element to the AndroidManifest.xml file.

Exercise 16-1   Review and modify the Task List app widget

In this exercise, you'll review the code for the Task List app widget. Then, you'll test this app widget to see how it works. Finally, you'll modify this app widget to improve how it works.

Review the app widget

1. Open the project named ch16_ex1_TaskList.

2. Open the app_widget_top3 layout. Then, view it in the graphical editor, and view its XML.

3. Open the AppWidgetTop3 class and view its code. This code should call the getTopTaskNames method from the TaskListDB class, and it should receive the TASK_MODIFIED action.

4. Open the TaskListDB class. Then, review the code for the getTopTaskNames method. Also, review the methods of the TaskListDB class that modify the data in the Task table. Note that some of these methods broadcast the TASK_MODIFIED action.

5. Open the app_widget_top3_info.xml file that's in the res\xml folder. This file should specify that the app widget should be 4 cells wide by 1 cell tall and that it should be updated every 6 hours.

6. Open the AndroidManifest.xml file and review the receiver element. This file should specify the actions that the app widget receives, and it should specify the info file for the app.

Test the app widget

7. Run the app. This should start the Task List app. If this app displays less than three tasks, add tasks until this app displays at least five tasks.

8. Add the app widget to the Home screen. To do that, you may need to remove other app icons or widgets to make room for the app widget. Note the three tasks that the Task List app widget displays.

9. Switch to the Task List app and delete one of the three tasks that's currently displayed by the app widget.

10. Switch back to the Home screen to make sure the app widget has been updated to reflect the changed data.

Modify the app widget

11. Open the TaskListDB class. Make sure that all of the methods that modify the Task table broadcast the TASK_MODIFIED action. This should include the methods that support the content provider.

12. Open the app_widget_top3_info.xml file that's in the res\xml folder.

13. Modify the info file so it specifies that the app widget should be 3 cells wide by 1 cell tall.

14. Modify the info file so Android does not update the app widget at a specified interval. This should help to conserve battery life.

15. Test the app widget to make sure it still works correctly. To do that, run the app again. Then, remove the old app widget from the Home screen and add the modified app widget to the Home screen.

Exercise 16-2   Add an app widget that displays a count of tasks

In this exercise, you'll add another app widget to the Task List app presented in this chapter. This app widget should be 1 cell wide by 1 cell tall, and it should display a count of the total number of incomplete tasks.

1. Open the project named ch16_ex2_TaskList.

2. Open the TaskListDB class and note that it includes a method named getTaskCount that gets a count of incomplete tasks.

3. Navigate to the res\drawable-xhdpi directory and view its contents. Note that it includes an icon file named ic_task_count.png.

4. Create a layout named app_widget_task_count that uses a TextView widget to display the count of incomplete tasks on the icon named ic_task_count. To accomplish this, you can set the background property of the TextView to the icon resource.

5. Create a class named AppWidgetTaskCount that displays the count of task widgets. This class should update the data on the layout when the widget is first added to the Home screen and when the TASK_MODIFIED action is received.

6. Create an info file for the app widget that specifies that the icon is 1 cell wide by 1 cell tall.

7. Open the Android manifest file and add a receiver element that registers the app widget.

8. Run the app on an emulator or device. Note the number of tasks that are displayed by the app.

9. Add the app widget to the Home screen and check the task count to make sure the app widget is working correctly. If the app displays 3 tasks, the app widget should look like this:

Section 5

Advanced Android skills

In sections 1 through 4, you learned the skills that you need to develop many types of Android apps. In this section, chapter 17 shows several ways to deploy most apps, including how to publish an app on the Google Play store.

Chapter 18 shows how to develop the Run Tracker app. This app tracks a device's location and displays it on a map. In this chapter, you'll learn how to use Android to work with the newest version of Google Maps. Since this chapter requires some understanding of deployment, I recommend reading chapter 17 before chapter 18.

17

How to deploy an app

So far in this book, you've learned how to develop, test, and debug an Android app. Once you finish thoroughly testing and debugging your app, you're ready to deploy your app. One way to deploy an app is to distribute it directly to your users by making it available from a website. Another way is to publish it on an application marketplace. In this chapter, you'll learn how to publish your app on the most popular application marketplace, Google Play.

An introduction to distribution and monetization

Before you distribute an app, you should examine the various distribution options. Similarly, if you want to make money from your app, you need to find a way to get paid for it. This is known as *monetization*.

How distribution works

Figure 17-1 begins by showing two ways to distribute an app. Typically, you distribute an app through an *application marketplace*. Today, Google Play (formerly known as the Android Market) is the most popular application marketplace. This figure, for example, shows the Google Play listing for a free Tip Calculator app. This listing includes descriptive text and images for the app.

If you want to use another application marketplace instead of or in addition to Google Play, that's possible too. Many other application marketplaces exist, and they all have their advantages. For example, the Amazon Appstore has some features that aren't available from Google Play.

Another distribution option is to make an app available directly to users via a website download or an email attachment. This option is especially appropriate for a proprietary app that's only used within an organization. However, it can also be a good way to promote an app, especially a free version of an app.

A free app displayed in Google Play

Two ways to distribute an app

- Through an application marketplace
- Directly to the user via a website, email, etc.

Two application marketplaces

- Google Play
- Amazon Appstore

Description

- You can distribute an app through an *application marketplace* such as Google Play, or you can distribute it directly to users via a website or email.
- Google Play, formerly known as the Android Market, is a digital marketplace for Android apps that's developed and maintained by Google.

Figure 17-1   How distribution works

How monetization works

Figure 17-2 begins by listing three ways to monetize an app. First, you can create a *paid app*. For this type of app, the customer pays a one-time fee before he or she can install the app.

Second, you can use *in-app billing*. This provides for *in-app products* that have a one-time purchase or *subscriptions* that have a recurring, automated billing. To learn how to add in-app billing to your app, you can start by searching for "in-app billing" in the Android documentation. This figure, for example, shows a Google Play listing for a Sports Tracker app that's free but includes in-app billing.

Third, you can add advertising to your app. For example, you can add a banner ad to your app. To do that, you need to work with a *mobile advertising network* that serves ads to mobile devices. AdMob is a popular mobile advertising network that's owned by Google, but other advertising networks are also available. To learn how to add advertising to your app, you can start by searching for "advertising" in the Android documentation.

If you publish a free app on Google Play, you can't change it to being a paid app. However, you can add in-app billing or subscriptions, and you can add advertising. On the other hand, if you publish a paid app, you can change it at any time to being a free app.

A common strategy for monetizing an app is to provide a free version of an app that includes advertising. Then, you can provide a paid version of that app that removes the advertising. Or, you can provide a free version of an app that doesn't include all features. Then, the paid version can provide additional features. If one of your goals is to make money from the apps that you develop, you should create a strategy for monetizing your app before you publish it.

An app with in-app billing displayed in Google Play

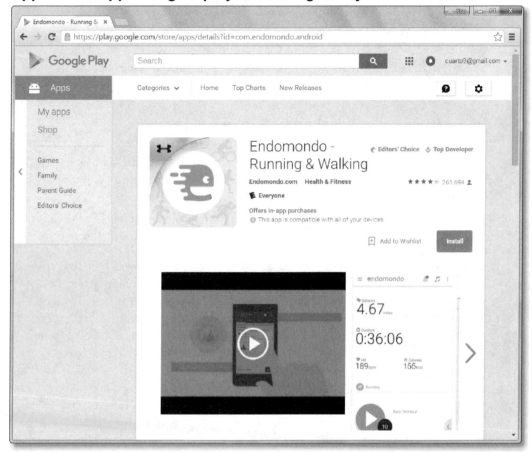

How to monetize an app

- Paid apps
- In-app billing
- Advertising

Description

- Getting paid for an app is referred to as *monetizing* an app.
- If you publish a free app, you can't change it to being a paid app. However, you can sell in-app products or subscriptions, and you can add advertising.
- If you publish a *paid app*, you can change it at any time to being a free app.
- *In-app billing* provides for *in-app products* that have a one-time purchase or *subscriptions* that have a recurring, automated billing.

Figure 17-2   How monetization works

How to create a release build

Now that you understand the options for distributing and monetizing an app, you're ready to create a release build for your app. Then, you can distribute this release build via a website or email. Or, you can publish this release build to a marketplace such as Google Play.

How to prepare an app for release

Figure 17-3 shows a Tip Calculator app that has been thoroughly tested and debugged in the development environment and is ready to be released. It also lists some tasks you should complete before creating the release build. Most of the time, you complete these tasks as you create, test, and debug the app. However, if you haven't already completed these tasks, you should complete them before releasing the app.

To start, you should make sure the package name for the app is unique and appropriate. That's because Android uses the package name to identify the app. As a result, no two apps can use the same package. If you plan to create free and paid versions of an app, you may want to adjust the package names accordingly.

You should make sure to set the final icon for the app. Most of the time, you set this icon when you create the app, but you may want to switch to a more professional app icon just before releasing the app.

You should make sure the version numbers are correct for the app and its components. To set the correct version for the app, you can edit the Android manifest file for the app. If the app uses another component such as a database, you can edit the version number for that component too. For example, you can edit the database class to set the version number for the database.

You should add any copyright info or *End User License Agreement* (*EULA*) you want. To do that, you can add copyright info to an About screen. Similarly, you can add a starting screen that displays the EULA and requires the user to accept it or decline it. If the user accepts the EULA, this screen should not be displayed on subsequent starts.

If you haven't already done it, you should finish localizing the app. This involves creating resource directories for various countries and adding strings. xml files to those directories that contain translations for other languages as described in chapter 3.

You should remove any old files from the project. During development, it's common to add files such as icon files that don't end up being used by the release version of that app. To keep the file size of the release build small, you should delete all unused files.

Finally, you should remove logging statements from the source code. These statements are useful during development, but you should remove them from a release build.

A Tip Calculator app

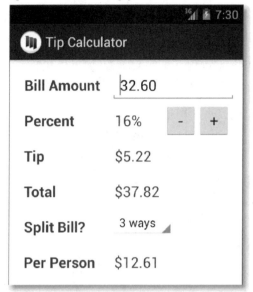

Final checklist

1. Set the final package name.
2. Set the final application icon.
3. Set the version numbers for the app and all of its components.
4. Add copyright info.
5. Add an *End User License Agreement* (*EULA*).
6. Finish localizing the app.
7. Remove any old files from the project.
8. Remove logging statements from the source code.

Description

*   After you have thoroughly tested and debugged your app for all devices and emulators in your development environment, you can prepare an app for release by performing some final checks on the app.

Figure 17-3   How to prepare an app for release

How to create the signed APK file

To run an app on a device or emulator, the app must be stored in an *APK file* (*Android package file*) that's *signed* with a *digital certificate*. During development, Android Studio automatically generates a *debug key* and uses it to sign the APK file. Before distributing an app, you must generate a *release key* and use it to sign the APK file.

The JDK includes command-line tools that you can use to generate a release key and sign an app. However, Android Studio provides an easier way to generate the release key and APK file for an app.

When you use the procedure shown in figure 17-4, you respond to several dialogs that are displayed by Android Studio. To start, in the first dialog, you can choose to create a new *key store*, which is a file that contains one or more digital certificates. Or, you can choose to use an existing key store. Either way, you need to specify the location and passwords for the key store and the release key. In this figure, the dialog shows that the key store is stored in a file named murach_release_keystore.jks and the release key has a name of murach_release_key.

Android uses the key store file to verify any updates to an app. If you lose this file, you won't be able to update your app. Conversely, if a hacker gains access to this file, he or she can update your app. For example, a hacker could replace your app with another app that contains malicious code. As a result, it's important to store this file in a secure location. And, of course, it's important to store this file in a location that you can remember.

When you create a new key store, you can use the dialog box shown in this figure to specify the information for a digitial certificate for a release key. When you do this, you typcially want to specify a number of years that's valid for the entire life of the app. Since it's hard to predict how long an app might live, you should specify a large value such as 25 or even 50 years. Another issue to consider is that Google Play requires the release key to be valid for at least 20 years.

After you finish the dialog shown in this figure, you can use another dialog to specify the location of the signed APK file. At this point, you can distribute this APK file directly to your users as described in the next figure, or you can publish it on Google Play as described after that.

The New Key Store dialog

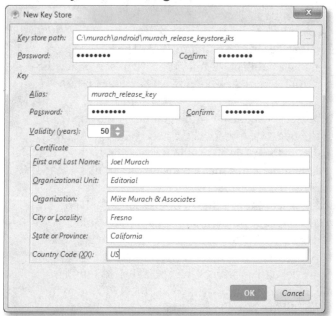

Procedure

1. Select the Build→Generate Signed APK item from the menu system.
2. Click the "Create new" button to create a new key store file, or click the "Choose existing" button to select an existing key store file.
3. Specify the location and password for the key store file.
4. Specify the alias and password for the release key.
5. If creating a new key store, specify the information for the certificate.
6. Specify the location and filename for the APK file.

Description

-   To run an app on a device or emulator, the app must be stored in an *APK file* (*Android package file*) that's *signed* with a *digital certificate*.
-   During development, the IDE automatically generates a *debug key* and uses it to sign the APK file. Before distributing an app, you must generate a *release key* and use it to sign the APK file.
-   A *key store* is a file that contains one or more digital certificates. If you lose this file, you won't be able to update your app. Conversely, if a hacker gains access to this file, he or she can update your app. As a result, it's important to store this file in a secure location.

Figure 17-4   How to create the signed APK file

How to distribute directly to the user

Once you have a signed APK file for an app, you can distribute it directly to your users via a website or email as described in figure 17-5. Then, your users can install the app without using an established Android marketplace. Since the user doesn't download the app from a marketplace, this is sometimes referred to as *side loading*.

If you are a developer, you need to delete any old versions of the app from your Android device that use a debug key. Otherwise, when you try to install the new version of the app that uses the release key, the installation will fail and Android will display a message that indicates that the installation failed due to different application signatures.

In addition, you may need to select the Unknown Sources item on your Android device to allow apps from sources other than Google Play. Some devices have this item selected by default. However, if this item isn't selected by default, you need to select it.

Once you have removed any debug versions of the app and selected the Unknown Sources item, you can install the app by copying the APK file to the device and clicking on it. Two of the easiest ways to copy the APK file to a device are to download it from a website or to receive it as an email attachment.

As you review the procedure shown in this figure, note that step 1 only applies to developers who have installed a version of the app that uses a debug key. In addition, on some devices, the Unknown Sources item described in step 2 is selected by default. As a result, some users only need to follow steps 3 and 4.

How to distribute via a website

To distribute via a website, you can create a mobile-friendly web page that includes a link to download the APK file for the app. Then, a user can download the app by using the browser on the Android device to view that web page and download the APK file.

How distribute via email

To distribute via email, you can attach the APK file to an email and send it to your users. Then, a user can click on the attachment to download the APK file. Although this works seamlessly when your users have a new Android device and a Gmail account, it might not work at all for older Android devices or other types of email accounts. In that case, your users need to find another way to copy the attachment to their device. One easy way to do that is to download the file from the web.

How to install an app from an APK file

1. If you are a developer, delete any old versions of the app from your Android device that use a debug key.
2. Make sure your Android device allows apps from sources other than Google Play.
   - For Android 4.0 and later, start the Settings app and select the Security→Unknown Sources item.
3. Copy the APK file to your Android device.
   - If the APK file is available from a web page, use the browser on your Android device to view that web page and download the APK file.
   - If the APK file has been sent to you as an email attachment, use your Android device to open the email and click on the attached APK file. This works seamlessly with newer Android devices and Gmail. However, this may not work as well with older devices or other types of email.
4. Click the APK file to install the app.

Description

- Once you have a signed APK file, you can distribute it directly to the user without using an Android marketplace.
- Installing an app from a source other than an established Android marketplace is sometimes referred to as *side loading*.

Figure 17-5   How to distribute directly to the user

How to publish on Google Play

Once you have a signed APK file for an app, you're just a few steps from making it available on the Google Play store. To do that, you need to create some online accounts. Then, you can use the Developer Console for Google Play to upload your app and manage its listing.

How to set up a publisher account

Before you can upload an APK file to Google Play, you must set up a publisher account as shown in figure 17-6. To start, you can sign in with the Google account that you want to use for publishing apps. A Google account is the same kind of account that you use to work with any Google product such as Gmail or the Google Calendar.

If you represent an organization, you may want to create a Google account for the organization. That way, several developers can use that account to publish the app. In this figure, for example, I am signed in as Joel Murach. However, to publish this app for an organization such as Mike Murach & Associates, I could sign in using the Google account for Mike Murach & Associates.

After you sign in with the correct Google account, you can accept the developer agreement, pay the registration fee ($25) for that account, and complete the details for that account. This creates a Google Play account that you can use to publish free apps. However, if you want to create paid apps or use in-app billing, you also need to set up a merchant account. To do that, you can follow one of the links that's displayed after you finish creating the Google Play account.

The Google Play signup page

`https://play.google.com/apps/publish/signup`

The Accept Developer Agreement page

Procedure

1. Go to the website shown above.
2. Sign in with your Google account.
3. Accept the developer agreement.
4. Pay the registration fee ($25). To do that, you need to create a Google Wallet account if you don't already have one.
5. Complete your account details.
6. If you want to create paid apps or use in-app billing, follow the links to set up a merchant account.

Description

- Before you can upload an APK file to the Google Play store, you must set up a publisher account.

Figure 17-6   How to set up a publisher account

How to use the Developer Console to publish an app

When you finish creating a Google Play account, there is a link for publishing an app on Google Play. You can use this link to navigate to a Developer Console like the one shown in figure 17-7. Or, you can navigate to the Developer Console by entering the URL shown at the top of this figure. Either way, once you display the Developer Console, you can use it to upload your app to Google Play and manage its listing.

To upload your app, you can click the APK link. Then, you can use the three tabs to upload the APK for alpha testing, beta testing, or production. If you upload the Tip Calculator app to the Production tab, this app will be released to the general public as soon as it is published.

The Developer Console won't let you publish your app until Google processes your registration fee, and that can take up to 48 hours. Before then, the app has a status of "Draft". This lets you enter the listing and pricing details for the app. However, the app isn't published until you select the "Publish app" item. After you do that, the status of the app changes to "Published".

To modify the listing for the app, you can click the Store Listing link. Then, you can specify the required text and images for the app. This can include text and images like the ones shown in the first two figures in this chapter. To create the required text, you may want to work with someone who specializes in marketing. Similarly, to create the required images, you may want to work with someone who specializes in graphic design.

To modify the pricing and distribution for the app, you can click the Pricing and Distribution link. Then, you can specify whether the app is a free or paid app. In addition, you can specify the countries where you want to distribute the app. Of course, you typically want to make sure your app provides for the primary language of each country.

If the app is a paid app, you need to specify the price for each country. This is necessary because every country uses a different currency. Fortunately, if you specify a default price in US dollars, the Developer Console can generate prices for other countries based on current exchange rates.

The Developer Console page

`https://play.google.com/apps/publish`

The Developer Console page for the Tip Calculator app

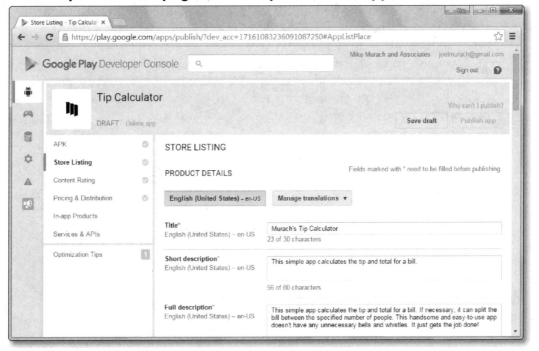

Procedure

1. Log in to the Developer Console.
2. Upload the APK file.
3. Edit the store listing. This includes specifying a title, a description, screenshots, and so on.
4. Edit the pricing and distribution (optional).
5. Edit the in-app products (optional).
6. Select the "Publish this app" item from the combo box to the right of the app's name.

Description

- You can use the Developer Console to publish your app and to manage its listing on Google Play.
- At a minimum, the listing for an app must include the name of the app, a brief description of the app, two screenshots of the app, and an image for the app's logo.
- It may take 48 hours for your registration fee to be processed. The Developer Console won't allow you to publish an app until that fee is processed.

Figure 17-7   How to use the Developer Console to publish an app

How to view the listing for an app

Once you publish an app on Google Play, you can view the listing for the app by going to Google Play and searching for the app. Figure 17-8, for example, shows the listing for the Tip Calculator app. I displayed this listing by using a browser to navigate to Google Play, searching for "murach's tip calculator", and clicking on the link for this app.

If you use the Developer Console to edit the listing for your app, the changes should appear when you reload the Google Play page for your app. For example, if I used the Developer Console shown earlier in this chapter to edit the text or images for this app, those changes should appear in this listing.

The Google Play listing

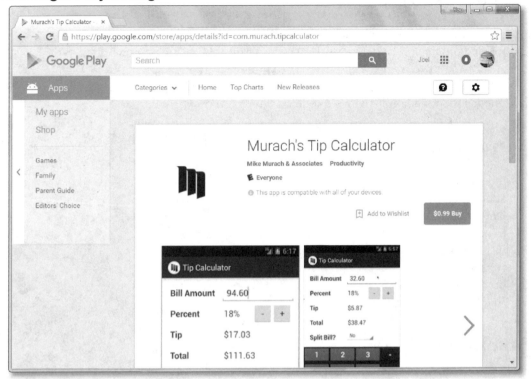

Description

- Once you publish an app on Google Play, you can view the listing for the app by going to Google Play and searching for the app.

Figure 17-8   How to view the listing for an app

Perspective

The skills presented in this chapter should be enough for you to publish a free or paid app on Google Play. However, if you want to add in-app billing or advertising to your app, you'll need to do more research to learn how to do that. Of course, the skills presented in this chapter and throughout this book should provide the foundation you need to learn how to accomplish these tasks.

Terms

monetization	APK file (Android package file)
application marketplace	Signed app
paid app	digital certificate
in-app billing	debug key
in-app products	release key
subscriptions	key store
mobile advertising network	side loading
End User License Agreement (EULA)	

Summary

- You can distribute an app through an *application marketplace* such as Google Play, or you can distribute it directly to users via a website or email.

- Google Play, formerly known as the Android Market, is a digital market-place for Android apps that's developed and maintained by Google.

- Getting paid for an app is referred to as *monetizing* an app.

- *In-app billing* provides for *in-app products* that have a one-time billing or *subscriptions* that have a recurring, automated billing.

- To run an app on a device or emulator, the app must be stored in an *APK file* (*Android package file*) that's *signed* with a *digital certificate*.

- During development, the IDE automatically generates a *debug key* and uses it to sign the APK file. Before distributing an app, you must generate a *release key* and use it to sign the APK file.

- A *key store* is a file that contains one or more digital certificates. This file is required to update your app. As a result, it's important to store this file in a secure location that you can remember.

- Installing an app from a source other than an established Android market-place is sometimes referred to as *side loading*.

Exercise 17-1   Install the Tip Calculator app on your device

In this exercise, you'll install the release build of the Tip Calculator app on an Android device.

1. Select the Android device that you want to use.

2. Uninstall any old versions of the Tip Calculator app that use the debug key.

3. Make sure the Android device allows apps from sources other than Google Play. To do that, you can start the Settings app and select the Unknown Sources item.

4. Copy the APK file that's in the ex_starts\ch17_APK directory to the device. One easy way to do this is to attach it to an email and send it to a Gmail account. Then, you can open Gmail on the device, open the email, and click on the attachment.

5. Click the APK file to install the Tip Calculator app.

6. Start the Tip Calculator app and make sure it's the app presented in this chapter.

Exercise 17-2   Publish an app on Google Play

In this exercise, you'll publish your first app on Google Play.

1. Design and develop an app that you want to publish on Google Play.

2. Test and debug that app on all devices and emulators in your development environment.

3. Prepare the app for release.

4. Create the signed APK file for the app.

5. Create a Google Play publisher account. To do that, you must pay the registration fee ($25).

6. Use the Developer Console to publish your app. To do that, you must upload the APK file and provide the required text and images for the app.

7. Use Google Play to install the app on one or more devices.

8. Test the installed app on one or more devices and make sure it works the way you want.

9. Check Google Play for feedback and continue to improve your app.

18

How to work with locations and maps

Most mobile devices include hardware that can provide the location of the device to an Android app. As a result, you can use the location of a device to customize your app so it's appropriate for the user's current location.

Often apps that use a device's location also use maps. For example, they might display a map and zoom in on the current location. To do that, the app can use any Android API for mapping. This chapter shows how to use the newest version of the most popular Android API for mapping, Google Maps version 2.

An introduction to locations and maps

This chapter begins by presenting the user interface for an app that tracks a device's location and displays it on a map. Then, it describes two competing Android APIs for working with maps, Google Maps and MapQuest. Finally, this topic describes two versions of the Android API for Google Maps.

Although it's possible to use an emulator to test locations and maps, it's easier to use an actual device. As a result, this chapter doesn't show how to use an emulator to test locations. Instead, it assumes that you have an actual device that you can use for testing. If you need to use an emulator for testing, you can search the Internet for information to learn how to send test location data to an emulator.

The user interface for the Run Tracker app

The skills presented in this chapter are necessary to create the Run Tracker app shown in figure 18-1. When the Run Tracker app starts, the Stopwatch activity displays a stopwatch that you can use to start, stop, and reset a run. Typically, you begin by clicking the Start button to start a run. Then, the Start button becomes a Stop button.

If you want to pause the run, you can click on the Stop button. Then, the Stop button becomes a Start button that you can use to continue the run. However, if you want to delete the current run, you can click the Reset button to do that. Then, you can click the Start button to start a new run.

At any time, you can view the map for the run by clicking the View Map button. This should display a red marker for the user's current location and a black line to indicate the route of the run. In addition, it should display zoom controls that you can use to zoom in or out. When you're ready to switch back to the Stopwatch activity, you can click on the View Stopwatch button. This button is transparent by default, which is usually what you want since it allows you to view the map and compass behind it.

When you start the stopwatch, the app starts a service that uses *GPS* (*Global Positioning System*) to get the location of the device. As a result, Android displays the GPS icon and a stopwatch icon in the notification area. In this figure, the user has started a run. As a result, it shows both icons in the notification area above both activities.

To remove these notifications, you can stop or reset the stopwatch. This stops the service that uses GPS. As a result, Android removes the GPS icon and the stopwatch icon from the notification area.

Although I designed this app for running or jogging, you can use it for walking, biking, or even driving. It should work for any activity where you are moving and your device can access GPS. However, GPS does not usually work indoors, so you typically need to be outside.

The Stopwatch activity

The Run Map activity

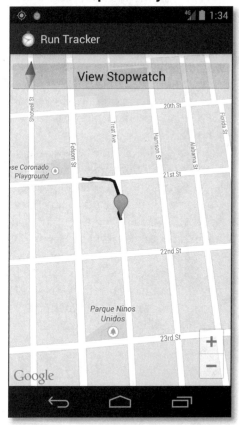

Description

- When the app starts, it displays the Stopwatch activity.
- To start or continue a run, the user can click on the Start button.
- To pause a run, the user can click on the Stop button.
- To delete a run, the user can click on the Reset button.
- To display the Run Map activity, the user can click on the View Map button.
- The Run Map activity displays a red marker for the user's current location and uses a black line to indicate the path of the run.
- When the user starts the stopwatch, the app displays a stopwatch notification icon and starts a service that uses *GPS* (*Global Positioning System*) to track the user's location. This causes Android to display a notification icon for GPS.
- To remove these notifications, the user can stop or reset the timer.
- To display the Stopwatch activity, the user can click on the View Stopwatch button.

Figure 18-1   The user interface for the Run Tracker app

An introduction to determining location

From an Android developer's point of view, there are three ways to determine the location of a device. Figure 18-2 describes their pros and cons.

First, you can determine a device's location by using the device's GPS receiver to check signals from GPS satellites. The advantage of this approach is that it's the most accurate. However, for GPS to work, the device must have a GPS receiver and an unobstructed line of sight to at least three GPS satellites. As a result, GPS doesn't work well or at all when the device is indoors. Also, GPS consumes a lot of battery power, significantly more than the other techniques. Finally, GPS returns the location more slowly than using cell or Wi-Fi signals.

Second, you can determine a device's location by using signals from cell towers or Wi-Fi hotspots to determine location. Although this technique isn't as accurate as GPS, it works outdoors and indoors, consumes less battery power, and returns the location more quickly.

Third, you can listen to location updates that Android is sending to other apps. The advantage of this approach is that it doesn't use any extra battery power. The disadvantage is that it doesn't let you control when your app gets updates. For example, if other apps don't request location updates, your app doesn't receive any updates at all.

GPS

- Uses signals from GPS (Global Positioning System) satellites to determine the location of the device

Pros
- Most accurate

Cons
- Only works outdoors, not indoors
- Consumes more battery power
- Returns the location slowly

Network

- Uses signals from the cell network or Wi-Fi hotspots to determine location

Pros
- Works outdoors and indoors
- Consumes less battery power
- Returns the location more quickly

Cons
- Less accurate

Passive

- Listens to location updates that are being sent to other apps

Pros
- Doesn't use any extra battery power

Cons
- Doesn't let you control when your app gets updates

Description

- You can use a device's GPS receiver or its network signals (cell or Wi-Fi) to find a device's current location.
- You can also find a device's current location by listening for location updates that are being sent to other apps.

Figure 18-2   An introduction to determining location

An introduction to maps

You've probably used more than one Google Map by now on a mobile device. If so, you know how easy one is to use. As you might expect, Google Maps is the most popular Android API for mapping. However, there are other Android APIs for mapping. One of the most prominent is MapQuest, one of the oldest mapping services. Figure 18-3 describes some of the pros and cons of these two APIs for mapping.

Google Maps is a complete mapping solution that provides (1) the Android API, (2) the base maps, and (3) the servers that serve the map tiles to the app. On the good side, this makes Google Maps easy for developers to use. On the bad side, this makes it more difficult to use one component of Google Maps without using another component.

Google Maps is available for free for most apps. At the time of this writing, for example, Google Maps allows 25,000 free map requests per day before it starts charging a small fee.

Google Maps also provides many features. It provides for *geocoding*, which is the process of converting latitude/longitude coordinates to street addresses and back. It provides for *routing*, which is the process of providing directions. And it provides for other features such as street view, traffic layers, and 3D buildings.

However, Google Maps isn't free for its heaviest users. As a result, if your app is going to generate more than 25,000 map requests per day, you might want to investigate alternatives to Google Maps. In addition, Google Maps isn't open-source. As a result, the developer doesn't have much control over the API or its pricing.

MapQuest works much like Google Maps, but it has recently embraced open-source mapping, allowing developers to choose between its proprietary maps or open maps that use data from the OpenStreetMap project. The goal of the OpenStreetMap project is to create a free map of the world that's available under the Open Database License.

Although MapQuest doesn't have as many features as Google Maps, the open version of its Android API is free for all users. As a result, if your app is going to generate more than 25,000 map requests per day, you may want to cut costs by using MapQuest. Similarly, if you want to have more control over the map API, or if you are an open-source enthusiast, you may want to try using MapQuest.

Google Maps

Website
`https://developers.google.com/maps/`

Description
- Google Maps includes (1) the Android API, (2) the base maps, and (3) the servers to serve map tiles.

Pros
- Popular and familiar
- Free for most users
- Many features:
  - Geocoding
  - Routing
  - Street view
  - Traffic
  - 3D Buildings

Cons
- Not free for the heaviest users
- Lack of control

MapQuest

Website
`http://developer.mapquest.com/`

Description
- MapQuest works much like Google Maps, but it has recently embraced open-source mapping, allowing developers to choose between its proprietary maps or open maps that use data from the OpenStreetMap project. The goal of the OpenStreetMap project is to create a free map of the world that's available under the Open Database License.

Pros
- Open version is free for all users

Cons
- Not as many features

Description
- Google Maps is the most popular Android API for working with maps. However, there are several alternatives to Google Maps. One alternative is MapQuest.
- *Geocoding* is the process of converting latitude/longitude coordinates to street addresses and back.
- *Routing* is the process of using map data to provide directions.

Figure 18-3   An introduction to maps

An introduction to the Google Maps Android API

Figure 18-4 describes two versions of the Google Maps Android API. Version 1 (v1) was widely used prior to March 2013 because it makes it easy to add powerful mapping capabilities to an app. As a result, many existing apps use this version of the API.

However, the Android documentation recommends version 2 (v2) for all new development. In fact, Google Maps no longer accepts requests for an API key for version 1. As a result, if you don't already have an API key for version 1, you must use version 2. In addition, as you might expect, version 2 has many new features that improve upon the features available in version 1. For example, version 2 uses fragments to provide more flexibility for working with different screen sizes, displays faster, uses less bandwidth, and supports 3D. For these reasons, this chapter only covers version 2 of the API.

If you need to work on an app that uses version 1, you shouldn't have much trouble finding tutorials and documentation for it. There are plenty of books and online tutorials available for version 1, and the official online documentation for version 1 is still available too.

Google Maps Android API versions

Version 1 (v1)

- Has been officially deprecated as of December 3rd, 2012.
- Requests for API keys are no longer accepted as of March 18th, 2013.
- New features are no longer being added.
- Continues to work for existing apps.
- Uses the MapActivity class to encapsulate maps. This doesn't provide much flexibility for displaying maps on both small and large screens.
- Makes it easy to add powerful mapping capabilities to your application.
- Provides built-in downloading, rendering, and caching of map tiles.
- Provides a variety of display options and controls.
- Uses a MapView widget to display a map with data obtained from the Google Maps service. This widget handles key and touch events to pan and zoom the map automatically, including handling network requests for additional map tiles.
- Allows several types of overlays on the map.
- Is not part of the standard Android library.
- Is a part of the Google APIs add-on.

Version 2 (v2)

- Is encouraged for all new app development.
- Is distributed as part of the Google Play services SDK.
- Uses the MapFragment class to encapsulate maps. This allows you more flexibility for displaying maps on both small and large screens.
- Uses vector tiles. This allows maps to display faster and use less bandwidth.
- Uses improved caching. This typically allows the map to display without showing empty areas.
- Supports 3D. This allows you to show the map with perspective by moving the user's viewpoint.

Description

- Version 1 of the Google Maps Android API was widely used prior to March 2013. Many existing apps use this version of the API.
- Version 2 of the Google Maps Android API is recommended for all new development as of March 2013.

Figure 18-4   An introduction to the Google Maps Android API

How to configure the Google Maps Android API v2

Now that you understand some of the options for working with maps, you're ready to learn how to configure version 2 of the Android API for Google Maps. To do that, you need to add the Google Play services library to your project, get an API key for your app, register that key, and set the required permissions for the app.

How to create a new Google Maps project

The easiest way to add the Google Play services library to your app is to use Android Studio to create a new project that's based on the Google Maps Activity template in figure 18-5. If you don't have an API key, this is also the easiest way to get one. Using this approach also creates a skeleton map activity that you can use as a starting point.

Sometimes, however, you might want to add a map to an app that already exists. To do that, you can manually add the Google Play services API to your project as shown in the next figure.

How to create a new Google Maps project

Procedure

1. Start Android Studio and create a new project as described in chapter 2.
2. Make sure to select the Google Maps Activity template.

Description

- The easiest way to add the Google Play services library and get a Google Maps API key is to create a new project that's based on the Google Maps Activity template.

Figure 18-5   How to create a new Google Maps project

How to add Google Play services to an existing project

Figure 18-6 shows how to add Google Play services to an existing project. To start, right-click on the app node of your project and select the Open Module Settings item to open the Project Structure dialog. In this dialog, click the Dependencies tab. If the row of tabs isn't visible, make sure you select the app module from the list of Modules on the left.

In the Dependencies tab, click the Add (+) button on the right side of the dialog, and select the Library Dependency item. This should display a dialog that allows you to add the play-services library to your project. When you add this library, Gradle should automatically rebuild your project.

How to add Google Play services to an existing project

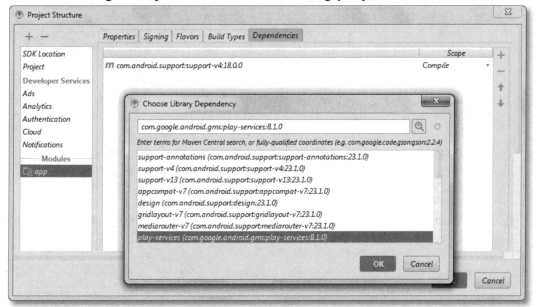

Procedure

1. In the Project window, right-click on the app node, and select the Open Module Settings item.
2. Select the Dependencies tab, click the Add (+) button on the right side of the dialog, and select the Library Dependency item from the resulting menu.
3. Select the play-services library from the dialog.

Description

*   You can add the Google Play services library to an existing project using the Project Structure dialog.

Figure 18-6   How to add Google Play services to an existing project

How to get a Google Maps API key

There are several ways you can get a Google Maps API key. If you don't already have one, the easiest way to get one is to use Android Studio to create a new project that's based on the Google Maps Activity template as described earlier in this chapter. Once you do that, Android Studio automatically opens a google_maps_api.xml file that contains a link and directions that make it easy to get an API key. These directions are summarized in figure 18-7.

To start, you copy the link provided in the google_maps_api.xml file and paste it into your browser. From here, you can follow the directions to create the API key. If you aren't logged into a Google account, you'll need to do that before you can continue. Once you are logged in, there will be a short pause as a project is created for you.

When the project has been created, your browser should display a message indicating that the API is enabled, along with a button that says "Go to credentials." If you click this button, it should display the web page shown in this figure.

From this web page, you can provide a different name for the key, as well as restrict usage of the key to certain android apps. For the purposes of this book, we recommend you don't restrict usage. That way, you'll be able to use the same key for all of your learning and testing purposes. To remove the restrictions on the key, click the X to the right side of the SHA-1 certificate fingerprint listing so that there are no packages and no SHA-1 fingerprints listed. Then, click the Create button.

After the website creates your new API key, it should display it in a dialog. This makes it easy for you to select the entire API key and copy it to the clipboard. Then, you can switch to the google_maps_api.xml file, delete the text that says "YOUR_KEY_HERE", and replace it with your key. This creates a string resource named google_maps_key that's used by the Android manifest file to set the API key as shown in the next figure.

To make sure the API key works, you can run the skeleton app that Android Studio created for you. If the key is working, you should see a map on the screen of your emulator or device after the app has started. Since the app has to download the map data from the Internet, it may take several seconds for the map to appear.

If you followed the procedure in this figure, this key isn't tied to any particular app. As a result, you can use it for all of the map apps you develop in this book, including the Run Tracker project presented at the end of the chapter.

Web page for creating an API key

A typical API key

`AIzaSyAPUmbZW6OuCJu5RuceQLD83PvZxR4zEKg`

Procedure

1. Create a new project based on the Google Maps Activity template.

2. Copy the link from the google_maps_api.xml file and paste it into your browser.

3. Follow the web pages to create the API key. To use the key for all of your applications (recommended), click the X to the right of the SHA-1 fingerprint box so that the list of package names and fingerprints is empty.

4. After you create the API key, copy it to the clipboard.

5. Switch to the google_maps_api.xml file, delete the text that says "YOUR_KEY_HERE", and paste the API key in its place. This creates a string resource named google_maps_key that the Android manifest file uses to set the API key for the app.

6. Test the key by running the new project. If a map appears on your emulator or device, the key is working correctly.

Description

- There are several ways to obtain a Google Maps API key. This chapter presents the easiest method.

- When learning how to use Google Maps, we recommend you don't tie the API key to any particular application. That way, you can use it for all of the test applications you build.

Figure 18-7   How to get a Google Maps API key

How to set permissions and features

For maps to work properly, your app will need certain permissions. For example, it will need permission to access the Internet so that it can download map tiles. Figure 18-8 shows how to use the AndroidManifest.xml file to set the permissions and features necessary to use the Google Maps Android API v2 service. To start, you can secure your app by defining a custom permission named MAPS_RECEIVE and setting its protection level to "signature". Then, you can request that permission to allow the app to access the Google Maps servers.

After defining the MAPS_RECEIVE permission, you must request permissions that allow the app to access the Internet, check the Internet connection, cache map tiles, and access Google web-based services. In addition, you typically need to enable the ACCESS_COARSE_LOCATION or ACCESS_FINE_LOCATION permission to allow your app to access the device's current location. Part 2 of this figure describes these permissions in more detail.

Finally, Google Maps Android API version 2 requires OpenGL ES version 2. As a result, you must add a uses-feature element to notify external services of this requirement. This prevents the Google Play store from displaying your app for devices that don't support OpenGL ES version 2.

Part of the AndroidManifest.xml file for the Run Tracker app

```
...
<!-- set up MAPS_RECEIVE permission -->
<permission
    android:name="com.murach.runtracker.permission.MAPS_RECEIVE"
    android:protectionLevel="signature" />
<uses-permission
    android:name="com.murach.runtracker.permission.MAPS_RECEIVE" />

<!-- set other permissions -->
<uses-permission
    android:name="android.permission.INTERNET" />
<uses-permission
    android:name="android.permission.ACCESS_NETWORK_STATE" />
<uses-permission
    android:name="android.permission.WRITE_EXTERNAL_STORAGE" />
<uses-permission
    android:name=
        "com.google.android.providers.gsf.permission.READ_GSERVICES" />
<uses-permission
    android:name="android.permission.ACCESS_FINE_LOCATION" />

<!-- Maps API version 2 requires OpenGL ES version 2 -->
<uses-feature
    android:glEsVersion="0x00020000"
    android:required="true" />

<application
    android:allowBackup="true"
    android:icon="@drawable/ic_launcher"
    android:label="@string/app_name"
    android:theme="@style/AppTheme" >

    <meta-data
        android:name="com.google.android.maps.v2.API_KEY"
        android:value="@string/google_maps_key" />
...
```

Description

- To secure your app, you can define a custom MAPS_RECEIVE permission for your app and set its protection level.

- When you use Google Maps, your app must define permissions that allow the app to use the Internet and access the Google Maps servers.

- Google Maps Android API version 2 requires OpenGL ES version 2. As a result, when you use Google Maps, you must add a uses-feature element to notify external services of this requirement. This prevents the Google Play store from displaying your app for devices that don't support OpenGL ES version 2.

Figure 18-8   How to set permissions and features (part 1 of 2)

Part 2 summarizes the elements and permissions used in part 1. For the most part, these permissions are boilerplate code that you can copy from one app that uses version 2 of the Google Maps Android API to another. However, if you want to learn more about how these elements and permissions work, you can read more about them.

In the permissions, the ACCESS_COARSE_LOCATION permission isn't used in part 1. That's because the Run Tracker app presented in this figure only uses GPS to determine a device's location. As a result, this app only needs the ACCESS_FINE_LOCATION permission. However, if you want to use Wi-Fi or mobile cell data (or both) to determine a device's location, you need to add the ACCESS_COARSE_LOCATION permission to your app.

The meta-data element

Attribute	Description
name	The type of API key.
value	The value of the API key.

The permission element

Attribute	Description
name	The name of the permission that corresponds with the uses-permission element.
protectionLevel	The level of protection for the permission. A value of "signature" allows the system to grant this permission without notifying the user only if the requesting app is signed with the same certificate as the app that declared the permission.

A summary of permissions

Permission	Description
MAPS_RECEIVE	Allows the API to securely contact your app. This prevents other apps from impersonating the API.
INTERNET	Allows the API to download map tiles from Google Maps servers.
ACCESS_NETWORK_STATE	Allows the API to check the connection status in order to determine whether map data can be downloaded.
WRITE_EXTERNAL_STORAGE	Allows the API to cache data such as map tiles in the device's external storage area.
READ_GSERVICES	Allows the API to access Google web-based services.
ACCESS_FINE_LOCATION	Allows the API to use GPS to determine the device's location.
ACCESS_COARSE_LOCATION	Allows the API to use cell tower or Wi-Fi signals to determine the device's location.

The uses-feature element

Attribute	Description
glEsVersion	The OpenGL ES version required by the application. To specify version 2.0, specify a value of "0x00020000".
required	Specifies whether the application requires the feature.

Description

- The meta-data element that stores the Maps API key must be declared within the application element.
- If your app needs to access the device's current location, you must enable the ACCESS_COARSE_LOCATION or ACCESS_FINE_LOCATION permissions.

Figure 18-8   How to set permissions and features (part 2 of 2)

How to work with locations

Now that you have configured version 2 of the Android API for Google Maps, you're ready to learn how to use it to work with locations. In particular, you're ready to learn how to get and track the device's location. To do that, you begin by connecting to Google Play services.

How to connect to Google Play services

Figure 18-9 shows how to connect to Google Play services. To start, you declare an activity that implements two interfaces: ConnectionCallbacks and OnConnectionFailedListener. In this figure, the methods for these interfaces are shown in part 2. At this point, these methods don't contain any code. However, you'll learn how to add code to these methods in the next few figures.

After implementing these interfaces, you can declare a GoogleApiClient object as an instance variable. Then, in the onCreate method, you can create a GoogleApiClient object. To do that, you can create a new Builder object for the GoogleApiClient class and use the this keyword to indicate that the client object belongs to this class.

After you create the client builder object, you can specify which Google APIs you want to add to the client. In addition, you can specifiy any listeners you want to add to the client. To do that, you can use method chaining as shown in this figure. In this figure, the code adds the LocationServices API. Then, it adds the callback and listener methods needed to work with a connection to the LocationServices API. Finally, it calls the build method to create the client object from the builder object.

The LocationViewerActivity class   **Page 1**

```java
package com.murach.locationviewer;

import android.location.Location;
import android.app.Activity;
import android.os.Bundle;
import android.widget.TextView;

import com.google.android.gms.common.ConnectionResult;
import com.google.android.gms.common.api.GoogleApiClient;
import com.google.android.gms.common.api.GoogleApiClient.ConnectionCallbacks;
import com.google.android.gms.common.api.GoogleApiClient.OnConnectionFailedListener;

public class LocationViewerActivity extends Activity implements
        ConnectionCallbacks, OnConnectionFailedListener {

    private GoogleApiClient googleApiClient;
    private TextView coordinatesTextView;

    @Override
    protected void onCreate(Bundle savedInstanceState) {
        super.onCreate(savedInstanceState);
        setContentView(R.layout.activity_location_viewer);
        coordinatesTextView =
            (TextView) findViewById(R.id.cooridinatesTextView);
        googleApiClient = new GoogleApiClient.Builder(this)
                .addApi(LocationServices.API)
                .addConnectionCallbacks(this)
                .addOnConnectionFailedListener(this)
                .build();
    }
```

Constructors and methods of the GoogleApiClient.Builder class

Constructor/Method	Description
Builder(ctx)	Creates a GoogleApiClient.Builder object with the specified context.
addApi(api)	Adds the specified API to the client.
addConnectionCallbacks(ctx)	Adds connection callbacks to the client.
addOnConnectionFailedListener(ctx)	Adds a connection failed listener to the client.
build()	Returns a GoogleApiClient object.

Figure 18-9   How to connect to Google Play services (part 1 of 2)

After you create the GoogleApiClient object, you typically call its connect method from within the onStart method as shown in part 2. That way, when the activity is started, it starts an asynchronous thread that opens a connection to Google Play services in the background. When the connection is opened, this calls the onConnected method.

When you aren't using a connection, you should close it. To do that, you can call the disconnect method of the GoogleApiClient object from within the onStop method. That way, when the activity is stopped, it closes the connection to Google Play services. This calls the onConnectionSuspended method.

In addition, part 2 shows the onConnectionFailed method of the OnConnectionFailedListener interface. This method is called when a connection to Google Play services fails. That doesn't usually happen, but it can happen when the network isn't available. It can happen when the client attempts to connect to the service but the user isn't signed in or is using an invalid account name. And it can happen when Google Play services is missing on a device or is out of date.

The LocationViewerActivity class
Page 2

```java
    @Override
    public void onStart() {
        super.onStart();
        googleApiClient.connect();
    }

    @Override
    public void onStop() {
        googleApiClient.disconnect();
        super.onStop();
    }

    //*************************************************************
    // Implement ConnectionCallbacks interface
    //*************************************************************
    @Override
    public void onConnected(Bundle dataBundle) {
        // Put code to run after connecting here
    }

    @Override
    public void onConnectionSuspended() {
        // Put code to run before disconnecting here
    }

    //*************************************************************
    // Implement OnConnectionFailedListener
    //*************************************************************
    @Override
    public void onConnectionFailed(ConnectionResult connectionResult) {
        // Put code to run if connection fails here
    }
}
```

Methods of the GoogleApiClient class

Method	Description
connect()	Starts an asynchronous thread that opens a connection to Google Play services.
disconnect()	Closes the connection to Google Play services.

Methods of the ConnectionCallbacks interface

Method	Description
onConnected(dataBundle)	Called when the connection is opened.
onConnectionSuspended()	Called when the connection is closed.

Method of the OnConnectionFailedListener interface

Method	Description
onConnectionFailed()	Called when the connection fails.

Figure 18-9   How to connect to Google Play services (part 2 of 2)

How to get the current location

Figure 18-10 shows code that you can add to the onConnected method shown in the previous figure to get the current location. To start, you can get a Location object by calling the static getLastLocation method of the FusedLocationApi field of the LocationServices class. This Location object contains detailed information about the best and most recent location of a device.

After you get a Location object, you should check to make sure that it is not null. If that's true, you can call the getLatitude and getLongitude methods of the Location object to get the degree of latitude and longitude for the current location. In this figure, for example, the code gets the latitude and longitude for the current location and displays them on a TextView widget. Later in this chapter, you'll learn how to use these coordinates to display the device's location on a map.

If necessary, you can use other methods of the Location object to get more information about the location. For example, you can use the getTime method to get the time of the location fix in milliseconds. This might be useful for any app that's keeping a historical record of a device's location.

Similarly, you can use the getAccuracy method to determine the accuracy of the location. This might be useful if you want to handle the location differently depending on its estimated accuracy.

Code that gets the current location

```
@Override
public void onConnected(Bundle bundle) {
    Location location =
            LocationServices.FusedLocationApi
                    .getLastLocation(googleApiClient);
    if (location != null) {
        coordinatesTextView.setText(
                location.getLatitude() + "|" + location.getLongitude());
    }
}
```

A method of the LocationServices.FusedLocationApi field

Method	Description
getLastLocation(client)	A static method that returns the best and most recent location currently available.

Some methods of the Location class

Method	Description
getLatitude()	Gets the degree of latitude.
getLongitude()	Gets the degree of longitude.
getTime()	Gets the time of this location fix in milliseconds.
getAccuracy()	Gets the estimated accuracy of this location in meters. If you draw a circle centered at this location's latitude and longitude with a radius equal to the accuracy, there is a 68% probability that the true location is inside the circle. If this location does not have an accuracy, this method returns 0.0.

Description

- You can use the LocationServices.FusedLocationApi field to get a Location object that contains detailed information about the best and most recent location of a device.

Figure 18-10   How to get the current location

How to handle a failed connection

Figure 18-11 shows code that you can add to the onConnectionFailed method shown earlier in this chapter to handle a failed connection. But first, you should define a constant for an integer code that uniquely identifies your request for a resolution. In this figure, the code sets the constant to a value of 9000, but you can use any integer value that doesn't conflict with other request codes.

After defining this constant, you can add the code to the onConnectionFailed method like the code shown in this figure. To start, this code calls the hasResolution method from the ConnectionResult parameter to check whether Google Play services knows of a possible solution to the problem that caused the failed connection.

If there is a possible solution, you can call the startResolutionForResult method to start any intents that require user interaction. When you call this method, you pass it the current activity and the constant that identifies your request.

If there is no solution, you can display an error to the user to indicate that the connection has failed. In this figure, for example, the code uses the AlertDialog. Builder class to display an error message that includes the code for the error that caused the connection to fail. To do that, this statement calls the getErrorCode method from the ConnectionResult parameter.

In some cases, you may want to do additional processing when the user returns from the activity that attempts to resolve the problem. To do that, you can add an onActivityResult method like the one shown in this figure. This method is a lifecycle method of the Activity class that's executed when the user returns from an activity that returns a result. In this figure, the onActivityResult method begins by checking if the request code parameter matches the constant for the request code. If so, it can do additional processing. If necessary, this processing can include checking the result code against the constants of the ConnectionResult class to determine the cause of the failed connection.

Code that handles a failed connection

The constant for the request code

```
private final static int CONNECTION_FAILURE_RESOLUTION_REQUEST = 9000;
```

The onConnectionFailed method

```
@Override
public void onConnectionFailed(ConnectionResult connectionResult) {
    if (connectionResult.hasResolution()) {
        try {
            connectionResult.startResolutionForResult(this,
                    CONNECTION_FAILURE_RESOLUTION_REQUEST);
        } catch (IntentSender.SendIntentException e) {
            e.printStackTrace();
        }
    } else {
        new AlertDialog.Builder(this)
                .setMessage("Connection failed. Error code: " +
                        connectionResult.getErrorCode())
                .show();
    }
}
```

The onActivityResult method

```
@Override
protected void onActivityResult(int requestCode, int resultCode,
        Intent data) {
    super.onActivityResult(requestCode, resultCode, data);

    if (requestCode == CONNECTION_FAILURE_RESOLUTION_REQUEST) {
        // perform additional processing here
    }
}
```

Methods of the ConnectionResult class

Method	Description
`hasResolution()`	Returns true if Google Play services can possibly resolve the error.
`startResolutionForResult(` `activity, code)`	Attempts to resolve the error by starting any intents requiring user interaction with the specified request code.
`getErrorCode()`	Gets the code for the error that caused the connection to fail.

Description

- In the onConnectionFailed method, you can use the ConnectionResult parameter to allow Google Play services to attempt to resolve the error that caused the connection to fail.

Figure 18-11   How to handle a failed connection

How to get location updates

Figure 18-12 shows how to get updates of a device's location at a specified interval of time. To start, you import the classes for working with locations. When you do that, it's important to import the correct LocationListener class. If you import the wrong one, which is a very easy mistake to make, your code won't compile.

After you import the classes for working with locations, you implement the LocationListener interface. This interface defines a single method, the onLocationChanged method.

After you implement the interface, you define the constants and instance variables for the class. Here, the UPDATE_INTERVAL constant sets the milliseconds for the requested update interval to 5000, which is 5 seconds. Then, the FASTEST_UPDATE_INTERVAL sets the milliseconds for the fastest rate at which the app can update the location to 2000, which is 2 seconds. Next, this code declares a variable for a LocationRequest object.

In the onCreate method, you can create the LocationRequest object. Then, you can use one of the constants described in part 2 of this figure to set its priority. In this figure, the code sets the priority to high accuracy. This uses GPS, which results in the highest accuracy, but also uses the most battery. In addition, you can set the update interval and the fastest update interval. In this figure, the code sets these intervals to the constants defined in step 3.

In the onConnected method, you can turn on location updates. To do that, you call the static requestLocationUpdates method on the LocationServices. FusedLocationApi field and pass it the GoogleApiClient object, the LocationRequest object, and the LocationListener object. In this figure, the current activity is the LocationListener object.

In the onConnectionSuspended method, you can turn off location updates. To do that, you call the static removeLocationUpdates method of the LocationServices.FusedLocationApi field and pass it the GoogleApiClient object, and the LocationListener object. But first, you should use the isConnected method of the GoogleApiClient object to make sure the connection is still available.

In the onLocationChanged method, you can use the Location parameter to process the current location. In this figure, the code just displays the latitude and longitude for the location on a TextView widget. Later in this chapter, you'll learn how to update the device's location on a map.

Step 1: Import the classes for working with locations

```
import com.google.android.gms.location.LocationListener;
import com.google.android.gms.location.LocationRequest;
import com.google.android.gms.location.LocationServices;
```

Step 2: Declare the LocationListener interface

```
public class LocationViewerActivity extends Activity
        implements ConnectionCallbacks, OnConnectionFailedListener,
            LocationListener {
```

Step 3: Declare the constants and instance variables

```
public static final int UPDATE_INTERVAL = 5000;            // 5 seconds
public static final int FASTEST_UPDATE_INTERVAL = 2000;    // 2 seconds
private LocationRequest locationRequest;
```

Step 4: Create the location request in the onCreate method

```
locationRequest = LocationRequest.create()
        .setPriority(LocationRequest.PRIORITY_HIGH_ACCURACY)
        .setInterval(UPDATE_INTERVAL)
        .setFastestInterval(FASTEST_UPDATE_INTERVAL);
```

Step 5: Request updates in the onConnected method

```
LocationServices.FusedLocationApi.requestLocationUpdates(
    googleApiClient, locationRequest, this);
```

Step 6: Remove updates in the onConnectionSuspended method

```
if (googleApiClient.isConnected()) {
    LocationServices.FusedLocationApi.removeLocationUpdates(
        googleApiClient, this);
}
```

Step 7: Implement the LocationListener interface

```
@Override
public void onLocationChanged(Location location) {
    coordinatesTextView.setText(
            location.getLatitude() + "|" + location.getLongitude());
}
```

Description

- You can use a LocationRequest object to configure a request for location updates from Google Play services.

- You can use a LocationListener object to listen for location updates from Google Play services.

- You can use the static methods of the FusedLocationApi field of the LocationServices class to turn location updates on and off.

Figure 18-12   How to get location updates (part 1 of 2)

Part 2 begins by showing some of the methods and constants that you can use to create and configure the LocationRequest object. In addition, part 2 shows some more methods of the FusedLocationApi field of the LocationServices class. If you want to learn more about how these methods and constants work, you can read more about them.

The setPriority method sets the accuracy parameter to one of the constants shown in this figure. If your main priority is accuracy, you should use the PRIORITY_HIGH_ACCURACY constant. This constant uses GPS to request the most accurate locations available. However, this also uses the most battery of all options. If your app only needs accuracy of about 100 meters, use the PRIORITY_BALANCED_POWER_ACCURACY constant instead as it consumes less battery.

Another option is to use the PRIORITY_NO_POWER constant. This requests the best accuracy possible with zero additional battery consumption. This only returns locations when another app is receiving location updates. In that case, your app listens to those locations.

The setInterval method sets the interval that your app prefers to receive location updates. If no other apps are receiving updates, your app receives updates at this rate.

However, if other apps are receiving updates at a faster rate, your app can receive updates at that faster rate too with no additional use of battery. Unfortunately, a faster rate may cause problems with your app. For example, it may cause UI (user interface) flicker or data overflow. That's why you need to use the setFastestInterval method to set the interval at which your app can handle location updates.

Methods of the LocationRequest class

Method	Description
`create()`	Creates and returns a LocationRequest object.
`setPriority(accuracy)`	Sets the accuracy parameter to one of the constants shown below.
`setInterval(millis)`	Sets the rate in milliseconds that your app prefers to receive location updates. If no other apps are receiving updates, your app receives updates at this rate.
`setFastestInterval(millis)`	Sets the fastest rate in milliseconds at which your app can handle location updates. This prevents problems with UI flicker or data overflow and helps to save power. If other apps have requested a faster rate, you get the benefit of a faster rate with no additional use of battery.

Priority constants from the LocationRequest class

Constant	Description
`PRIORITY_HIGH_ACCURACY`	Requests the most accurate locations available. This uses the most battery of all options and requires the ACCESS_FINE_LOCATION permission.
`PRIORITY_BALANCED_POWER_ACCURACY`	Requests accuracy of about 100 meter accuracy. This consumes less battery and requires the ACCESS_COARSE_LOCATION permission.
`PRIORITY_NO_POWER`	Requests the best accuracy possible with zero additional battery consumption. This only returns locations when another app is receiving location updates. In that case, your app listens to those locations.

Methods of the LocationServices.FusedLocationApi field

Method	Description

`requestLocationUpdates(`
    `client, request,`
    `listener)`
	A static method that requests location updates using the specified GoogleApiClient, LocationRequest, and LocationListener objects.

`removeLocationUpdates(` `client, listener)`	A static method that removes location updates from the specified GoogleApiClient and LocationListener objects.

Figure 18-12 How to get location updates (part 2 of 2)

How to make sure GPS is enabled

If your app relies on GPS to get the location, you typically want to make sure it's enabled on the user's device before you start your app. To do that, you can use the code shown in figure 18-13.

The code presented in this figure is typically coded within the onCreate, onStart, or onResume method of an activity. That way, it's executed when the activity is first created or anytime it's displayed. Here, the first statement uses the getSystemService method to get a LocationManager object. Then, an if statement uses the LocationManager object to check whether GPS is enabled. If not, this code displays a toast that asks the user to enable GPS. Then, it creates an intent for the activity of the Settings app that allows the user to enable and disable GPS. Finally, this code calls the startActivity method and passes it that intent.

Code that makes sure GPS is enabled

```
LocationManager locationManager =
        (LocationManager) getSystemService(LOCATION_SERVICE);

if (!locationManager.isProviderEnabled(LocationManager.GPS_PROVIDER)){
    Toast.makeText(this, "Please enable GPS!",
            Toast.LENGTH_LONG).show();
    Intent intent = new Intent(
            Settings.ACTION_LOCATION_SOURCE_SETTINGS);
    startActivity(intent);
}
```

Method of the Activity class

Method	Description
getSystemService(serviceType)	Gets the specified system service. You can use the LOCATION_SERVICE constant to get a Location-Manager object.

Constants and methods of the LocationManager class

Constant/Method	Description
GPS_PROVIDER	The provider that uses GPS to determine location.
NETWORK_PROVIDER	The provider that uses Wi-Fi or cell networks to determine location.
isProviderEnabled(providerType)	You can use this to check if a type of location provider is available.

Description

- You can use the LocationManager object to check if GPS is enabled. If it isn't, you can display an error message, start the Settings app, and display the activity that allows the user to enable GPS.

Figure 18-13   How to make sure GPS is enabled

How to work with Google Maps

Now that you know how to track a device's location, you're ready to learn how to use that information with a Google Map. To do that, the first step is to add a Google Map widget to your user interface.

How to add a map fragment to a layout

The easiest way to add a Google Map widget to your user interface is to add a fragment element to a layout as shown in figure 18-14. To do that, you can add the XML code for the fragment to the layout. This fragment should include a class attribute that specifies the SupportMapFragment class.

If you switch to the graphical layout editor, Android Studio doesn't display the map. However, it displays a grayed out area where the map will be displayed when you run the app.

A layout that includes a map fragment

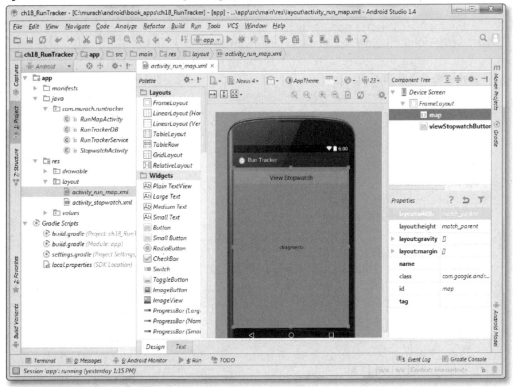

The XML for map fragment

```
<fragment
    android:id="@+id/map"
    android:layout_width="match_parent"
    android:layout_height="match_parent"
    class="com.google.android.gms.maps.SupportMapFragment" />
```

Description

- To add a map to a layout, add a fragment element for the SupportMapFragment class.

Figure 18-14   How to add a map fragment to a layout

How to display a map

Figure 18-15 shows how to display a map. First, you create an activity that extends the FragmentActivity class. In this figure, for example, the activity named LocationViewerActivity extends the FragmentActivity class.

Second, you declare an instance variable for the GoogleMap object. This is the primary object for working with a Google Map. In this figure, for example, the code declares an instance variable for a GoogleMap object named map.

Third, you add code to the onStart method of the activity to get the GoogleMap object. That way, when the activity is displayed, the code gets the map and displays it. In this figure, the code begins by checking whether the map is null. If so, it gets a FragmentManager object, and uses it to get the SupportMapFragment object for the map. Then, it calls the getMap method from that object to get the GoogleMap object.

Once you have a GoogleMap object, you can call its methods to change its type or its UI (user interface) settings. In this figure, for example, the code checks to make sure the GoogleMap object exists. Then, it calls the getUiSettings method from the GoogleMap object to return a UiSettings object. From that object, this code calls the setZoomControlsEnabled method to enable the zoom controls. As a result, the map should include zoom controls (plus and minus buttons) like the ones shown on the Run Activity in figure 18-1.

In this figure, the code uses the default map type (the normal type). This displays a map like the one shown on the Run Map activity in figure 18-1. However, if you want to use a different map type, you can call the setMapType method from the GoogleMap object to change the map type to one of the constants shown in this figure. For example, you can change the map type to a satellite map, a hybrid map, or a terrain map.

Similarly, the code in this figure uses most of the default UI settings for the map. The only UI setting change this code makes is to enable the zoom controls. However, if you want, you can call methods from the GoogleMap object to customize most UI settings of the map so they're appropriate for your app. For example, you can use the methods of the UiSettings object shown in this figure to disable the compass or zoom gestures.

At this point, the code should display a map of the world with zoom controls. In the next figure, you'll learn how to zoom in on a particular location.

Step 1: Extend the FragmentActivity class

```
public class LocationViewerActivity extends FragmentActivity
        implements ConnectionCallbacks, OnConnectionFailedListener,
                LocationListener {
```

Step 2: Declare the instance variable for the GoogleMap object

```
        private GoogleMap map;
```

Step 3: Get the GoogleMap object in the onStart method

```
            // if GoogleMap object is not already available, get it
            if (map == null) {
                FragmentManager manager = getSupportFragmentManager();
                SupportMapFragment fragment = (SupportMapFragment)
                    manager.findFragmentById(R.id.map);
                map = fragment.getMap();
            }

            // if GoogleMap object is available, configure it
            if (map != null) {
                map.getUiSettings().setZoomControlsEnabled(true);
            }
```

A method of the SupportMapFragment class

Method	Description
getMap()	Gets a GoogleMap object.

Some constants and methods of the GoogleMap class

Constant/Method	Description
MAP_TYPE_NORMAL	The default map.
MAP_TYPE_HYBRID	A satellite map with a transparent layer of major streets.
MAP_TYPE_SATELLITE	A satellite map with no labels.
MAP_TYPE_TERRAIN	A terrain map.
setMapType(mapType)	Sets the type of map tiles that should be displayed.
getUiSettings()	Gets a UiSettings object for the user interface settings for the map.

Some methods of the UiSettings class

Method	Description
setCompassEnabled(bool)	Enables or disables the compass.
setZoomControlsEnabled(bool)	Enables or disables the zoom controls.
setZoomGesturesEnabled(bool)	Enables or disables zoom gestures.

Figure 18-15   How to display a map

How to zoom in on a location

Figure 18-16 shows how to use a GoogleMap object to zoom in on a location. In this figure, the code begins by checking to make sure the GoogleMap object exists before calling any methods from that object. This is a good practice as it prevents a NullPointerException from being thrown.

If the GoogleMap object exists, this code calls the animateCamera method from it to zoom in on the specified target. To help visualize how this works, you can think of a virtual camera pointing at the specified target from the specified position. Here, the target is the latitude and longitude that are stored in the Location object named location. The zoom level is 16.5, which zooms in so the map displays a few city blocks. The bearing is 0, which means the camera is pointing due north. And the tilt is 25, which means that the camera is tilted 25 degrees towards the bearing (as opposed to the default setting of 0, which points straight down). This displays a map like the one shown in the Run Map activity in figure 18-1.

When you use the animateCamera method to zoom in on the current location, the API uses an animation to zoom in as if the camera is flying. If you don't like that, you can use the moveCamera method. It works the same as the animateCamera method, but it just displays the map from the camera position without using any animations to get there.

You may need to experiment a bit to find the best zoom level, bearing, and tilt for your app. The maximum zoom level is limited by the underlying map, and varies for the different locations around the world. For example, you can zoom in more closely on populated areas and can't zoom in as closely on other areas such as the north pole. Also, the maximum zoom level varies depending on the map type. If you're curious, you can use the getMaxZoomLevel to return the maximum zoom level for the current camera position and map type.

The maximum tilt value depends on the zoom level. However, it's never more than 67.5. Fortunately, if you specify a value that's greater than the maximum, it doesn't cause an error. Instead, the API just uses the maximum value.

How to zoom in on a map

```
if (map != null) {
    map.animateCamera(
        CameraUpdateFactory.newCameraPosition(
            new CameraPosition.Builder()
                .target(new LatLng(location.getLatitude(),
                                   location.getLongitude())))
                .zoom(16.5f)
                .bearing(0)
                .tilt(25)
                .build()));
}
```

More methods of the GoogleMap class

Method	Description
animateCamera(cameraPosition)	Animates a move to the specified latitude and longitude, zooms in to the specified amount, and sets the bearing and the tilt of the camera.
moveCamera(cameraPosition)	Works like the animateCamera method, but doesn't animate the move.
getMaxZoomLevel()	Returns the maximum zoom level for the current camera position and map type.

Some methods of the CameraPosition.Builder class

Method	Description
target(latLng)	Sets the location to the specified LatLng object.
zoom(level)	Sets the zoom level to the specified float value. At level 0, a map that's 256dp wide can display the whole world. At level 16.5, a map that's 256dp wide can only display a few blocks.
bearing(degrees)	Sets the direction that the camera is pointing, in degrees clockwise from north where 0 is north and 180 is south.
tilt(degrees)	Sets the angle of the camera in degrees from directly facing the Earth. The minimum degrees value is 0 (no tilt), and the maximum degrees value varies depending on the zoom level but is never more than 67.5.
build()	Returns the CameraPosition object.

Constructor of the LatLng class

Constructor	Description
LatLng(latitude, longitude)	Creates a LatLng object with the specified latitude and longitude.

Description

- You can use a GoogleMap object to zoom in on a location.

Figure 18-16   How to zoom in on a location

How to add markers

Figure 18-17 shows how to add one or more markers to a map. To do that, you can use the addMarker method of the GoogleMap class. If necessary, you can use the clear method of the GoogleMap class to clear any existing markers.

In this figure, the code calls the clear method to clear any old markers from the map. Then, it uses the addMarker method to add a marker for the device's current location. To do that, this code creates a new MarkerOptions object, sets its position to the latitude and longitude that are stored in the Location object named location, and sets its title to a value of "You are here". Then, this code passes the newly created MarkerOptions object to the addMarker method. This displays a default red marker icon like the one shown in the Run Map activity in figure 18-1.

If you don't want to use the default marker icon, you can use the other methods of the MarkerOptions class to customize it. For example, you can use the icon method to change the default icon. To do that, you need to add an appropriate icon to the project's resources. Then, you can use the BitmapDescriptorFactory to get a BitmapDescriptor object from that resource. In this figure, the second code example gets a BitmapDescriptor object from a resource named ic_runner.png that's stored in the project's res\drawable directory.

How to add a marker for the current location

```
if (map != null) {
    map.clear();    // clear any old markers
    map.addMarker(
        new MarkerOptions()
            .position(new LatLng(location.getLatitude(),
                                 location.getLongitude()))
            .title("You are here"));
}
```

An example that uses the icon method to set a new icon for a marker

```
.icon(BitmapDescriptorFactory.fromResource(R.drawable.ic_runner))
```

More methods of the GoogleMap class

Method	Description
clear()	Clears all markers from the map.
addMarker(marker)	Adds a marker to the map at the specified latitude and longitude.

Some methods of the MarkerOptions class

Method	Description
position(latLng)	Sets the location for the marker.
title(title)	Sets the title for the marker.
icon(bitmapDescriptor)	Sets the icon for the marker. If you don't set an icon, Google Maps uses the default marker icon.

Description

- You can use a GoogleMap object to add one or more markers to a map.

Figure 18-17   How to add markers

How to add lines

Figure 18-18 shows how to display a series of connected lines on a map. To do that, you can call the addPolyline method from the GoogleMap object and pass it a PolylineOptions object that contains a series of points. Then the API draws a line that connects these points in the order in which they were added.

In this figure, the code begins by creating a new PolylineOptions object. Then, an if statement checks whether an ArrayList<Location> object named locationsList contains any locations. If so, this code loops through this list, converts each Location object to a LatLng object, and adds each LatLng object to the PolylineOptions object. After the loop, this code uses the addPolyline method to add this polyline to the map. This displays a crooked black line that connects a series of points like the line shown in the Run Map activity in figure 18-1.

If you don't want to use the default settings for the line, you can use the other methods of the PolylineOptions class to customize it. For example, you can use the width method to make the line thicker or thinner. In this figure, the second code example sets the width of a line to 10 pixels. Similarly, you can use the color method to change the color of the line. In this figure, the third example sets the color of the line to red. To do that, it uses the RED constant that's available from the Color class of the android.graphics package.

How to add lines to a map

```
if (map != null) {
    PolylineOptions polyline = new PolylineOptions();
    if (locationList.size() > 0) {
        for (Location l : locationList) {
            LatLng point = new LatLng(
                l.getLatitude(), l.getLongitude());
            polyline.add(point);
        }
    }
    map.addPolyline(polyline);
}
```

An example that sets the width of a line

```
polyline.width(10);
```

An example that sets the color of a line

```
polyline.color(Color.RED);
```

More methods of the GoogleMap class

Method	Description
addPolyline(polyline)	Adds a line that connects the points in the specified PolylineOptions object.

Constructor/methods of the PolylineOptions class

Constructor/Method	Description
PolylineOptions()	Creates a PolylineOptions object.
add(latLng)	Adds the point specified by the LatLng object.
width(pixels)	Specifies the width of the line to the specified number of pixels.
color(argbColor)	Sets the color of the polyline as a 32-bit ARGB color. The default color is black.

Description

- You can use a GoogleMap object to display a series of connected lines on a map.

Figure 18-18   How to add lines to a map

The Run Tracker app

At this point, you have all the skills you need to understand the code for the Run Tracker app. Since the Stopwatch activity for this app doesn't contain much code that works with locations or maps, this chapter doesn't present this activity. Instead, this chapter presents the layout and class for the Run Map activity. In addition, it presents the class for the Run Tracker service. These files contain most of the code that works with locations and maps. Of course, if you want to see the complete code, you can open the project for this app and view its code.

The activity_run_map layout

Figure 18-19 shows the XML for the activity_run_map layout. This layout displays the View Stopwatch button over a map. To do that, this XML uses a frame layout. Within the frame layout, the fragment element adds the map. Then, the Button element adds a button across the top of the map.

The activity_run_map layout

```xml
<?xml version="1.0" encoding="utf-8"?>
<FrameLayout xmlns:android="http://schemas.android.com/apk/res/android"
    android:layout_width="match_parent"
    android:layout_height="match_parent" >

    <fragment
        android:id="@+id/map"
        android:layout_width="match_parent"
        android:layout_height="match_parent"
        class="com.google.android.gms.maps.SupportMapFragment" />

    <Button
        android:id="@+id/viewStopwatchButton"
        android:layout_width="match_parent"
        android:layout_height="wrap_content"
        android:layout_marginTop="10dp"
        android:gravity="center"
        android:text="@string/stopwatch"
        android:textSize="20sp" />

</FrameLayout>
```

Figure 18-19   The activity_run_map layout

The RunMapActivity class

Figure 18-20 shows the code for the RunMapActivity class. To start, this class imports some classes from the Java API. Then, it imports some classes from the standard Android API. Finally, this class imports classes from the Google Maps Android API v2. By now, you should be familiar with all of these classes.

The RunMapActivity class extends the FragmentActivity class and implements three interfaces. The first interface is necessary to execute code when the user clicks on the View Stopwatch button. The next two interfaces are necessary to connect to Google Play services and get the current location.

After the declaration for the the RunMapActivity class, this code defines some constants and instance variables that are needed by the class. This includes a constant named INTERVAL_REFRESH that specifies that the activity should refresh itself every 10 seconds. In addition, it includes a RunTrackerDB object named db that can be used to read and write data from a database. Although the RunTrackerDB class isn't shown in this chapter, it works similarly to the other database classes presented in this book.

The onCreate method displays the layout for the activity. Then, it initializes the View Stopwatch button, its intent, and the RunTrackerDB object. Next, it checks whether GPS is enabled. If not, it displays a toast asking the user to enable GPS and an activity within the Settings app that allows the user to enable GPS. Finally, this code initializes a GoogleApiClient object.

The onStart method begins by getting the GoogleMap object if it isn't already available. Then, it enables the zoom controls on this map object. Finally, it attempts to connect to Google Play services. If successful, this executes the onConnected method shown in part 4. In other words, the app should connect to Google Play services when the user displays the activity.

The onStop method disconnects from Google Play services. This executes the onConnectionSuspended method shown in part 4. In other words, the app should disconnect from Google Play services if the user navigates away from this activity.

The updateMap method begins by checking whether the app is connected to Google Play services. If so, it calls another method in this class to set the marker for the current location. Then, it calls another method in this class to display the route for the run.

The setCurrentLocationMarker method gets the current location, zooms in on that location, and adds a marker to identify the location.

The displayRun method gets a list of locations from the database and adds a line that connects those locations to the map.

The RunMapActivity class

```java
package com.murach.runtracker;

import java.util.List;
import java.util.Timer;
import java.util.TimerTask;

import android.app.AlertDialog;
import android.content.Intent;
import android.content.IntentSender;
import android.location.Location;
import android.location.LocationManager;
import android.os.Bundle;
import android.provider.Settings;
import android.support.v4.app.FragmentActivity;
import android.support.v4.app.FragmentManager;
import android.view.View;
import android.view.View.OnClickListener;
import android.widget.Button;
import android.widget.Toast;

import com.google.android.gms.common.api.GoogleApiClient;
import com.google.android.gms.common.api.GoogleApiClient.ConnectionCallbacks;
import com.google.android.gms.common.api.GoogleApiClient.OnConnectionFailedListener;
import com.google.android.gms.common.ConnectionResult;
import com.google.android.gms.location.LocationServices;
import com.google.android.gms.maps.GoogleMap;
import com.google.android.gms.maps.SupportMapFragment;
import com.google.android.gms.maps.model.CameraPosition;
import com.google.android.gms.maps.model.LatLng;
import com.google.android.gms.maps.model.MarkerOptions;
import com.google.android.gms.maps.CameraUpdateFactory;
import com.google.android.gms.maps.model.PolylineOptions;

public class RunMapActivity extends FragmentActivity
        implements OnClickListener, ConnectionCallbacks,
                   OnConnectionFailedListener {

    private final static int CONNECTION_FAILURE_RESOLUTION_REQUEST = 9000;
    private static final int INTERVAL_REFRESH = 10 * 1000;   // 10 seconds

    private GoogleMap map;
    private GoogleApiClient googleApiClient;
    private List<Location> locationList;

    private RunTrackerDB db;

    private Button stopwatchButton;
    private Intent stopwatchIntent;

    private Timer timer;
```

Figure 18-20   The RunMapActivity class (part 1 of 4)

The RunMapActivity class

```java
@Override
protected void onCreate(Bundle savedInstanceState) {
    super.onCreate(savedInstanceState);
    setContentView(R.layout.activity_run_map);

    stopwatchButton = (Button) findViewById(R.id.viewStopwatchButton);
    stopwatchButton.setOnClickListener(this);
    stopwatchIntent = new Intent(getApplicationContext(),
            StopwatchActivity.class).addFlags(
                    Intent.FLAG_ACTIVITY_CLEAR_TOP);

    db = new RunTrackerDB(this);
    googleApiClient = new GoogleApiClient.Builder(this)
            .addApi(LocationServices.API)
            .addConnectionCallbacks(this)
            .addOnConnectionFailedListener(this).build();

    // if GPS is not enabled, start GPS settings activity
    LocationManager locationManager =
            (LocationManager) getSystemService(LOCATION_SERVICE);
    if (!locationManager.isProviderEnabled(LocationManager.GPS_PROVIDER)){
        Toast.makeText(this, "Please enable GPS!",
                Toast.LENGTH_LONG).show();
        Intent intent =
                new Intent(Settings.ACTION_LOCATION_SOURCE_SETTINGS);
        startActivity(intent);
    }
}

@Override
protected void onStart() {
    super.onStart();

    // if GoogleMap object is not already available, get it
    if (map == null) {
        FragmentManager manager = getSupportFragmentManager();
        SupportMapFragment fragment =
                (SupportMapFragment) manager.findFragmentById(R.id.map);
        map = fragment.getMap();
    }

    // if GoogleMap object is available, configure it
    if (map != null) {
        map.getUiSettings().setZoomControlsEnabled(true);
    }

    googleApiClient.connect();
}

@Override
protected void onStop() {
    googleApiClient.disconnect();
    super.onStop();
}
```

Figure 18-20   The RunMapActivity class (part 2 of 4)

The RunMapActivity class   **Page 3**

```
private void updateMap() {
    if (googleApiClient.isConnected()) {
        setCurrentLocationMarker();
    }
    displayRun();
}

private void setCurrentLocationMarker() {
    if (map != null) {
        // get current location
        Location location = LocationServices.FusedLocationApi
                .getLastLocation(googleApiClient);

        if (location != null) {
            // zoom in on current location
            map.animateCamera(
                    CameraUpdateFactory.newCameraPosition(
                        new CameraPosition.Builder()
                            .target(new LatLng(location.getLatitude(),
                                                location.getLongitude()))
                            .zoom(16.5f)
                            .bearing(0)
                            .tilt(25)
                            .build()));

            // add a marker for the current location
            map.clear();       // clear old marker(s)
            map.addMarker(     // add new marker
                    new MarkerOptions()
                            .position(new LatLng(location.getLatitude(),
                                    location.getLongitude()))
                            .title("You are here"));
        }
    }
}

private void displayRun() {
    if (map != null) {
        locationList = db.getLocations();
        PolylineOptions polyline = new PolylineOptions();
        if (locationList.size() > 0) {
            for (Location l : locationList) {
                LatLng point = new LatLng(
                        l.getLatitude(), l.getLongitude());
                polyline.add(point);
            }
        }
        map.addPolyline(polyline);
    }
}
```

Figure 18-20   The RunMapActivity class (part 3 of 4)

The setMapToRefresh method starts a timer that updates the map at the specified refresh interval, which was set to 10 seconds earlier in the class. To do that, this method uses the Timer and TimerTask classes that were described in chapter 10. As a result, you shouldn't have much trouble understanding how this works.

The onConnected method calls the updateMap method to update the map. Then, it calls the setMapToRefresh method so that map will update every 10 seconds.

The onConnectionSuspended method cancels the timer. This stops the thread for the timer when this activity isn't displayed, which is what you want.

The onConnectionFailed method attempts to resolve the problem that caused the connection to fail. If it can't resolve the problem, it displays a dialog box that displays an error message that includes the code for the error that caused the connection to fail.

The onClick method is executed when the user clicks on the View Stopwatch button. The code for this method starts the Stopwatch activity.

The RunMapActivity class   Page 4

```
    private void setMapToRefresh() {
        timer = new Timer();
        TimerTask task = new TimerTask() {
            @Override
            public void run() {
                RunMapActivity.this.runOnUiThread(new Runnable() {
                    @Override
                    public void run() {
                        updateMap();
                    }
                });
            }
        };
        timer.schedule(task, INTERVAL_REFRESH, INTERVAL_REFRESH);
    }

    @Override
    public void onConnected(Bundle dataBundle) {
        updateMap();
        setMapToRefresh();
    }

    @Override
    public void onConnectionSuspended(int i) {
        timer.cancel();
        Toast.makeText(this, "Disconnected", Toast.LENGTH_SHORT).show();
    }

    @Override
    public void onConnectionFailed(ConnectionResult connectionResult) {
        // if Google Play services can resolve the error, display activity
        if (connectionResult.hasResolution()) {
            try {
                // start an Activity that tries to resolve the error
                connectionResult.startResolutionForResult(this,
                        CONNECTION_FAILURE_RESOLUTION_REQUEST);
            } catch (IntentSender.SendIntentException e) {
                e.printStackTrace();
            }
        } else {
            new AlertDialog.Builder(this)
                    .setMessage("Connection failed. Error code: "
                            + connectionResult.getErrorCode())
                    .show();
        }
    }

    @Override
    public void onClick(View v) {
        startActivity(stopwatchIntent);
    }
}
```

Figure 18-20   The RunMapActivity class (part 4 of 4)

The RunTrackerService class

Figure 18-21 shows the code for the RunTrackerService class. This class defines a service that continues running as long as the stopwatch is running, even if the user navigates away from the Stopwatch activity. However, the user can stop this service from running by clicking on the Stop or Reset buttons.

To start, this class imports some classes from the Android API. Then, this class imports classes from the Google Maps Android API v2. If you read chapter 10 and this chapter, you should understand how these classes work.

The RunTrackerService class implements three interfaces. The first two are necessary to connect to Google Play services and get the current location, and the third is necessary to track the device's location.

After declaring the RunTrackerService class, this code defines some constants and instance variables that are needed by the class. To start, this code defines the constant that's used to request location updates every 5 seconds. Then, it defines the constant that indicates that the app can handle location updates every 2 seconds if necessary. Next, it defines three instance variables including a RunTrackerDB object named db that can be used to work with a database. Although the RunTrackerDB class isn't shown in this chapter, it works similarly to the other database classes presented in this book. To see the code for this class, you can view the downloadable source code for this app.

The onCreate method initializes the RunTrackerDB object, the GoogleApiClient object, and the LocationRequest object. Here, the LocationRequest object specifies that its priorty is high accuracy. As a result, this app uses GPS to determine the location of the device. In addition, the LocationRequest object uses the constants defined earlier in this class to specify the requested update interval and the fastest allowable update interval.

The RunTrackerService class   **Page 1**

```java
package com.murach.runtracker;

import android.app.Service;
import android.content.Intent;
import android.location.Location;
import android.os.Bundle;
import android.os.IBinder;
import android.widget.Toast;

import com.google.android.gms.common.ConnectionResult;
import com.google.android.gms.common.api.GoogleApiClient;
import com.google.android.gms.common.api.GoogleApiClient.ConnectionCallbacks;
import com.google.android.gms.common.api.GoogleApiClient.OnConnectionFailedListener;
import com.google.android.gms.location.LocationListener;
import com.google.android.gms.location.LocationRequest;
import com.google.android.gms.location.LocationServices;

public class RunTrackerService extends Service
        implements ConnectionCallbacks, OnConnectionFailedListener,
                   LocationListener {

    public static final int UPDATE_INTERVAL = 5000;        // 5 seconds
    public static final int FASTEST_UPDATE_INTERVAL = 2000; // 2 seconds

    private GoogleApiClient googleApiClient;
    private LocationRequest locationRequest;

    private RunTrackerDB db;

    @Override
    public void onCreate() {
        super.onCreate();

        // get database
        db = new RunTrackerDB(getApplicationContext());

        googleApiClient = new GoogleApiClient.Builder(this)
                .addApi(LocationServices.API)
                .addConnectionCallbacks(this)
                .addOnConnectionFailedListener(this).build();

        // get location request and set it up
        locationRequest = LocationRequest.create()
                .setPriority(LocationRequest.PRIORITY_HIGH_ACCURACY)
                .setInterval(UPDATE_INTERVAL)
                .setFastestInterval(FASTEST_UPDATE_INTERVAL);
    }

    @Override
    public IBinder onBind(Intent intent) {
        return null;
    }
```

Figure 18-21   The RunTrackerService class (part 1 of 2)

The onStartCommand method is executed when the service is started. The first statement in this method attempts to connect to Google Play services. This causes the onConnected method shown later in this class to be executed.

The onDestroy method is executed when the service is stopped. The statements in this method attempt to disconnect from Google Play services. This causes the onConnectionSuspended method shown later in this class to be executed.

The onConnected method is executed when the device connects to Google Play services. This method begins by getting the current location for the user. Then, it writes that location to the database. Next, it requests location updates.

The onConnectionSuspended method is executed when the device disconnects from Google Play services. This method removes its request for location updates, which is what you want when you disconnect.

The onConnectionFailed method is executed if the connection fails. Since it's difficult to display a dialog box from a service, this method displays a toast that indicates that the connection has failed. This works differently than the code for the same method that's in the Run Map activity where it's easy to display a dialog box.

The onLocationChanged method is executed every time Android gets a fix on a new location. The code within this method writes some of the data that's stored in the Location object to the database for this app. That way, the Run Map activity can read this data from the database.

The RunTrackerService class   **Page 2**

```java
    @Override
    public int onStartCommand(Intent intent, int flags, int startId) {
        googleApiClient.connect();
        return super.onStartCommand(intent, flags, startId);
    }

    @Override
    public void onDestroy() {
        if (googleApiClient.isConnected()) {
            googleApiClient.disconnect();
        }
        super.onDestroy();
    }

    @Override
    public void onConnected(Bundle dataBundle) {
        Location location =
            LocationServices.FusedLocationApi.getLastLocation(googleApiClient);
        if (location != null){
            db.insertLocation(location);
        }
        LocationServices.FusedLocationApi
                .requestLocationUpdates(googleApiClient,
                    locationRequest, this);
    }

    @Override
    public void onConnectionSuspended(int i) {
        if (googleApiClient.isConnected()) {
            LocationServices.FusedLocationApi
                .removeLocationUpdates(googleApiClient, this);
        }
    }

    @Override
    public void onConnectionFailed(ConnectionResult connectionResult) {
        Toast.makeText(this, "Connection failed! " +
                        "Please check your settings and try again.",
                Toast.LENGTH_SHORT).show();
    }

    @Override
    public void onLocationChanged(Location location) {
        if (location != null){
            db.insertLocation(location);
        }
    }
}
```

Figure 18-21   The RunTrackerService class (part 2 of 2)

Perspective

The skills presented in this chapter should be enough to get you started with locations and maps. However, this is a huge topic, and there's plenty more to learn about it. For example, you may want to learn more about the exciting features that are available from version 2 of the Android API for Google Maps.

One feature is geofencing. This feature lets your app specify a boundary around a location. Then, your app can receive notifications when users enter or leave that boundary.

Another feature is activity recognition. This feature lets you determine the user's current activity. For example, you can determine whether the user is walking, cycling, or riding in a vehicle.

On the other hand, you may need to learn more about less exciting features that have been available for years. For example, you may need to use geocoding to convert latitude and longitude to a street address. You may need to use routing to get directions for your users. Or, you may need to learn how to test an app that uses locations in an emulator. To do that, you can search the Internet for more information.

Terms

GPS (Global Positioning System)
Geocoding
Routing

Summary

- You can use a device's *GPS* (*Global Positioning System*) receiver or its network signals (cell or Wi-Fi) to find a device's current location.

- Google Maps is the most popular Android API for working with maps. One alternative is MapQuest.

- *Geocoding* is the process of converting latitude/longitude coordinates to street addresses and back.

- *Routing* is the process of using map data to provide directions.

- Version 2 of the Google Maps Android API is recommended for all new development as of March 2013.

- To add the Google Play services library to your project, you can import it into your workspace, and add a reference to it from your project.

- A Maps API key is necessary to access the Google Maps servers.

- You can use a GoogleApiClient object to open and close a connection to Google Play services and to get a Location object that contains detailed information about the best and most recent location of a device.

- You can use a LocationRequest object to configure a request for location updates from Google Play services.

- You can use a LocationListener object to listen for location updates from Google Play services.

- You can use the LocationManager object to check if GPS is enabled.

- You can use a GoogleMap object to zoom in on a location, add one or more markers to a map, and to display a series of connected lines on a map.

Exercise 18-1   Test and modify the Run Tracker app

In this exercise, you'll test the Run Tracker app on a physical device. To do that, you'll need to get your own Google Maps API key so you can display a map. Then, you'll modify some of the settings for the map to change its appearance.

This exercise only describes how to test on an actual device, not on an emulator. In addition, this device must have a valid version of Google Play installed, and it must support OpenGL ES version 2.

Test the Run Tracker app on your device

1. Start Android Studio and import the project named ch18_ex1_RunTracker.

2. Open the Project Structure dialog and make sure that the play-services library is a dependency. If it's not, add it.

3. Connect the device you want to use for testing. If you have already installed the Run Tracker app on this device, uninstall it now.

4. Run the Run Tracker app on this device. Then, click on the View Map button. This should display a blank map, and the LogCat view should display an error message for an authentication error. That's because the app hasn't registered a valid API key for Google Maps.

5. Get an API key for Google Maps.

6. In Android Studio, open the google_maps_api.xml file and replace the YOUR_KEY_HERE text with your key.

7. Open the Android manifest file and note that it uses the string resource in the google_maps_api.xml file to set the API key for the app.

8. Run the Run Tracker app with the valid API key on your device. Then, click on the View Map button. This time, the app should display a map and zoom in on the current location. If it doesn't, check for error messages in the LogCat view and do your best to troubleshoot the problem.

9. Click the View Stopwatch button. Then, click on the Start button to start a run. Go outside and walk for a while. Switch back and forth between the Stopwatch and Run Map activities. The map should display a marker for your current location and track your route (or at least somewhat close to your route).

10. Click the Stop button to stop the run. At this point, the map should still display the route of the run.

11. Click the Reset button to delete the run. At this point, the map should only display your current location, not your route.

Modify the settings for the map

12. Open the RunMapActivity class and review its code.

13. Change the map type from the default type to a hybrid map. Then, run the app. Note how this changes the appearance of the map.

14. Use the MAP_TYPE_NORMAL constant to change the map type back to the default type.

15. Use the moveCamera method to change the position of the camera instead of using the animateCamera method. This should display the current location without using any animation to zoom in on it.

16. Change zoom level to 18 and change the tilt to 10 degrees. This should zoom in even further with less tilt.

17. Modify the code so the marker uses the icon named ic_runner that's in the res\drawable directories instead of using the default icon for the marker. This should display a blue runner.

18. Change the width of the line that marks the route to 10 pixels.

19. Change the color of the line that marks the route to blue.

Appendix A

How to set up Windows for this book

This appendix shows how to install and configure the software that we recommend for developing Android apps on the Windows operating system. This appendix also shows how to install the source code for this book.

As you read this appendix, please remember that most websites and install programs are continually updated. As a result, some of the procedures in this appendix may have changed since this book was published. Nevertheless, these procedures should still be good guides to installing the software. And if there are significant changes to these setup instructions, we will post updates on our website (www.murach.com).

How to install the source code for this book

Figure A-1 shows how to download and install the source code for this book. This includes the source code for the apps presented in this book, the starting points for the chapter exercises, and the solutions to the exercises.

When you finish this procedure, the book apps as well as the starting points and solutions for the exercises should be in the directories shown in this figure. Then, you can review the apps presented in this book, and you're ready to do the exercises in this book.

The Murach website

```
www.murach.com
```

The directories for the book apps and exercise starts

```
C:\murach\android\book_apps
C:\murach\android\ex_starts
C:\murach\android\ex_solutions
```

How to download and install the files for this book

1. Go to www.murach.com,

2. Find the page for *Murach's Android Programming (2nd Edition)*.

3. If necessary, scroll down to the FREE downloads tab.

4. Click the FREE downloads tab.

5. Click the Download Now button for the exe file, and respond to the resulting pages and dialog boxes. This should download an installer file named and2_allfiles.exe to your hard drive.

6. Use the Windows Explorer to find the exe file on your C drive.

7. Double-click this file and respond to the dialog boxes that follow. This installs the files for this book in directories that start with C:\murach\android.

How to use a zip file instead of a self-extracting zip file

- Although we recommend using the self-extracting zip (exe) file to install the downloadable files as described above, some systems won't allow self-extracting zip files to run. In that case, you can download a regular zip file (and2_allfiles.zip) from our website. Then, you can extract the files stored in this zip file into the C:\murach directory. If the C:\murach directory doesn't already exist, you will need to create it.

Description

- You can install the source code for this book by downloading it from murach.com.

Figure A-1   How to install the source code for this book

How to install the Java SE JDK

For Android development, you need to have the *Java Development Kit* (*JDK*) for *Java SE* (*Standard Edition*) installed on your computer. If you've already done some Java development, you probably already have the JDK installed on your system.

If not, figure A-2 shows how to install the JDK. To start, you download the file for the installer program for the most recent version of the JDK from the Java website. Then, you navigate to the directory that holds the JDK, run the installer file, and respond to the resulting dialog boxes.

Since the Java website may change after this book is printed, we've kept the procedure shown in this figure somewhat general. As a result, you may have to do some searching to find the current version of the JDK. In general, you can start by searching the Internet for the download for Java SE. Then, you can find the most current version of the JDK for your operating system.

The Android apps in this book only use Java SE 6 features. As a result, you can use JDK 1.6 or later. For example, you can use JDK 1.8, which corresponds with Java SE 8, for this book. This works because Java has a good record of being backwards compatible.

The download page for Java SE

www.oracle.com/technetwork/java/javase/downloads

Procedure

1. Go to the download page for Java SE. If necessary, you can search the Internet to find this page.
2. Follow the instructions for downloading the installer program for the JDK for your operating system.
3. Save the installer program to your hard disk.
4. Run the installer program by double-clicking on it.
5. Respond to the resulting dialog boxes. When you're prompted for the JDK folder, use the default folder.

The default folder for the JDK for Java SE 8

C:\Program Files\Java\jdk1.8.0_45

Description

- For Android development, you need to install *Java SE* (*Standard Edition*). To do that, you can download and install the *JDK* (*Java Development Kit*).
- For more information about installing the JDK, you can refer to the Oracle website.

Figure A-2   How to install Java

How to install Android Studio

Before you can develop Android apps, you need to install the *Android SDK* (*Software Development Kit*). Fortunately, this SDK is installed automatically when you install Android Studio as shown in figure A-3.

The download page for Android Studio

`http://developer.android.com/sdk`

How to install Android Studio

1. Go to the download page for Android Studio. The easiest way to find this page is to search the Internet for "Android Studio download".

2. Click on the link to download Android Studio and follow the instructions on the website. This should download an exe file.

3. Double-click on the exe file to start the installation.

4. Follow the dialogs to install Android Studio.

Description

*   For Android development, you need to install Android Studio. This also installs the *Android SDK* (*Software Development Kit*).

Figure A-3   How to install Android Studio

How to use the Android SDK Manager

Android Studio includes a starting set of Android tools and platforms. However, we recommend following the procedure presented in figure A-4 to make sure that you have the tools and platforms that you'll need as you work through this book.

When you use the Android SDK Manager, it displays "Installed" in the Status column if the most current version of a tool or platform has been installed. Otherwise, it displays a status of "Not Installed" or "Update Available".

All of the apps that you can download for this book use version 23.0.1 of the Android SDK build tools. As a result, we recommend that you install this version of the build tools.

Throughout this book, we have used Android 6.0 (API 23) for testing new devices and Android 4.1 (API 16) for testing old devices. As a result, we recommend installing these platforms to get started.

If you prefer to use other versions of Android, you can do that too. For example, you may want to use newer or older versions than the ones we use in this book. However, it's usually easier to follow along with the book if you use the same versions we use in this book. That way, this book will match your system, and you will encounter fewer bugs since we have tested the apps in this book on these versions of Android.

We also recommend installing two more items. First, we recommend installing the Google USB Driver item because you may need it to configure Android devices for testing and debugging. Second, we recommend installing the Android Support Library item because it's necessary to support some newer features on older devices.

This figure shows how to install the tools and platforms necessary to work with this book. If you want to install more tools or platforms, you can use the Android SDK Manager to do that. Of course, additional tools and platforms require more disk space and take longer to download. As a result, if you want to get started as fast as possible, you can follow the instructions in this figure.

The Android SDK Manager

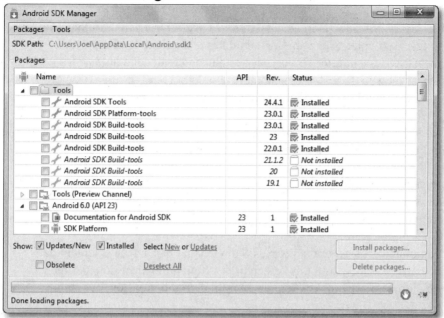

Procedure

1. Start Android Studio.

2. If this displays the Welcome page, select the Configure→SDK Manager option. Otherwise, click the toolbar button for the SDK Manager. Then, click the Launch Standalone SDK Manager link.

3. To view the tools and platforms, expand or collapse the category nodes.

4. To install more tools or platforms, select the tools or platforms. For this book, we recommend installing all of the defaults including the following items:
   - Tools→Android SDK Build-tools Rev. 23.0.1
   - Android 6.0 (API 23)→SDK Platform
   - Android 6.0 (API 23)→Intel x86 Atom System Image
   - Android 4.1.2 (API 16)→SDK Platform
   - Android 4.1.2 (API 16)→Intel x86 Atom System Image
   - Extras→Android Support Library
   - Extras→Google USB driver

5. Click the Install button. This should display another dialog box.

6. Select the Accept License radio button. Then, click the Install button. On some systems, this may take an hour or longer.

Description

- You can use the Android SDK Manager to install Android tools and platforms.

Figure A-4   How to use the Android SDK Manager

How to create an emulator

Figure A-5 shows how to create an *Android Virtual Device* (*AVD*) that you can use to test your apps. Since an AVD emulates a physical device, an AVD can also be called an *emulator*. If necessary, you can create an emulator for each platform that you want to test. Unfortunately, emulators tend to run extremely slowly on all but the most powerful computers. As a result, we recommend avoiding using emulators whenever possible.

If you do need to use an emulator, we recommend creating one that uses a low resolution such as hdpi or mdpi and an older version of the Android API. That way, the emulator should use fewer resources on your system, which should allow it to start and run more quickly.

When you create an emulator, you must provide a name for the emulator and the target platform. In this figure, I created an emulator named "Nexus S API 16".

This emulator is based upon a Nexus S device that has a 4-inch screen that's 480 by 800 pixels. As a result, this screen uses a high number of dots per inch (hdpi), which runs more quickly than an extra high number of dots per inch (xhdpi).

This emulator uses a system image for the Android 4.1 (API 16) platform. Since this API doesn't have as many features as later APIs, this also helps the emulator to run more quickly. In addition, this system image uses an x86 processor, which runs faster on a Windows system than an ARM processor. Most smartphones use ARM processors, so this emulation isn't completely accurate, but the improved performance usually makes this tradeoff worthwhile. For this to work, you must download the x86 system image for the corresponding platform as described in the previous figure.

The dialog box for creating an emulator

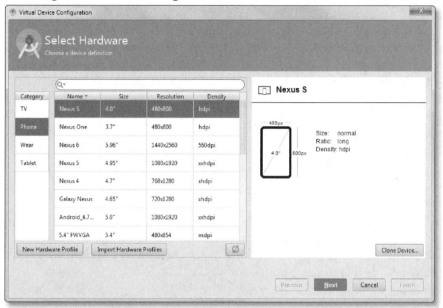

Procedure

1. Start Android Studio.
2. If this displays the Welcome page, open an existing project as described in the start of figure A-7.
3. Click the toolbar button for the AVD Manager. This should start the Android Virtual Device Manager and show all existing virtual devices.
4. Click the Create Virtual Device button. This should display the dialog box shown above. Follow the prompts to create a virtual device. For this book, we recommend creating a virtual device named "Nexus S API 16" with these properties:
   - Category: Phone
   - Hardware profile: Nexus S
   - System image: Jelly Bean (API 16) x86
   - Emulator name: Nexus S API 16

Description

- To test your Android apps, you can create an *Android Virtual Device* (*AVD*) for each platform that you wish to test. An Android Virtual Device can also be referred to as an *emulator*.

Figure A-5   How to create an emulator

How to configure a device for development

Figure A-6 shows how to configure a physical device so that you can test Android apps on it. Although you can also test Android apps on an emulator, we recommend testing on a physical device for two reasons. First, apps install and run much more quickly on a physical device than on an emulator. Second, testing on a physical device is the only way to truly see how an app works on that device.

To configure a physical device for development, you must connect your device to the computer with a USB cable and turn on the "USB debugging" option on your device. The procedure for doing this varies depending on the version of Android that's running on the device. As a result, you may need to search the Internet to learn how to enable this option for your device.

With Android 4.2 (API 17) and later, the developer options are hidden. This makes sense because most users aren't developers and don't need these options. As a result, if your device uses API 17 or later, you must start by enabling developer options on your device.

In addition, you must install a driver for the device. Again, the procedure for doing this varies depending on the device. For some devices, you can install the *Google USB Driver*. For other devices, you need to install an *OEM (Original Equipment Manufacturer) driver*. To determine what type of driver you need for your device, you can search the Internet. Or, you can consult the OEM USB Drivers document by going to the link shown at the bottom of this figure.

How to configure a device for development

1. Enable developer options on your device. For Android 4.2 (API 17) and later, you can do it like this:
   - Settings→More→About Device→Build Number (tap on it 7 times).
2. Turn on USB Debugging. For Android 4.2 (API 17) and later, you can do it like this:
   - Settings→More→Developer Options→USB Debugging.
3. Connect your Android device to your computer's USB port.
4. Download a driver for the device.
   - For an Android Developer Phone, a Nexus One, or a Nexus S, you can use the *Google USB Driver*. If you followed the steps in figure A-5, you should have already downloaded this driver.
   - For the Galaxy Nexus, go to the Samsung website and download the Google USB Driver for that phone, which is listed as model SCH-I515.
   - For most other devices, install an *OEM (Original Equipment Manufacturer) driver* as described below.

How to download the Google USB Driver

- Start the Android SDK Manager as described in figure A-5 and use it to download the Google USB Driver. To do that, expand the Extras category, select the Google USB Driver check box, and click the Install button. By default, this downloads the driver to this directory:

      C:\Users\YourName\android-sdks\extras\google\usb_driver

How to install an OEM driver

- Search the Internet or use the OEM USB Drivers document to find an OEM driver for your device. Then, download the driver to your computer.

The link to the OEM USB Drivers document

      developer.android.com/tools/extras/oem-usb.html

Description

- Since an actual device runs more quickly than an emulator, we recommend using an actual device whenever possible.

Figure A-6   How to configure a device for development

How to verify that your system is set up correctly

At this point, you have installed all of the software and source code that you need to work with this book, but you haven't tested it to make sure it's working correctly. To do that, you can run the Tip Calculator app for chapter 3 as shown by the procedure in figure A-7.

To begin, you can start Android Studio and open the project for the Tip Calculator app as shown in part 1 of the figure. At this point, the name of the project appears in Android Studio. However, the emulator doesn't appear on your screen until you've gone through the steps in part 2.

Android Studio and an emulator after an app has been run

Open an existing project

1. Start Android Studio.
2. If this displays the Welcome page, select the "Open an existing Android Studio project" item. Otherwise, click the Open button in the toolbar.
3. Use the resulting dialog to open the Tip Calculator app for chapter 3 that's stored in this directory:

    ```
    C:\murach\android\book_apps\ch03_TipCalculator
    ```

Description

*   To verify that you have set up your system correctly for this book, you can run the Tip Calculator app for chapter 3 on an Android device. Or, if you don't have an Android device available, you can run this app in an emulator.
*   If you get an error that says, "failed to find Build Tools revision 23.01", click on the link that says, "Install Build Tools and sync project" to install these build tools.

Figure A-7   How to verify that your system is set up correctly (part 1 of 2)

After opening the project in Android Studio, you can run the app as shown in part 2. This verifies that your system is set up correctly.

If you didn't download version 23.0.1 of the Android SDK build tools as described in figure A-4, you may get an error when you attempt to run the Tip Calculator app that says:

```
failed to find Build Tools revision 23.0.1
```

To fix this issue, you can install version 23.0.1 of the Android SDK build tools. Fortunately, Android Studio typically provides a link just below the error message that you can use to install the correct version of the build tools.

If you have an Android device that you have configured for development, you can run the app on that device. When you do this, you should display the Choose Device dialog before you connect your device to the computer. Otherwise, this dialog might not display your device.

Also, the first time you use a device with your computer, you need to authorize that device to work with your computer. To do that, you can use the authorization dialog that's displayed after you connect your device and unlock its screen. After you authorize your device, you should be able to keep your device connected to the computer, and the Choose Device dialog should display your device every time you run the app.

If you don't have access to an Android device, you'll need to test the app on an emulator. This provides a way for you to test your apps on devices even if you don't have that type of device. However, on most systems, emulators take a long time to start, consume a lot of system resources, and run slowly. As a result, we recommend testing on an actual device whenever possible.

If you are able to complete either of these tasks, you have set up your system correctly for this book. Congratulations!

The Choose Device dialog

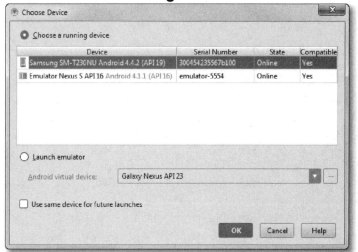

Test on a device

4. In the toolbar, click the Run button. This should display the Choose Device dialog.

5. Connect the device you configured in figure A-6 to the computer with a USB cable. This should display the device in the Choose Device dialog, though it may indicate that the device is unauthorized.

6. Unlock the screen on the device. If necessary, use the dialog that's displayed on the device to authorize the device for the computer.

7. Use the Choose Device dialog to select the device and click the OK button. This should run the app on the device.

8. If the Tip Calculator app displays correctly and you can use it to calculate a tip, your device is configured correctly!

Test on an emulator (optional)

9. Start the emulator. To do that, click on the AVD Manager button in the toolbar. Then, click on the Run button for the emulator described in figure A-5. Depending on your system, it may take a long time for the emulator to launch, so be patient!

10. Unlock the emulator by dragging the lock icon to the right.

11. In the toolbar, click the Run button. This should display the Choose Device dialog.

12. Use the Choose Device dialog to select the emulator. This should run the app on that emulator.

13. If the Tip Calculator app displays correctly and you can use it to calculate a tip, your emulator is set up correctly!

14. If the soft keyboard doesn't display when you click in the bill amount text box, use the emulator to start the Settings app and turn off the Language and Input→Keyboard and Input→Default→Hardware (Physical Keyboard) option. This should allow you to use the soft keyboard or your computer's keyboard to enter the amount.

Figure A-7   How to verify that your system is set up correctly (part 2 of 2)

Appendix B

How to set up Mac OS X for this book

This appendix shows how to install and configure the software that we recommend for developing Android apps on Mac OS X. This appendix also shows how to install the source code for this book.

As you read this appendix, please remember that most websites are continually updated. As a result, some of the procedures in this appendix may have changed since this book was published. Nevertheless, these procedures should still be good guides to installing the software. And if there are significant changes to these setup instructions, we will post updates on our website (www.murach.com).

How to install the source code for this book

Figure B-1 shows how to download and install the source code for this book. This includes the source code for the apps presented in this book, the starting points for the exercises, and the solutions to the exercises.

When you finish this procedure, the book apps as well as the starting points and solutions for the exercises should be in the directories shown in this figure. Then, you can review the apps presented in this book, and you're ready to do the exercises in this book.

As you read this book, you'll notice that it often instructs you to right-click, which is a common technique on Windows. On Mac OS X, right-clicking is not enabled by default. Instead, you can use the Ctrl-click instead of the right-click. Or, if you prefer, you can enable right-clicking by editing the system preferences for your mouse. Then, you can follow the instructions in this book more closely.

The Murach website

`www.murach.com`

The folders for the book apps and exercise starts

```
/murach/android/book_apps
/murach/android/ex_starts
/murach/android/ex_solutions
```

How to download and install the files for this book

1. Go to www.murach.com.
2. Find the page for *Murach's Android Programming (2nd Edition)*.
3. If necessary, scroll down to the FREE downloads tab.
4. Click the "FREE Downloads" tab.
5. Click the Download Now button for the zip file. This should download a zip file named and2_allfiles.zip onto your hard drive.
6. Use the Finder to browse to this file and double-click on it to unzip it. This creates the android folder and its subfolders.
7. Use the Finder to create the murach folder directly on the Mac hard drive.
8. Use the Finder to move the android folder into the murach folder.

A note about right-clicking

- This book often instructs you to right-click, because that's common on PCs. On a Mac, right-clicking is not enabled by default. Instead, you can use the Ctrl-click instead of the right-click. Or, if you prefer, you can enable right-clicking by editing the system preferences for your mouse.

Description

- You can install the source code for this book by downloading it from murach.com.

Figure B-1   How to install the source code for this book

How to install the Java SE JDK

For Android development, you need to have the *Java Development Kit* (*JDK*) for *Java SE* (*Standard Edition*) installed on your computer. If you've already done some Java development, you probably already have the JDK installed on your system.

If not, figure B-2 shows how to install the JDK. To start, you download the file for the installer program for the most recent version of the JDK from the Java website. Then, you navigate to the directory that holds the JDK, run the installer file, and respond to the resulting dialog boxes.

Since the Java website may change after this book is printed, we've kept the procedure shown in this figure somewhat general. As a result, you may have to do some searching to find the current version of the JDK. In general, you can start by searching the Internet for the download for Java SE. Then, you can find the most current version of the JDK for your operating system.

The Android apps in this book only use Java SE 6 features. As a result, you can use JDK 1.6 or later. For example, you can use JDK 1.8, which corresponds with Java SE 8, for this book. This works because Java has a good record of being backwards compatible.

The download page for Java SE

www.oracle.com/technetwork/java/javase/downloads

Procedure

1. Go to the download page for Java SE. If necessary, you can search the Internet to find this page.

2. Follow the instructions for downloading the installer program for the JDK for your operating system.

3. Save the installer program to your hard disk.

4. Run the installer program by double-clicking on it.

5. Respond to the resulting dialog boxes. When you're prompted for the JDK folder, use the default folder.

The default folder for the JDK for Java SE 8

/Library/Java/JavaVirtualMachines/jdk1.8.0_45

Description

- For Android development, you need to install *Java SE* (*Standard Edition*). To do that, you can download and install the *JDK* (*Java Development Kit*).

- For more information about installing the JDK, you can refer to the Oracle website.

Figure B-2   How to install Java

How to install Android Studio

Before you can develop Android apps, you need to install the *Android SDK* (*Software Development Kit*). Fortunately, this SDK is installed automatically when you install Android Studio as shown in figure B-3.

You may want to create an easier way to start Android Studio for subsequent sessions. For example, you can dock Android Studio by right-clicking on the app file for Android Studio and selecting the Options→Keep in Dock item. Or, you can create an alias for your desktop, by right-clicking on the app file for Android Studio and selecting the Make Alias item. Then, drag the alias to the desktop.

The download page for Android Studio

`http://developer.android.com/sdk`

How to install the Android Studio

1. Go to the download page for Android Studio. The easiest way to find this page is to search the Internet for "Android Studio download".

2. Click on the link to download Android Studio and follow the instructions on the website. This should download a dmg file.

3. Use the Finder to browse to this file and double-click on it to launch it.

4. Drag the Android Studio icon into the Apps folder.

A typical folder for the Android SDK

`User/name/Library/Android/sdk`

Description

- For Android development, you need to install Android Studio. This also installs the *Android SDK* (*Software Development Kit*).

Figure B-3   How to install Android Studio

How to use the Android SDK Manager

Android Studio includes a starting set of Android tools and platforms. However, we recommend following the procedure presented in figure B-4 to make sure that you have the tools and platforms that you'll need as you work through this book.

When you use the Android SDK Manager, it displays "Installed" in the Status column if the most current version of a tool or platform has been installed. Otherwise, it displays a status of "Not Installed" or "Update Available".

All of the apps that you can download for this book use version 23.0.1 of the Android SDK build tools. As a result, we recommend that you install this version of the build tools.

Throughout this book, we have used Android 6.0 (API 23) for testing new devices and Android 4.1 (API 16) for testing old devices. As a result, we recommend installing these platforms to get started.

If you prefer to use other versions of Android, you can do that too. For example, you may want to use newer or older versions than the ones we use in this book. However, it's usually easier to follow along with the book if you use the same versions we use in this book. That way, this book will match your system, and you will encounter fewer bugs since we have tested the apps in this book on these versions of Android.

We also recommend installing the Android Support Library item because it's necessary to support some newer features on older devices.

This figure shows how to install the tools and platforms necessary to work with this book. If you want to install more tools or platforms, you can use the Android SDK Manager to do that. Of course, additional tools and platforms require more disk space and take longer to download. As a result, if you want to get started as fast as possible, you can follow the instructions in this figure.

The Android SDK Manager

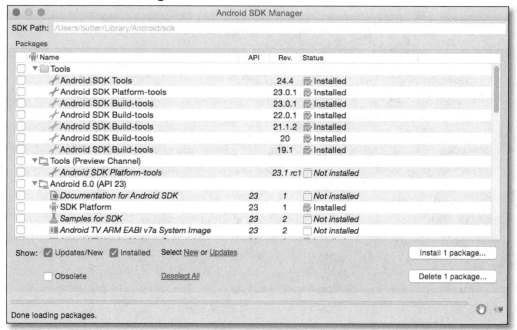

Procedure

1. Start Android Studio.
2. If this displays the Welcome page, select the Configure→SDK Manager option. Otherwise, click the toolbar button for the SDK Manager. Then, click the Launch Standalone SDK Manager link at the bottom of the window.
3. To view the tools and platforms, expand or collapse the category nodes.
4. To install more tools or platforms, select the tools or platforms. For this book, we recommend installing all of the defaults including the following items:
    - Tools→Android SDK Build-tools Rev. 23.0.1
    - Android 6.0 (API 23)→SDK Platform
    - Android 6.0 (API 23)→Intel x86 Atom System Image
    - Android 4.1.2 (API 16)→SDK Platform
    - Android 4.1.2 (API 16)→Intel x86 Atom System Image
    - Extras→Android Support Library
5. Click the Install button. This should display another dialog box.
6. Select the Accept License radio button. Then, click the Install button. On some systems, this may take an hour or longer.

Description

*   You can use the Android SDK Manager to install Android tools and platforms.

Figure B-4   How to use the Android SDK Manager

How to create an emulator

Figure B-5 shows how to create an *Android Virtual Device* (*AVD*) that you can use to test your apps. Since an AVD emulates a physical device, an AVD can also be called an *emulator*. If necessary, you can create an emulator for each platform that you want to test. Unfortunately, emulators tend to run extremely slowly on all but the most powerful computers. As a result, we recommend avoiding using emulators whenever possible.

If you do need to use an emulator, we recommend creating one that uses a low resolution such as hdpi or mdpi and an older version of the Android API. That way, the emulator should use fewer resources on your system, which should allow it to start and run more quickly.

When you create an emulator, you must provide a name for the emulator and the target platform. In this figure, I created an emulator named "Nexus S API 16".

This emulator is based upon a Nexus S device that has a 4-inch screen that's 480 by 800 pixels. As a result, this screen uses a high number of dots per inch (hdpi), which runs more quickly than an extra high number of dots per inch (xhdpi).

This emulator uses a system image for the Android 4.1 (API 16) platform. Since this API doesn't have as many features as later APIs, this also helps the emulator to run more quickly. In addition, this system image uses an x86 processor, which runs faster on a Mac computer than an ARM processor. Most smartphones use ARM processors, so this emulation isn't completely accurate, but the improved performance usually makes this tradeoff worthwhile. For this to work, you must download the x86 system image for the corresponding platform as described in the previous figure.

The dialog box for creating an emulator

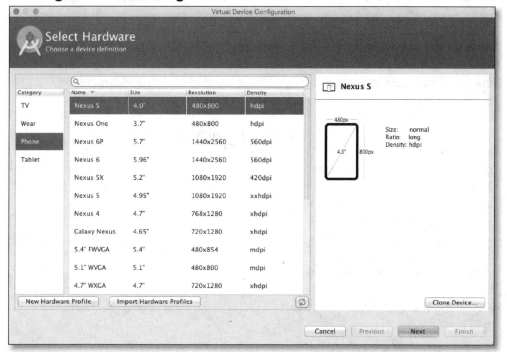

Procedure

1. Start Android Studio.
2. If this displays the Welcome page, open an existing project as described in the start of figure B-7.
3. Click the toolbar button for the AVD Manager. This should start the Android Virtual Device Manager and show all existing virtual devices.
4. Click the Create Virtual Device button. This should display the dialog box shown above. Follow the prompts to create a virtual device. For this book, we recommend creating a virtual device named "Nexus S API 16" with these properties:
    - Category: Phone
    - Hardware profile: Nexus S
    - System image: Jelly Bean (API 16) x86
    - Emulator name: Nexus S API 16

Description

- To test your Android apps, you can create an *Android Virtual Device* (*AVD*) for each platform that you wish to test. An Android Virtual Device can also be referred to as an *emulator*.

Figure B-5   How to create an emulator

How to configure a device for development

Figure B-6 shows how to configure a physical device so that you can test Android apps on it. Although you can also test Android apps on an emulator, we recommend testing on a physical device for two reasons. First, apps install and run much more quickly on a physical device than on an emulator. Second, testing on a physical device is the only way to truly see how an app works on that device.

To configure a device for development, you must connect your device to the computer with a USB cable and turn on the "USB debugging" option on your device. The procedure for doing this varies depending on the version Android that's running on the device. As a result, you may need to search the Internet to learn how to enable this option for your device.

With Android 4.2 (API 17) and later, the developer options are hidden. This makes sense because most users aren't developers and don't need these options. As a result, if your device uses API 17 or later, you must start by enabling developer options on your device.

How to configure a device for development

1. Enable developer options on your device. For Android 4.2 (API 17) and later, you can do it like this:
    -   Settings→More→About Device→Build Number (tap on it 7 times).

2. Turn on USB Debugging. For Android 4.2 (API 17) and later, you can do it like this:
    -   Settings→More→Developer Options→USB Debugging.

3. Connect your Android device to your computer's USB port.

Description

-   Since a physical device runs more quickly than an emulator, we recommend using a physical device whenever possible.

Figure B-6   How to configure a device for development

How to verify that your system is set up correctly

At this point, you have installed all of the software and source code that you need to work with this book, but you haven't tested it to make sure it's working correctly. To do that, you can run the Tip Calculator app for chapter 3 as shown by the procedure in figure B-7.

To begin, you can start Android Studio and open the project for the Tip Calculator app as shown in part 1 of the figure. At this point, the name of the project appears in Android Studio. However, the emulator won't appear on your screen until you've gone through the steps in part 2.

Android Studio and an emulator after an app has been run

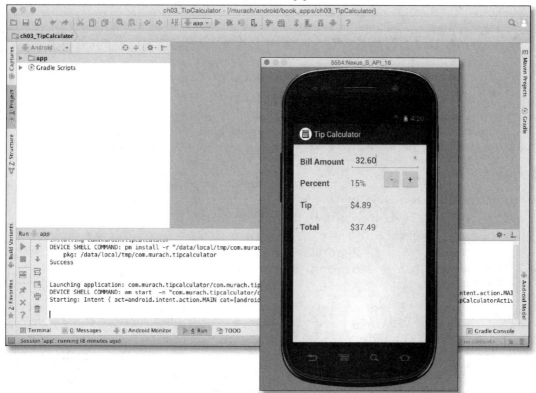

Open an existing project

1. Start Android Studio.

2. If this displays the Welcome page, select the "Open an existing Android Studio project" item. Otherwise, click the Open button in the toolbar.

3. Use the resulting dialog to open the Tip Calculator app for chapter 3 that's stored in this folder:

    murach/android/book_apps/ch03_TipCalculator

Description

- To verify that you have set up your system correctly for this book, you can run the Tip Calculator app for chapter 3 on an Android device. Or, if you don't have an Android device available, you can run this app in an emulator.

- If you get an error that says, "failed to find Build Tools revision 23.0.1", click on the link that says, "Install Build Tools and sync project" to install these build tools.

Figure B-7   How to verify that your Mac is set up correctly (part 1 of 2)

After opening the project in Android Studio, you can run the app as shown in part 2. This verifies that your system is set up correctly.

If you didn't download version 23.0.1 of the Android SDK build tools as described in figure B-4, you may get an error when you attempt to run the Tip Calculator app that says:

```
failed to find Build Tools revision 23.0.1
```

To fix this issue, you can install version 23.0.1 of the Android SDK build tools. Fortunately, Android Studio typically provides a link just below the error message that you can use to install the correct version of the build tools.

If you have an Android device that you have configured for development, you can run the app on that device. When you do this, you should display the Device Chooser dialog before you connect your device to the computer. Otherwise, this dialog might not display your device.

Also, the first time you use a device with your computer, you need to authorize that device to work with your computer. To do that, you can use the authorization dialog that's displayed after you connect your device and unlock its screen. After you authorize your device, you should be able to keep your device connected to the computer, and the Device Chooser dialog should display your device every time you run the app.

If you don't have access to an Android device, you'll need to test the app on an emulator. This provides a way for you to test your apps on devices even if you don't have that type of device. However, on most systems, emulators take a long time to start, consume a lot of system resources, and run slowly. As a result, we recommend testing on an actual device whenever possible.

If you are able to complete either of these tasks, you have set up your system correctly for this book. Congratulations!

The Device Chooser dialog

Test on a device

1. In the toolbar, click the Run button. This should display the Device Chooser dialog.

2. Connect the device you configured in figure B-6 to the computer with a USB cable. This should display the device in the Device Chooser dialog, though it may indicate that the device is unauthorized.

3. Unlock the screen on the device. If necessary, use the dialog that's displayed on the device to authorize the device for the computer.

4. Use the Device Chooser dialog to select the device and click the OK button. This should run the app on the device.

5. If the Tip Calculator app displays correctly and you can use it to calculate a tip, your device is configured correctly!

Test on an emulator (optional)

1. Start the emulator. To do that, click on the AVD Manager button in the toolbar. Then, click on the Run button for the emulator described in figure B-5. Depending on your system, it may take a long time for the emulator to launch, so be patient!

2. Unlock the emulator by dragging the lock icon to the right.

3. In the toolbar, click the Run button. This should display the Device Chooser dialog.

4. Use the Device Chooser dialog to select the emulator.

5. If the Tip Calculator app displays correctly and you can use it to calculate a tip, your emulator is set up correctly!

6. If the soft keyboard doesn't display when you click in the bill amount text box, use the emulator to start the Settings app and turn off the Language and Input→Keyboard and Input→Default→Hardware (Physical Keyboard) option. This should allow you to use the soft keyboard or your computer's keyboard to enter the amount.

Figure B-7   How to verify that your Mac is set up correctly (part 2 of 2)

Appendix C

How to set up Linux for this book

This appendix shows how to install and configure the software that we recommend for developing Android apps on the Linux operating system. This appendix also shows how to install the source code for this book.

Because there are many different distributions of Linux, it isn't possible for us to provide instructions to cover all of them. Instead, we have chosen to focus on two of the most popular: Ubuntu, and Fedora. Because most other popular distributions are based on one of these two, these instructions should work for most Linux distributions. If they don't, you can search the Internet for more information.

As you read this appendix, please remember that most websites and install programs are continually updated. As a result, some of the procedures in this appendix may have changed since this book was published. Nevertheless, these procedures should still be good guides to installing the software. And if there are significant changes to these setup instructions, we will post updates on our website (www.murach.com).

How to install the source code for this book

Figure C-1 shows how to download and install the source code for this book. This includes the source code for the apps presented in this book, the starting points for the chapter exercises, and the solutions to the exercises.

When you finish this procedure, the book apps as well as the starting points and solutions for the exercises should be in the directories shown in this figure. Then, you can review the apps presented in this book, and you're ready to do the exercises in this book.

The Murach website

www.murach.com

The directories for the book apps and exercise starts

```
/home/username/murach/android/book_apps
/home/username/murach/android/ex_starts
/home/username/murach/android/ex_solutions
```

How to download and install the files for this book

1. Go to www.murach.com.

2. Find the page for *Murach's Android Programming (2nd Edition)*.

3. If necessary, scroll down to the FREE downloads tab.

4. Click the FREE downloads tab.

5. Click the Download Now button for the zip file, and respond to the resulting pages and dialog boxes. This should download an archive file named and2_allfiles.zip to your hard drive.

6. Open a terminal window to access the Linux command line.

7. Enter the following commands to unarchive the file. Note the period at the end of the first command. You must include it.
   ```
   mv Downloads/and2_allfiles.zip .
   unzip and2_allfiles.zip
   ```

8. Verify that the directories listed above have been created. Then, you can close the terminal window.

Description

- You can install the source code for this book by downloading it from murach.com.

Figure C-1   How to install the source code for this book

How to install the Java SE JDK and required native libraries

For Android development, you need to have the *Java Development Kit* (*JDK*) for *Java SE* (*Standard Edition*) installed on your computer. If you've already done some Java development, you probably already have the JDK installed on your system. In addition, you'll need to install some native libraries that Android Studio requires to work correctly. Figure C-2 shows how to do this for both Ubuntu and Fedora.

Procedure for Ubuntu

1. Open a terminal window to access the Linux command line.

2. Enter the following commands to install the JDK and required libraries. You will probably be prompted for your root password after entering the first command. Go ahead and enter it.

```
sudo apt-get update
sudo apt-get install default-jdk
sudo apt-get install lib32z1 lib32ncurses5 lib32bz2-1.0 lib32stdc++6
```

3. To verify that the JDK has installed correctly, type "javac" at the command line and press enter. If Java has been properly installed, this should provide a usage message. If not, it will inform you that the command could not be found.

Procedure for Fedora

1. Open a terminal window to access the Linux command line.

2. Enter the following commands to install the JDK and required libraries. You will probably be prompted for your root password after entering the first command. Go ahead and enter it.

```
sudo dnf install java-1.8.0-openjdk-devel
sudo dnf install glibc.i686 glibc-devel.i686 zlib-devel.i686
sudo dnf install ncurses-devel.i686 libX11-devel.i686
sudo dnf install libXrender.i686 libXrandr.i686
```

3. To verify that the JDK has installed correctly, type "javac" at the command line and press enter. If Java has been properly installed, this should provide a usage message. If not, it will inform you that the command could not be found.

Description

- Android development requires the *Java SE* (*Standard Edition*) development kit.

- You need to install some native libraries that Android Studio requires. If you don't, Android Studio won't be able to install the Android SDKs correctly.

- If you get an error during the installation that says "Unable to run mksdcard SDK tool", it probably means you are missing one or more of the native libraries. Install them as shown above.

- If you get an error that indicates that the dnf command isn't available, you can use the yum command instead of the dnf command.

Figure C-2   How to install the JDK and required native libraries

How to get Android Studio

Before you can develop Android apps, you need to install the *Android SDK* (*Software Development Kit*). Fortunately, this SDK is installed automatically when you install Android Studio as shown in figure C-3.

The download page for Android Studio
http://developer.android.com/sdk

How to get Android Studio

1. Go to the download page for Android Studio. The easiest way to find this page is to search the Internet for "Android Studio download".

2. Click on the link to download Android Studio for Linux and follow the instructions on the website. This should download a zip file. Make a note of the zip file name, as the version number might be slightly different from the one below.

3. If necessary, open a terminal window to access the Linux command line.

4. Enter the following commands to unzip Android Studio. Note that in most cases, you only need to type the first part of the file name, followed by the Tab key. Linux will automatically complete the rest of the file name. Make sure to include the two periods at the end of the first command.

```
mv Downloads/android-studio-ide-141.2288178-linux.zip ..
unzip android-studio-ide-141.2288178-linux.zip
```

How to start Android Studio

1. If necessary, open a terminal window to access the Linux command line.

2. Type the following command:

```
android-studio/bin/studio.sh &
```

Description

*   For Android development, you need to install Android Studio. This also installs the *Android SDK* (*Software Development Kit*).

*   Once Android Studio has been installed, you can start it from the command line.

*   If you want an icon to start Android Studio, you can search the Internet for how to launch shell scripts from the desktop for your particular Linux distribution.

Figure C-3   How to get Android Studio

How to use the Android SDK Manager

Android Studio includes a starting set of Android tools and platforms. However, we recommend following the procedure presented in figure C-4 to make sure that you have the tools and platforms that you'll need as you work through this book.

When you use the Android SDK Manager, it displays "Installed" in the Status column if the most current version of a tool or platform has been installed. Otherwise, it displays a status of "Not Installed" or "Update Available".

All of the apps that you can download for this book use version 23.0.1 of the Android SDK build tools. As a result, we recommend that you install this version of the build tools.

Throughout this book, we have used Android 6.0 (API 23) for testing new devices and Android 4.1 (API 16) for testing old devices. As a result, we recommend installing these platforms to get started.

If you prefer to use other versions of Android, you can do that too. For example, you may want to use newer or older versions than the ones we use in this book. However, it's usually easier to follow along with the book if you use the same versions we use in this book. That way, this book will match your system, and you will encounter fewer bugs since we have tested the apps in this book on these versions of Android.

We also recommend installing the Android Support Library item because it's necessary to support some newer features on older devices.

This figure shows how to install the tools and platforms necessary to work with this book. If you want to install more tools or platforms, you can use the Android SDK Manager to do that. Of course, additional tools and platforms require more disk space and take longer to download. As a result, if you want to get started as fast as possible, you can follow the instructions in this figure.

The Android SDK Manager

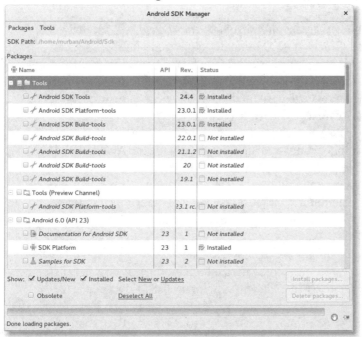

Procedure

1. Start Android Studio.

2. If this displays the Welcome page, select the Configure→SDK Manager option. Otherwise, click the toolbar button for the SDK Manager. Then, click the Launch Standalone SDK Manager link.

3. To view the tools and platforms, expand or collapse the category nodes.

4. To install more tools or platforms, select the tools or platforms. For this book, we recommend installing all of the defaults including the following items:
   - Tools→Android SDK Build-tools Rev. 23.0.1
   - Android 6.0 (API 23)→SDK Platform
   - Android 6.0 (API 23)→Intel x86 Atom System Image
   - Android 4.1.2 (API 16)→SDK Platform
   - Android 4.1.2 (API 16)→Intel x86 Atom System Image
   - Extras→Android Support Library

5. Click the Install button. This should display another dialog box.

6. Select the Accept License radio button. Then, click the Install button. On some systems, this may take an hour or longer.

Description

- You can use the Android SDK Manager to install Android tools and platforms.

Figure C-4   How to use the Android SDK Manager

How to create an emulator

Figure C-5 shows how to create an *Android Virtual Device* (*AVD*) that you can use to test your apps. Since an AVD emulates a physical device, an AVD can also be called an *emulator*. If necessary, you can create an emulator for each platform that you want to test. Unfortunately, emulators tend to run extremely slowly on all but the most powerful computers. As a result, we recommend avoiding using emulators whenever possible.

If you do need to use an emulator, we recommend creating one that uses a low resolution such as hdpi or mdpi and an older version of the Android API. That way, the emulator should use fewer resources on your system, which should allow it to start and run more quickly.

When you create an emulator, you must provide a name for the emulator and the target platform. In this figure, I created an emulator named "Nexus S API 16".

This emulator is based upon a Nexus S device that has a 4-inch screen that's 480 by 800 pixels. As a result, this screen uses a high number of dots per inch (hdpi), which runs more quickly than an extra high number of dots per inch (xhdpi).

This emulator uses a system image for the Android 4.1 (API 16) platform. Since this API doesn't have as many features as later APIs, this also helps the emulator to run more quickly. In addition, this system image uses an x86 processor, which runs faster on a Linux system than an ARM processor. Most smartphones use ARM processors, so this emulation isn't completely accurate, but the improved performance usually makes this tradeoff worthwhile. For this to work, you must download the x86 system image for the corresponding platform as described in the previous figure.

In some cases, you might get an error that says /dev/kvm could not be found. In that case, you need to install kvm virtual machine support on your Linux distribution. Most popular Linux distributions include kvm support by default, so this step isn't usually required. However, if your distribution doesn't come with kvm support out of the box, you can search the Internet for instructions on how to install it in your distribution.

The dialog box for creating an emulator

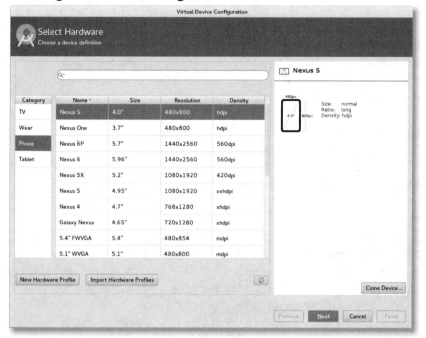

Procedure

1. Start Android Studio.

2. If this displays the Welcome page, open an existing project as described in the start of figure C-7.

3. Click the toolbar button for the AVD Manager. This should start the Android Virtual Device Manager and show all existing virtual devices.

4. Click the Create Virtual Device button. This should display the dialog box shown above. Follow the prompts to create a virtual device. For this book, we recommend creating a virtual device named "Nexus S API 16" with these properties:
   - Category: Phone
   - Hardware profile: Nexus S
   - System image: Jelly Bean (API 16) x86
   - Emulator name: Nexus S API 16

Description

- To test your Android apps, you can create an *Android Virtual Device* (*AVD*) for each platform that you wish to test. An Android Virtual Device can also be referred to as an *emulator*.

- If you get an error that /dev/kvm cannot be found when trying to launch the emulator, you need to install kvm support in your Linux distribution. Most popular Linux distributions already come with this installed.

Figure C-5   How to create an emulator

How to configure a device for development

Figure C-6 shows how to configure a physical device so that you can test Android apps on it. Although you can also test Android apps on an emulator, we recommend testing on a physical device for two reasons. First, apps install and run much more quickly on a physical device than on an emulator. Second, testing on a physical device is the only way to truly see how an app works on that device.

To configure a physical device for development, you must connect your device to the computer with a USB cable and turn on the "USB debugging" option on your device. The procedure for doing this varies depending on the version of Android that's running on the device. As a result, you may need to search the Internet to learn how to enable this option for your device.

With Android 4.2 (API 17) and later, the developer options are hidden. This makes sense because most users aren't developers and don't need these options. As a result, if your device uses API 17 or later, you must start by enabling developer options on your device.

In some cases, you may need to add a udev entry for your device to work properly, although this step is not required for most popular Linux distributions. However, if your device doesn't show up in the device list, you may need to do this. If you need to add udev entries, you can find instructions under the Using Hardware Devices section of the android developer documentation. At the time of this writing, these instructions could be found here:

`http://developer.android.com/tools/device.html`

How to configure a device for development

1. Enable developer options on your device. For Android 4.2 (API 17) and later, you can do it like this:
    -   Settings→More→About Device→Build Number (tap on it 7 times).
2. Turn on USB Debugging. For Android 4.2 (API 17) and later, you can do it like this:
    -   Settings→More→Developer Options→USB Debugging.
3. Connect your Android device to your computer's USB port.

Description

-   Since an actual device runs more quickly than an emulator, we recommend using an actual device whenever possible.
-   On some Linux distributions, you may need to add a udev rule for any physical device you want to support, but this isn't necessary on the most popular distributions. If you need to add udev rules, you can search the Internet for more information.

Figure C-6   How to configure a device for development

How to verify that your system is set up correctly

At this point, you have installed all of the software and source code that you need to work with this book, but you haven't tested it to make sure it's working correctly. To do that, you can run the Tip Calculator app for chapter 3 as shown by the procedure in figure C-7.

To begin, you can start Android Studio and open the project for the Tip Calculator app as shown in part 1 of the figure. At this point, the name of the project appears in Android Studio. However, the emulator doesn't appear on your screen until you've gone through the steps in part 2.

Android Studio and an emulator after an app has been run

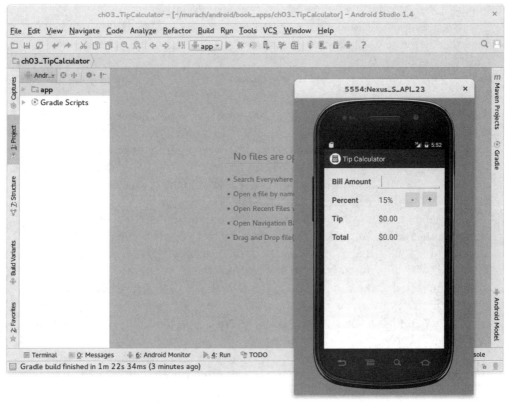

Open an existing project

1. Start Android Studio.
2. If this displays the Welcome page, select the "Open an existing Android Studio project" item. Otherwise, click the Open button in the toolbar.
3. Use the resulting dialog to open the Tip Calculator app for chapter 3 that's stored in this directory:

   ```
   /home/username/murach/android/book_apps/ch03_TipCalculator
   ```

Description

- To verify that you have set up your system correctly for this book, you can run the Tip Calculator app for chapter 3 on an Android device. Or, if you don't have an Android device available, you can run this app in an emulator.
- If you get an error that says, "failed to find Build Tools revision 23.0.1", click on the link that says, "Install Build Tools and sync project" to install these build tools.

Figure C-7   How to verify that your system is set up correctly (part 1 of 2)

After opening the project in Android Studio, you can run the app as shown in part 2. This verifies that your system is set up correctly.

If you didn't download version 23.0.1 of the Android SDK build tools as described in figure C-4, you may get an error when you attempt to run the Tip Calculator app that says:

```
failed to find Build Tools revision 23.0.1
```

To fix this issue, you can install version 23.0.1 of the Android SDK build tools. Fortunately, Android Studio typically provides a link just below the error message that you can use to install the correct version of the build tools.

If you have an Android device that you have configured for development, you can run the app on that device. When you do this, you should display the Device Chooser dialog before you connect your device to the computer. Otherwise, this dialog might not display your device.

Also, the first time you use a device with your computer, you need to authorize that device to work with your computer. To do that, you can use the authorization dialog that's displayed after you connect your device and unlock its screen. After you authorize your device, you should be able to keep your device connected to the computer, and the Device Chooser dialog should display your device every time you run the app.

If you don't have access to an Android device, you'll need to test the app on an emulator. This provides a way for you to test your apps on devices even if you don't have that type of device. However, on most systems, emulators take a long time to start, consume a lot of system resources, and run slowly. As a result, we recommend testing on an actual device whenever possible.

If you are able to complete either of these tasks, you have set up your system correctly for this book. Congratulations!

The Device Chooser dialog box

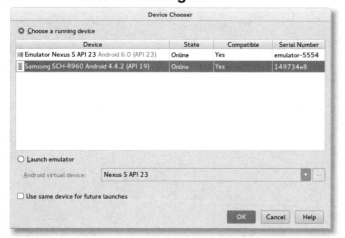

Test on a device

1. In the toolbar, click the Run button. This should display the Device Chooser dialog.

2. Connect the device you configured in figure C-6 to the computer with a USB cable. This should display the device in the Device Chooser dialog, though it may indicate that the device is unauthorized.

3. Unlock the screen on the device. If necessary, use the dialog that's displayed on the device to authorize the device for the computer.

4. Use the Device Chooser dialog to select the device and click the OK button. This should run the app on the device.

5. If the Tip Calculator app displays correctly and you can use it to calculate a tip, your device is configured correctly!

Test on an emulator (optional)

1. Start the emulator. To do that, click on the AVD Manager button in the toolbar. Then, click on the Run button for the emulator described in figure C-5. Depending on your system, it may take a long time for the emulator to launch, so be patient!

2. Unlock the emulator by dragging the lock icon to the right.

3. In the toolbar, click the Run button. This should display the Device Chooser dialog.

4. Use the Device Chooser dialog to select the emulator. This should run the app on that emulator.

5. If the Tip Calculator app displays correctly and you can use it to calculate a tip, your emulator is set up correctly!

6. If the soft keyboard doesn't display when you click in the bill amount text box, use the emulator to start the Settings app and turn off the Language and Input→Keyboard and Input→Default→Hardware (Physical Keyboard) option. This should allow you to use the soft keyboard or your computer's keyboard to enter the amount.

Figure C-7   How to verify that your system is set up correctly (part 2 of 2)

Index

U

V

W

X, Y, Z

The software you need for this book

- Java SE 6 (JDK 1.6) or later.
- Android Studio 1.4 or later.
- You can download this software for free and install it as described in appendix A (Windows), B (Mac OS X), or C (Linux).

The downloadable source code for this book

- Complete source code for the apps presented in this book so you can view the code and run the apps as you read each chapter.
- Starting source code for the exercises presented at the end of each chapter so you can get more practice in less time.
- Source code for the solutions to the exercises that you can use to check your exercise solutions.

How to download the source code for this book

- Go to www.murach.com.
- Navigate to the page for *Murach's Android Programming (2nd Edition)*.
- Follow the instructions there to download the file that contains the source code for the book apps.
- For more details, see appendix A (Windows), B (Mac OS X), or C (Linux).

www.murach.com